D0854063

An Introduction to Clay Colloid Chemistry

An Introduction to

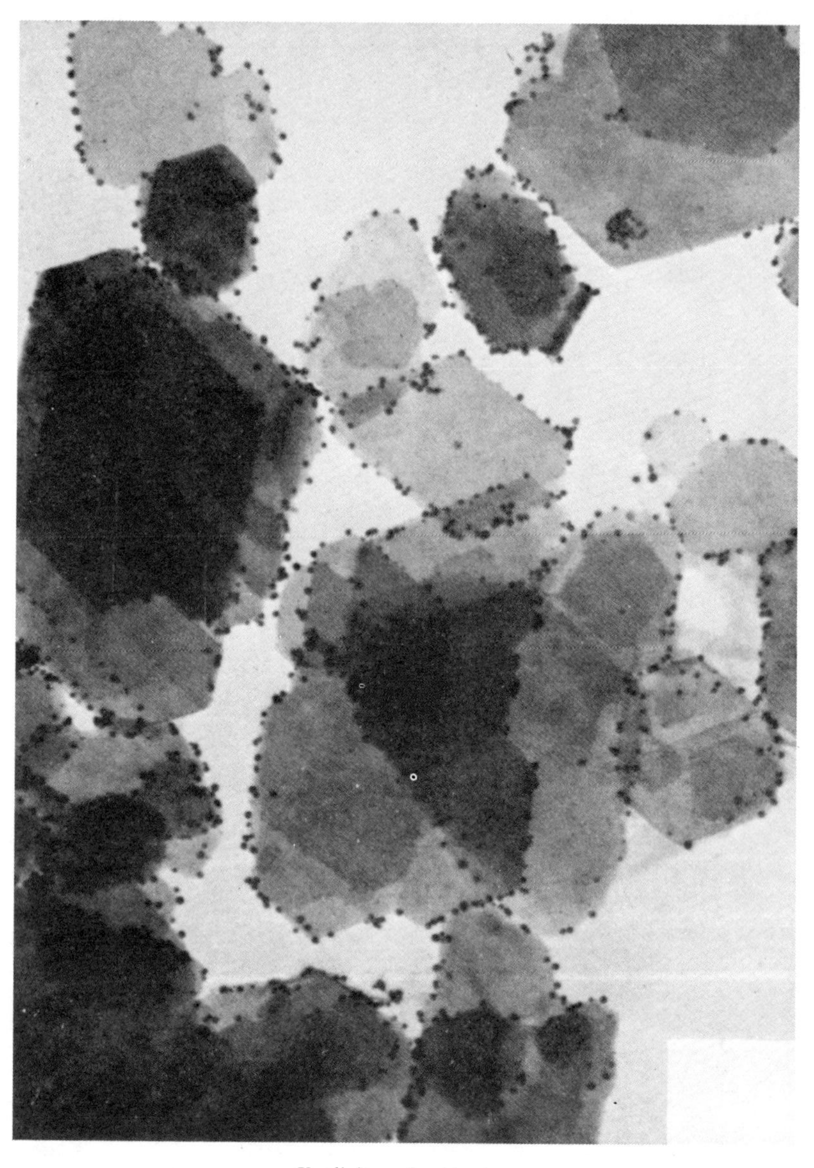

Kaolinite and gold sol.

Clay Colloid Chemistry

For Clay Technologists, Geologists, and Soil Scientists

H. van Olphen
National Academy of Sciences
Washington D.C.

Second Edition

A Wiley-Interscience Publication
JOHN WILEY & SONS
New York • London • Sydney • Toronto

Library of Congress Cataloging in Publication Data:

Van Olphen, H 1912–
An introduction to clay colloid chemistry.

"A Wiley-Interscience publication."
Includes bibliographic references.
1. Clay. 2. Colloids. I. Title.
QD54.9.V27 1977 541′.3451 77-400
ISBN 0-471-01463-X

Printed in the United States of America

10 9 8 7 6

PREFACE

It is not too surprising that the total research effort on clay systems is and has been tremendous: We live through the courtesy of plants grown on clayey soils; we eat our food from ceramic dinnerware; we live in buildings partly made of clay bricks which rest on clay-containing soils into which they sometimes tend to disappear. Part of our energy supply comes from petroleum, which often originates in clay-containing sedimentary rocks. Clays are used in many industrial products and processes. Outside the ceramic industry, clays are an essential part of paper, drilling fluids, and certain lubricating greases; they are applied in formulations of insecticides, adhesives, ointments, and rubber or synthetic plastics; they act as catalysts or catalyst supports in many processes, and they are used for clearing wine.

In almost every field of clay study, one has to deal at one time or another with dispersions of clay in water or in another fluid. Such dispersions, which are characterized by the large interfacial area between the extremely small clay particles and the surrounding liquid, are colloidal systems. Colloid chemistry therefore enters to some degree every technological problem involving clays and liquids, such as problems of soil consolidation, plant nutrition, molding of ceramic objects, and the circulation of drilling fluids in an oil well.

The purpose of this book is to familiarize those engaged in some phase of clay technology, in sedimentary geology, or in soil science, with the modern views of colloid science and its application to clay systems.

Colloid chemistry is a rather specialized branch of physical chemistry, and the colloid chemistry of clay systems is indeed a specialty within a specialty. Communication between the colloid chemist and the clay technologist or geologist is often hampered by differences in terminology. Although the technologist and the geologist frequently use terms not especially appealing to the colloid chemist, the latter must admit that he too uses a rather specialized jargon which is often misunderstood by the noncolloid chemist and which has led to many misconceptions. In this book, therefore, an attempt has been made to discuss the usage of terms very carefully.

Another communication barrier between the colloid chemist and technologist stems from a difference in point of view. The colloid chemist looks primarily at a system on a microscopic scale. He deals with the arrangement of ions and molecules on the surface of the submicroscopic particles and attempts to analyze the forces which act between these particles in a suspension. The technologist, on the other hand, is primarily interested in the bulk physical and mechanical properties of his systems.

The clay technologist is well aware that the bulk properties of the clay systems depend on the concentration as well as on the type of clay; but above all, he is familiar with—and often puzzled by—the remarkably strong dependence of these properties on the composition of the fluid phase. Comparatively minor changes in the composition of the liquid often have surprisingly large effects on the behavior of the system. Here is exactly the area where the colloid chemist and the technologist should get together, since the effects of small amounts of dissolved chemicals in the clay system are governed by the rules of colloid chemistry. A knowledge and an explanation of these rules should enable the technologist to understand the behavior of the systems and consequently to handle them more efficiently. Actually, in many applications he has a free hand to tailor the properties of the clay systems by the incorporation of comparatively small amounts of soluble additives.

The principal themes of this book are therefore a discussion of (*1*) the effect of changes in fluid composition on the forces acting between suspended clay particles, and (*2*) the consequences of such changes on the bulk physical and mechanical properties of the suspensions which are important in clay technology.

Item (*2*) will be illustrated with several examples from the field of clay and soil technology. Although in the choice of the examples emphasis will be on how to solve the problems in the manipulation of drilling fluids (with which the author is most familiar), problems of the same nature frequently arise in other fields where the same solutions can often be applied.

In the arrangement of the chapters, the specialized discussion of the colloid chemistry of clay systems will be preceded by the presentation of the general facts and modern theories of colloid chemistry. In the subsequent treatment of the clays, certain complications must be introduced which are inherent in the unusual structural features of clay particles.

To keep this presentation within the scope of an introduction, it will be elementary, and the mathematical formulations of modern colloid science will be avoided in the main text. However, for the benefit of those readers who plan to engage in physicochemical clay research a brief survey of electric double-layer computations as they apply to clay systems is given in an appendix. In the main text, only the gist of the theoretical approach is discussed, and a sufficiently detailed account is given of the practical results to supply the technologist with a workable background knowledge. The highlights of each chapter have been included in a relatively extensive *Synopsis* following Chapter Twelve.

Selected references at the end of each chapter furnish a good entry to the literature. In addition, a bibliography of books and publications on clay literature is presented in Appendix V. References to this bibliography are prefixed by the letter B to indicate books or by the letter P to indicate periodicals.

Since the appearance of the first edition of this book in 1963, developments in clay research have been substantial, particularly through intensive application of X-ray and electron diffraction, nuclear magnetic resonance, electron spin resonance, infrared absorption, and other tools of physics. The scanning electron microscope has provided detailed pictures of clay agglomerates. Various physicochemical studies have improved our understanding of the mobility and position of ions and adsorbed molecules on clay surfaces. Considerable advances have been made in understanding the reactivity of clay surfaces and catalysis—a subject that is, however, outside the scope of this book. In general, research of the last decade has not made our previous basic concepts obsolete, but our knowledge of clay structure and clay behavior has become more refined. The same is true for the relevant concepts of colloid science, but in both fields some new phenomena were discovered and some new concepts were developed. In colloid science, discreteness of charge effects in electrical particle interaction, and the so-called "hydrophobic bonding" mechanism of stability have received much attention. In clay science, the phenomenon of expansion of the kaolinite structure by potassium acetate and certain other salts ("intersalation"), as well as by hydrogen bonding compounds, (briefly mentioned in the first edition) has been intensively studied in the past decade.

In the present edition both the text and the literature references have been updated to recognize these new developments. Furthermore, recent international recommendations on terminology for both colloid science (IUPAC) and clay mineralogy (AIPEA) have been adopted in this edition (see Appendix V). Also, SI units are used with some convenient exceptions—with due apology. Thus, equivalents were not altogether discarded: the Angstrom unit has been maintained, and electrical double-layer calculations are primarily presented in terms of the electrostatic unit system (esu), rather than the rationalized four quantity system which is part of the SI system. The reason for the latter is that the use of the esu system provides easier access to the now classic literature on the electrical double layer. However, in a few cases, formulas and calculations are presented in both rationalized units and in electrostatic units, to demonstrate the differences between the two systems.

H. van Olphen

Reston, Virginia
1976

ACKNOWLEDGMENT

The author wishes to thank Professor J. Th. G. Overbeek of the University of Utrecht, and Professor G. H. Bolt of the Agricultural University of Wageningen, The Netherlands, as well as his colleagues of the Exploration and Production Research Division of Shell Development Company, Houston, Texas, for their stimulating criticism. The late Dr. H. P. Studer provided the excellent electron micrographs. It is a pleasure to acknowledge the careful editing of the manuscript by Mrs. J. G. Breeding and her staff of this laboratory.

The author thanks the Management of Shell Development Company for having given him the opportunity to publish this book.

H. van Olphen

CONTENTS

CHAPTER ONE

Clay Suspensions and Colloidal Systems in General

I. THE COLLOIDAL SOLUTION OF A CLAY IN WATER

Clays as found in nature are often light to dark grey and are sometimes greenish or bluish owing to organic and inorganic impurities, although certain ion constituents of the clays themselves may cause some coloration. In the purified form, most clays are white. We shall consider a pure white clay powder, for example, a bentonitic clay, and describe what happens when a small amount of the powder is stirred with a large volume of water.

A. Observations with the Naked Eye

The powdered clay seems to dissolve in water, just like common salt. However, some unusual optical effects are observed which are not displayed by a salt solution. When one looks *through* the clay "solution" against the light, the solution appears to be perfectly transparent, but it is brownish in color. When one looks *at* the illuminated clay "solution" from the side, against a dark background, the solution appears bluish white and seems to be slightly turbid. This optical effect is known as the *Tyndall effect*. It is explained as follows:

The clay "solution" is actually a dispersion of very small clay particles. These tiny particles scatter part of the incident light in all directions. According to the theory of light scattering, the intensity of the scattered light increases with decreasing wavelength. Therefore, when the incident light is white, the scattered light has a bluish color, and the transmitted light, which has lost a relatively larger portion of the smaller wavelengths, shows a complementary yellowish to brownish color. Of course, for a non-white clay, certain colors are preferentially absorbed by the clay particles, and a different color distribution between scattered and transmitted light is observed.

Light is also scattered by pure liquids because of local refractive index fluctuations owing to the thermal motion of the molecules of the liquid, but the intensity of the scattered light is small compared with that from dispersed particles. In true solutions, light scattering is also caused by local concentration fluctuations.

According to Rayleigh's theoretical treatment (1871), the intensity of the scattered light is inversely proportional to the fourth power of the wavelength for molecules of a gas and for dispersed particles with a size of up to a few tenths of the wavelength of light. The classical explanation of the deep blue appearance of a clear sky and the yellow to red color of the sun when viewed through the atmosphere is based on Rayleigh's theory.

For larger particles, theory predicts that the dependence of the scattering intensity on the wavelength becomes less pronounced. Indeed, with increasing particle size the color of the Tyndall light becomes less bluish and it becomes white for a coarse suspension. Cigar smoke first appears blue, but it becomes grey when the smoke particles grow by the absorption of water from the atmosphere. The scattered light from clouds of tiny ice crystals is white.

The following microscopic observations show that very small clay particles are indeed present in the clay solution and are responsible for the Tyndall effect.

B. Observations with the Ordinary Light Microscope

When a drop of the clay solution is observed under ordinary lighting conditions in the microscope, the drop appears as transparent as a drop of water, although in some clay solutions particles of microscopic size may be seen. The picture changes completely when the drop is observed in *dark-field illumination* with the *ultramicroscopical* technique.

C. Observations with the Ultramicroscope

In the ultramicroscopical arrangement, light is admitted to the sample under the microscope in such a way that no light can enter the objective lens directly; hence the field of observation is dark. When the light beam meets a particle, it scatters the light in all directions, and part of the scattered light enters the objective lens. In this way, the presence of a particle is revealed by a light speck on a dark background. With this technique, numerous particles are observed in the clay solution (Figure 1). Since the majority of these particles are not observed under normal lighting conditions in the microscope, the disturbances in the light beam caused by the particles apparently are too small to be discernible against a light background. This fact usually indicates that the particles are smaller than the wavelength of light. Yet, it is possible that particles of microscopic size are present which are too transparent to be observed against a light background. If the particles are indeed smaller than the wavelength of light, the

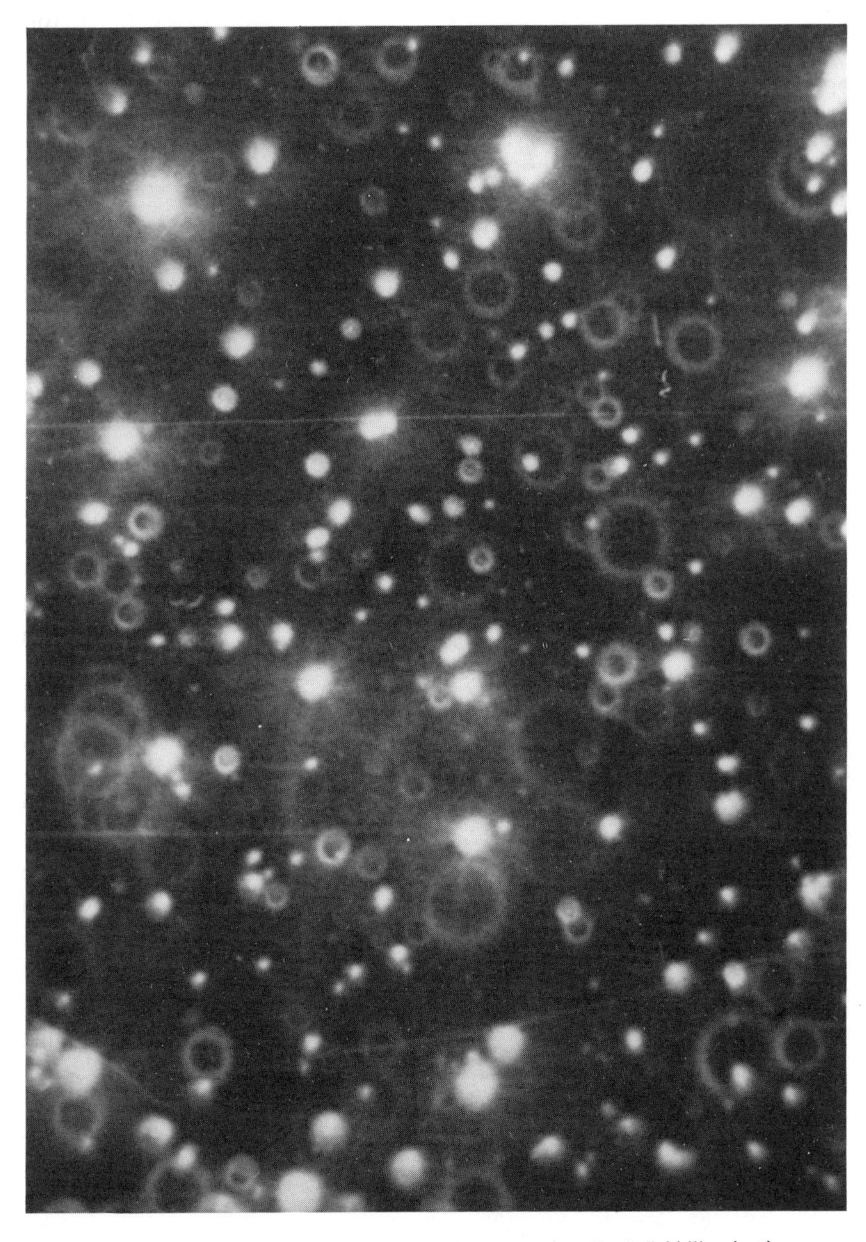

Figure 1. Photomicrograph of a stable clay suspension. Dark-field illumination.

light speck observed in the ultramicroscope cannot be an image of the particle and does not reveal its dimensions. Such particles have been called *ultramicrons*. Extremely small particles, that is, those in the millimicron range, are not individually discernible in the ultramicroscope, but their presence is shown by an undifferentiated diffuse light. In the classical literature, these particles are called *amicrons*.

In the ultramicroscope, the clay particles in a drop of a solution display a vivid irregular motion in all directions. This phenomenon is called the *Brownian motion* of the particles, since it was first described by the botanist R. Brown in 1827. The Brownian motion is a result of the thermal motion of the water molecules surrounding the particles. The water molecules continuously collide with the particles; however, at a certain instant the collisions happen to be more numerous on one side of a particle than on the other side, resulting in a net force in one direction. The next instant the collisions may be more numerous on the other side, and the particle moves in the opposite direction. In the kinetic equilibrium, the average translational kinetic energy of the particles is equal to that of the water molecules:

$$[\tfrac{1}{2}\, m\overline{v^2}]\ \text{particle} = [\tfrac{1}{2}\, m\overline{v^2}]\ \text{water molecules}$$

Therefore, the average velocity of the particles decreases with increasing mass, and the Brownian motion becomes less vivid with increasing particle size.

In addition to a translational Brownian motion, the particles also obtain a rotational motion. Owing to this rotation, anisodimensional particles, such as plates or rods, continuously change the position of their axes relative to the optical axis of the microscope. Since the intensity of the scattered light in different directions from the axes of the particle varies, the scattered light reaching the objective lens from the rotating particle continuously varies in intensity. Such fluctuations are clearly observed in clay suspensions, indicating that the particles are nonspherical. If one of the dimensions of the particle is in the microscopic range, the light speck vaguely reveals the shape of the particle. For example, in certain clay suspensions one may observe disc-shaped light specks showing a tumbling motion.

From the ultramicroscopic observations, it must be concluded that the clay "solution" is actually a homogeneous dispersion of very small particles. Such a dispersion is called a *colloidal solution*, or *sol*, if the dimensions of the particles are such that they do not settle within a reasonable time. When the dispersed particles are large and settle comparatively rapidly, the dispersion is called a *suspension*. The distinction between a sol and a suspension is, however, entirely arbitrary; there is no difference in principle. The borderline between sol and suspension is usually chosen at an

Table 1

Units, Dimensions, and Resolution Limits

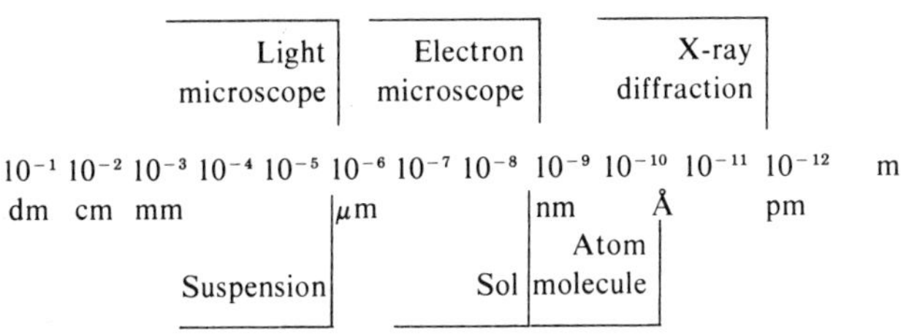

equivalent spherical radius, or *Stokes radius,* of 1 μm. The equivalent spherical radius of a particle of any shape is computed from its settling velocity by applying Stokes law for spherical particles. Particles with a Stokes radius smaller than 1 μm are considered to fall in the colloidal size range.

The particle size in natural clay dispersions usually varies widely and often covers a range which is typical of both sols and suspensions. Since the rapid settling of the larger particles is most obvious to the observer, it has become customary to refer to clay dispersions as *clay suspensions.* However, for specially prepared small-particle-size fractions, the term *clay sol* is more appropriate.

More information about the dispersed clay particles in sols and suspensions is obtained when we take the next step in magnifying power and study the clays by means of electron-microscopic techniques. (For a review of the resolution limits of various observation techniques and the notations for units of length, the reader is referred to Table 1.)

D. Observations with the Electron Microscope

Electron micrographs of dried clay suspensions and sols, as well as scanning electron microscope pictures, show that the particles are shaped either as flat plates or as needles or laths depending on the type of clay (Figures 2, 3, 4, and 5).

Since the electron-microscopic observation technique requires a high vacuum, only dry particles can be observed. There is no guarantee, therefore, that the electron micrograph reflects the shape and size of the particles when they are in a dispersed condition. In most cases, however, there is little doubt that plates remain plates and needles remain needles when suspended in water. This opinion is supported by various unusual optical and other effects shown by the suspensions and sols which are indicative of the anisodimensional shape of the dispersed particles. As will

Figure 2. Electron micrograph of Wyoming bentonite clay, shadowed. Micrograph by H. P. Studer.

be discussed later, the quantitative evaluation of such effects enables us to determine the average particle dimensions in suspension.

The regular shapes of the particles of some clays, as observed in electron micrographs—for example, the hexagonal shape of plates of kaolinite in Figure 5*a*—suggest that the particles are crystalline.

This supposition is indeed confirmed by the ultimate resolution of atom arrangements in the clay particles by means of electron diffraction and X-ray diffraction techniques.

E. X-Ray Diffraction Patterns of Clays

The Debye-Scherrer X-ray diffraction patterns obtained for dry clay powders as well as for concentrated clay suspensions show that the clay particles are crystalline. During the last decades, the atomic structures of the small clay crystallites have been investigated intensively. Since the results of these studies have contributed greatly to the understanding of the colloidal phenomena in clay sols and suspensions, they will be discussed in detail in Chapter Six.

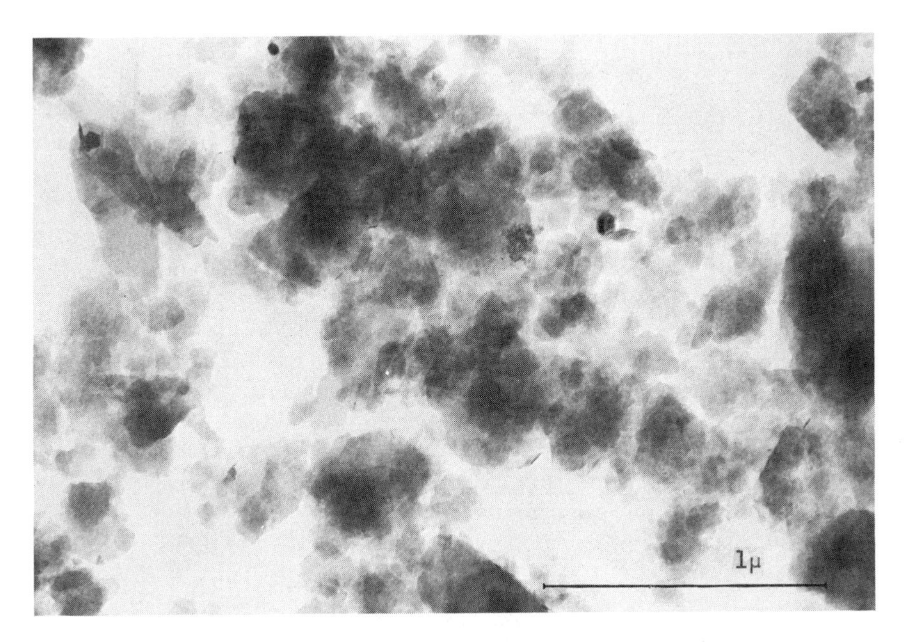

Figure 3. Electron micrograph of Fithian illite. Micrograph by H. P. Studer.

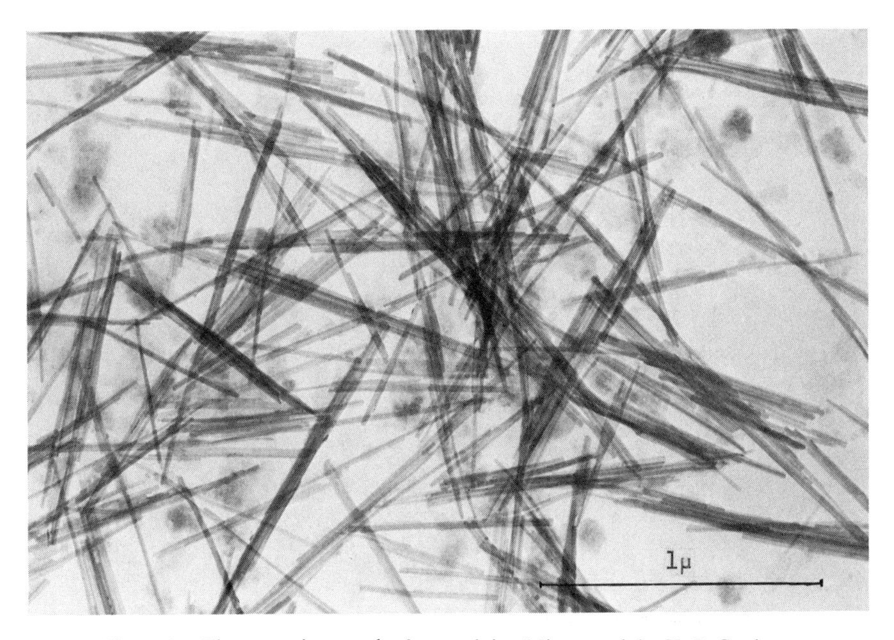

Figure 4. Electron micrograph of attapulgite. Micrograph by H. P. Studer.

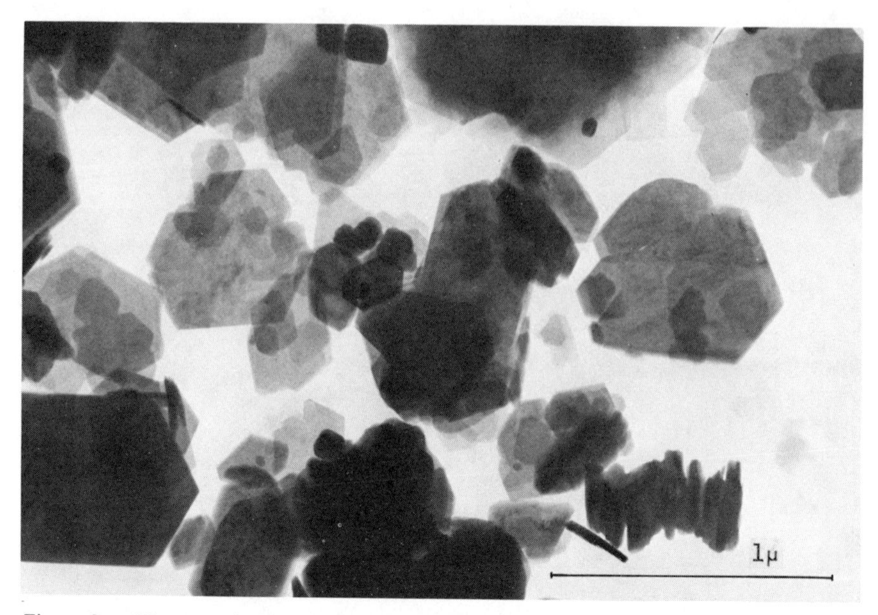

Figure 5a. Electron micrograph of kaolinite; a stacked pack of plates is shown in the lower right corner. Micrograph by H. P. Studer.

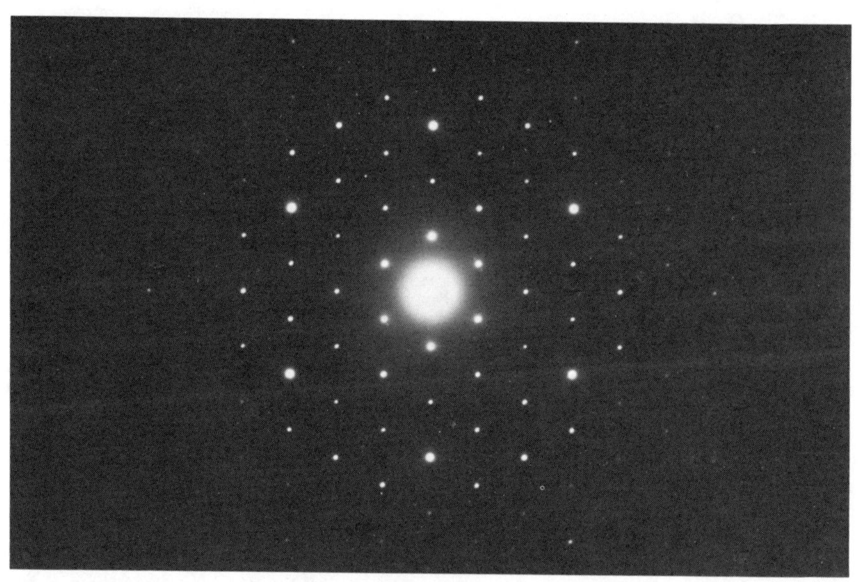

Figure 5b. Selected-area electron diffraction (SAD) pattern of a kaolinite flake. Micrograph by H. P. Studer.

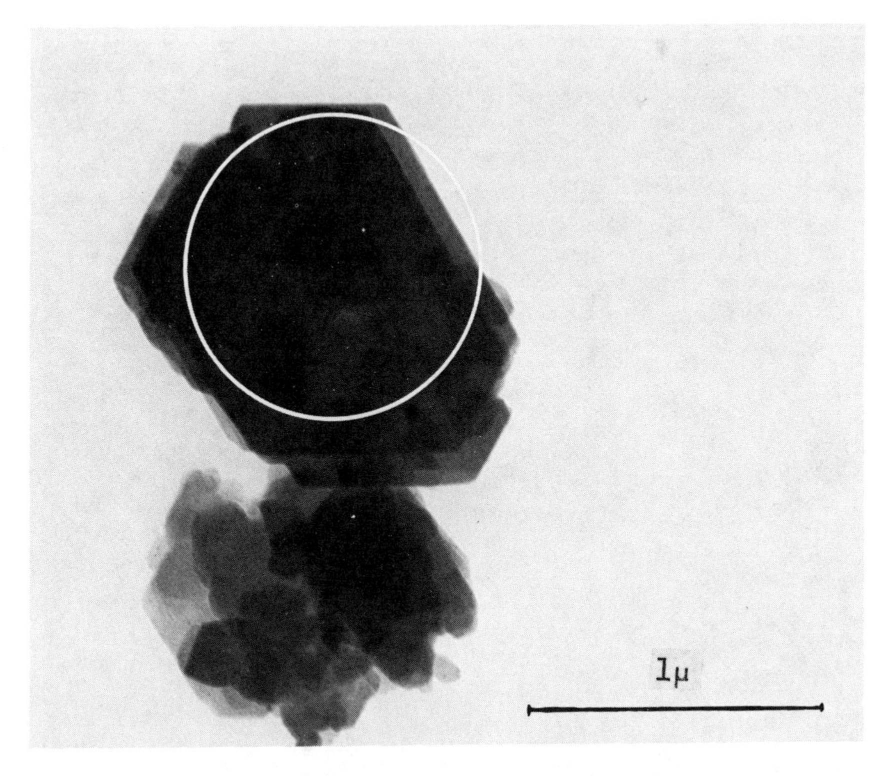

Figure 5c. Electron micrograph of a diffracted crystal in corresponding orientation; diffracted area indicated by circle. Micrograph by H. P. Studer.

F. Electron Diffraction Patterns of Clay Particles

With most electron microscopes it is possible to obtain an electron diffraction pattern of a preselected single clay particle. Such a pattern is shown in Figure 5*b* (selected area diffraction, SAD).

II. PARTICLE INTERACTION

The dispersed clay particles collide frequently because of their Brownian motion. After the collision they separate again, as can be observed in the ultramicroscope. The picture changes completely when a small amount of salt—a few tenths of one percent—is added to the clay dispersion. The particles begin to stick together upon collision, and agglomerates grow in the suspension, as shown by Figure 6. When the salt-containing suspension is observed with the naked eye, the agglomerates appear as flocs which set-

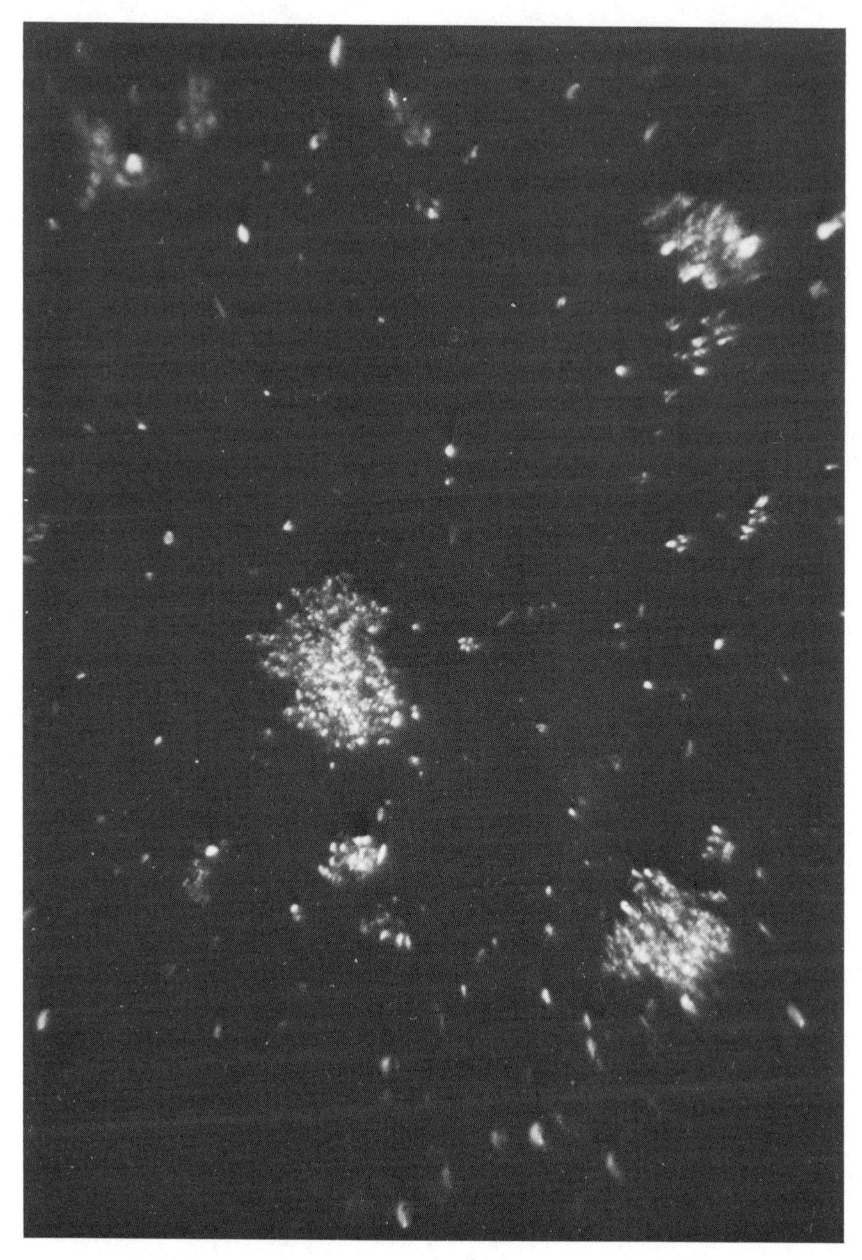

Figure 6. Photomicrograph of a flocculated clay suspension. Dark-field illumination.

tle rather quickly. The turbid-looking suspension is finally separated in a bottom sediment and a clear, particle-free supernatant liquid. This phenomenon is called *flocculation* or *coagulation.* In this condition, the sol or suspension is called *flocculated, coagulated,* or *colloidally unstable.* Upon removal of the salt and stirring of the suspension, it is usually possible to restore the original dispersion of individual particles. This restoration of the sol is called *deflocculation, peptization,* or *stabilization.* The second term was coined by Graham in analogy with peptic digestion. The original and the restored sols are called *deflocculated, peptized,* or *colloidally stable* (or briefly, *stable*). There terms are used interchangeably.

The flocculating effect of small amounts of a salt on a colloidal dispersion is one of the basic phenomena of colloid chemistry, and, at the same time, it is of great importance in the technology of colloidal systems. From the ultramicroscopic observation of the sticking together of the particles upon collision, it is obvious that attractive forces exist between the particles. What is the nature of these forces, and why are they seemingly not effective in the stable sol? These questions have been answered by the modern theory of the stability and flocculation of colloidal dispersions. We shall discuss this theory in some detail later, but at this point a qualitative description of the results of the analysis of particle interaction will be presented.

The attractive force between dispersed particles is attributed to the general van der Waals attraction forces between all the atoms of one particle and all the atoms of another particle. (It may be recalled that the van der Waals forces between the molecules of a gas explain certain deviations from ideal gas behavior as shown by real gases.) The total attractive force between the particles is the sum of the forces between all the atom pairs. The magnitude of this total force depends on the size and shape of the particle and, to some extent, on the character of the dispersion medium. However, in hydrous dispersions, the total force is little dependent on the salt content of the system.

In a stable, salt-free sol, the particle attraction is practically the same as in a salt-flocculated sol; however, in a stable sol, the particle attraction is successfully counteracted by a repulsive force between the particles. This repulsive force is of an electrical nature. It has been known for a long time that the suspended particles in a sol move in an electric field, indicating that the particles carry an electric charge. The magnitude of the electrical repulsion due to the particle charge does depend on the presence or absence of ionized salt in the sol. The repulsion becomes less effective when the ion concentration increases. Therefore, in the presence of salt, the repulsive force is no longer able to counteract the van der Waals attraction, and the sol flocculates.

To summarize, in both fresh and salt water, interparticle attraction and repulsion operate simultaneously. The attraction is independent of the salt concentration, but the repulsion decreases with increasing salt concentration. In fresh water the repulsion predominates, and the sol is stable; in salt water the repulsion is reduced to the point at which the attraction begins to dominate, and the sol flocculates.

Prior to the discussion of the quantitative and more precise formulation of the stability theory, we shall review some of the major empirical facts and rules of classical colloid chemistry. We shall digress for a while from the discussion of clay suspensions and deal with colloidal systems in general.

III. TERMINOLOGY IN COLLOID CHEMISTRY

Materials which can be brought into colloidal solution, such as clays, are often referred to as "colloids." This terminology is misleading, since it suggests that the colloidal state is a different state of matter, to be distinguished from the crystalline state. This view was indeed held in the early days of the development of colloid chemistry. Thomas Graham (1), one of the founders of colloid chemistry, wrote in 1861,

". . . As gelatine appears to be its type, it is proposed to designate substances of the class as colloids, and to speak of their particular form of aggregation as the colloidal condition of matter. Opposed to the colloidal is the crystalline condition of matter. Substances affecting the latter form will be classed as crystalloids. The distinction is, no doubt, one of intimate molecular constitution."

The word "colloids" was coined by Graham from the Greek "κωλλα," which means "glue." Accordingly, colloid chemistry literally means "glue chemistry." This term has persisted, although a substantial part of colloid chemistry does not deal with glue-type materials.

We now know that many colloidally dispersed materials are in the crystalline state, for example, clays. Therefore, it is more accurate to use the word colloid with reference to a colloidal *system*. The difference between the right and the wrong usage becomes at once clear when an emulsion of water in oil is considered. The emulsion, which is a dispersion of very fine water droplets in oil, is a colloidal system, but nobody would refer to water as a colloid.

In modern terms, a colloidal dispersion is defined as a system in which particles of colloidal dimensions (i.e., roughly between 1 nm and 1 μm in at least one direction) are dispersed in a continuous phase of a different composition. The dispersed particles or "kinetic units" in colloidal systems may be small solid particles, macromolecules, small droplets of liquids, or small gas bubbles. The continuous phase in which these units are dispersed may be a solid, a liquid, or a gas. Thus, in addition to sols, suspensions, and emulsions, which have been mentioned already, colloid science also deals with such systems as mists, smokes (aerosols), and foams. We shall confine the discussion of colloidal systems mainly to sols and suspensions, although emulsions will be dealt with occasionally.

The reader of colloid chemical literature faces a confusing abundance of terms which are sometimes only vaguely defined or are used interchangeably. During the rapid growth of this branch of physical chemistry, many terms have been coined on the basis of phenomenological developments. When theoretical progress was made, several of these terms were found to be misleading, but they still persist, such as the term "colloid" itself. In this book, an attempt is made to describe the proper usage of each term in accordance with internationally accepted terminology. (See Appendix V.)

IV. CLASSIFICATION OF COLLOIDAL SYSTEMS

The two major classes of hydrous colloidal systems are the *hydrophobic* and the *hydrophilic colloids.* They were formerly called *suspensoids* and *emulsoids,* respectively. A typical example of a hydrophobic colloid is a gold sol, which is a dispersion of submicroscopic gold particles in water. Typical hydrophilic colloids are colloidal solutions of gums.

The terminology "hydrophobic" colloid is somewhat misleading. The adjective hydrophobic (or, in general, lyophobic) means that a material repels water (or, in general, the solvent). A material which displays an affinity for water (or solvent) is described as hydrophilic (lyophilic). At present, these terms are in common use to describe the wetting properties of surfaces. For example, a solid surface which is preferentially wet by water in competition with oil is called a hydrophilic surface and, at the same time, an oleophobic surface. If a surface is preferentially wet by oil in competition with water, it is called oleophilic and hydrophobic.

Obviously, the particles in a hydrophobic sol are not hydrophobic at all—they are certainly wet by water, and usually one or two monolayers of water are more or less strongly adsorbed on the particle surface. Historically, the term "hydrophobic colloids" was introduced to distinguish these colloids from those which were called "hydrophilic colloids." The latter comprise the colloidal solutions of gums, which display a spectacular greediness for water. When a dry gum is contacted with water, a colloidal solution is spontaneously formed. If this spontaneous creation of the hydrophilic sol is compared with the great efforts required to make a colloidal dispersion of a material such as gold, one can see the point in calling the latter systems hydrophobic, relatively speaking. Still, the term hydrophobic colloid is not a happy choice.

With the growing knowledge of colloidal systems, present day classification and nomenclature are based on more fundamental characteristics of colloidal systems. It has been recognized that the hydrophilic sols which are prepared from organic macromolecular substances such as natural and synthetic gums should be considered true solutions of macromolecules or macro-ions. Consequently, the hydrophilic colloids are now called *macromolecular colloids* or *polyelectrolyte solutions*. Their "colloidal" properties are a result of the large size of the dispersed molecules with respect to the size of the molecules of the liquid medium.

The hydrophobic colloids are liquid dispersions of small solid particles, each consisting of a large number of atoms or molecules. They are suitably considered as two-phase systems with a large total interfacial area. In these systems, the properties of the particle surfaces play a dominant part. The historical, widely accepted term hydrophobic colloids is still in common use, despite its misleading connotation.

A very characteristic difference between hydrophobic and macromolecular sols may be mentioned at this point. They are distinguished by a great difference in sensitivity toward the addition of salt. The hydrophobic sols flocculate in the presence of rather small amounts of salt, as mentioned previously in the discussion of clay suspensions and sols. The macromolecular sols, which are true solutions, are rather insensitive toward salt. They do not flocculate in the sense that a hydrophobic sol flocculates, but large amounts of salt may adversely affect the solubility of the macro-ions and may cause the precipitation of the macromolecular compound. Many macromolecular compounds remain dissolved in highly concentrated, even saturated, salt solutions.

At present, a third class of colloidal systems is separated from the original group of lyophilic colloids. They are called *association colloids*, examples of which are soap solutions. In these systems an equilibrium exists between truly dissolved simple molecules and large kinetic units consisting of associated molecules. These units are called *micelles*, and they appear only beyond a certain minimum concentration of the soap, which is called the *critical micellization concentration* (c.m.c.). The term "micelle" also has been used more generally for the kinetic units in hydrophobic sols, and the equilibrium liquid in which the micelles are dispersed has been called the *intermicellar liquid* in the older literature.

It would carry us too far to go deeper into the differences between the three classes of colloidal systems. Certain colloids have typical features in common with more than one class of colloids. In fact, there is some difficulty in classifying the colloidal clay systems. In some respects they behave like macromolecular colloids and in some respects like association colloids. However, the liquid dispersions of small clay crystallites with a very large surface area behave in most respects like hydrophobic colloids, and they are most suitably treated as such.

Reference

1. Graham, Thomas (1861), Liquid diffusion applied to analysis, *Phil. Trans. Royal Soc. (London)*, **151,** 183–224.

CHAPTER TWO

Properties of Hydrophobic Sols

I. SETTLING, AGING, AND FLOCCULATION

We have seen that in a colloidally stable hydrophobic sol, such as a clay sol in fresh water, the individual particles are homogeneously dispersed in the liquid. Such a dispersion may remain virtually unchanged for a long period of time. Still, several changes do occur in such a system, although at a very slow rate, indicating that the system is not really stable in the thermodynamic sense, as is a true solution.

In the first place, if the particles are heavier than the liquid, they may settle slowly, and the homogeneity of the sol is disturbed. Furthermore, sometimes slow recrystallization of the particles takes place (aging). Finally, even in a "stable" sol, a very slow particle agglomeration (flocculation) does occur.

A. Settling

The rate of settling of the particles in a stable colloidal dispersion depends on the size and shape of the particles and on the difference between the density of the particles and that of the liquid. We have already discussed the distinction between a sol and a suspension, which is based on the different settling rates of the particles. The large particles of suspensions and of the coarsest sols settle to a bottom sediment, leaving a particle-free supernatant liquid. In finely dispersed sols, settling may be effectively counteracted by diffusion forces. Barring convection currents, an equilibrium state is attained in which the particles are distributed in the same way as the gas molecules in the earth's atmosphere. As will be discussed in Chapter Four (Section IV-A), settling is sometimes also counteracted by particle-interaction forces.

B. Aging

The large interface between particle and liquid is the seat of interfacial free energy. Since a system tends to lower its free energy, there is a tendency for a spontaneous reduction of the surface area of the particles. Small particles, which are more soluble than large particles, disappear, and the larger ones grow at the cost of the smaller ones (*coarsening* of sols). Irregular surfaces of individual particles tend to become more regular. This process, called "aging," may be comparatively rapid if the solubility of the solid is not too small, but in many systems the solubility is very small, and thus aging in these systems is practically negligible.

C. Flocculation (Particle Agglomeration)

As mentioned before, flocculation, or the agglomeration of particles, occurs when salt is added to the stable sol. However, it is often observed that in a "stable," salt-free sol, particle agglomeration does occur, though at an extremely slow rate. Eventually, therefore, the stability of the sol will be destroyed. The theoretical explanation of this process will be given when the quantitative theory of the stability of hydrophobic sols is discussed.

The agglomeration of particles is an irreversible process; the particles are unable to disengage spontaneously as fast as or faster than they associate. For this reason, the hydrophobic sol is called an irreversible system. The macromolecular sols, on the other hand, are true solutions, and they are reversible systems. The same is true for the association colloids. Hence in modern treatises of colloid science, the colloidal systems are classified as *irreversible systems* (covering the hydrophobic colloids) and *reversible systems* (comprising the macromolecular and the association colloids, i.e., the hydrophilic colloids). (See Appendix V, reference B-5.)

II. THE ORIGIN OF THE ELECTRIC CHARGE OF THE PARTICLES

When an electric field is applied to a hydrophobic sol, ultramicroscopical observation shows that the particles move toward one of the electrodes. When the polarity of the field is reversed, the particles immediately change direction and move to the other electrode. In some types of sols, the particles move toward the positive electrode; in other sols, they move toward the negative electrode. Apparently, both negative and positive sols exist. For example, the particles in clay sols are negatively charged, since they move toward the positive electrode. Other negative sols are silica and quartz sols, sulfur sols, and sols of several sulfides. Examples of positive sols are ferric hydroxide sols and other metal hydroxide sols.

The transport of the colloidal particles in an electric field is called *electrophoresis*. This term has replaced the formerly used, more restricted term *cataphoresis*.

Like an ionic solution, the hydrophobic sols does not have a net electric charge; therefore, the particle charge must be internally compensated in the sol. The internal balance of charges in a sol is incorporated in the concept of the *electrical double layer.* The double layer consists of the particle charge and an equivalent amount of ionic charge which is accumulated in the liquid near the particle surface. The accumulated ions of opposite sign are called the *counter-ions* or *gegenions*. The latter term is borrowed from the German literature.

An important question in colloid chemistry is how the particle obtains its charge. It appears that the particle charge may be created in two different ways:

1. Imperfections of the crystal structure of the particle may be the cause of a net positive or a net negative charge. Such a net charge will be compensated by the accumulation of an equivalent amount of ions of opposite sign in the liquid immediately surrounding the particle, keeping the whole assembly electroneutral. This situation is schematically sketched in Figure 7. This origin of an electric double layer is comparatively rare in hydrophobic colloids, but, actually, the clay particle is an example in which an electric double layer originates from crystal imperfections. This fact is responsible for certain unusual properties of clay sols and suspensions which are not generally encountered in hydrophobic colloids.

2. In most hydrophobic sols, the particle charge is created by the

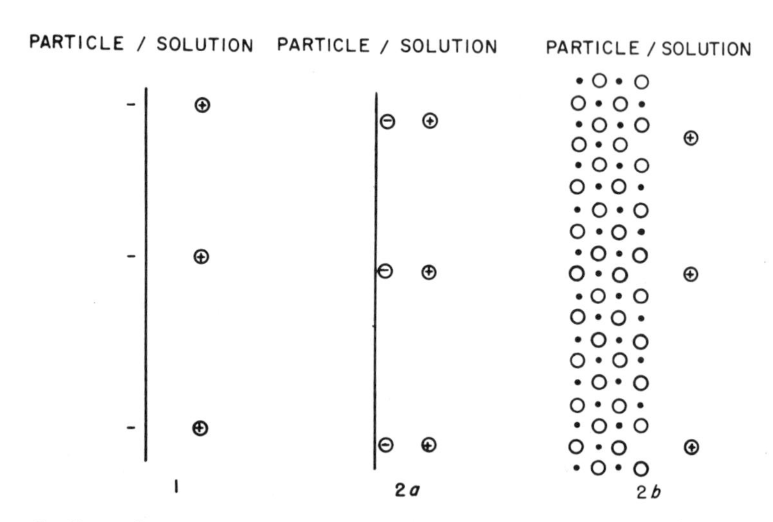

Figure 7. Formation of an electric double layer. (1) Charge originating in the interior of the crystal. (2*a*) Surface charge originating from specific adsorption of ions. (2*b*) Charge originating from the adsorption of ions constituting the crystal (potential determining ions).

preferential adsorption of certain specific ions on the particle surface. Such ions are called *peptizing ions* for the sol, since they create the stabilizing or peptizing charge. The adsorbed peptizing ions constitute the *inner coating* of the electric double layer. An equivalent amount of counter-ions is accumulated near the charged surface, constituting the *outer coating* of the double layer. The charging process requires the presence of a sufficient amount of an electrolyte containing the kind of ions which are specifically adsorbed on the particle surface. Such an electrolyte is called a peptizing electrolyte. The other ions of the electrolyte supply the counter-ion charge in the outer coating of the double layer. For those sols which acquire their particle charge by peptizing ion adsorption, the previous statement that a sol in pure water is stable must be amended; a certain amount of peptizing electrolyte must be present. However, it is not always necessary to add a peptizing electrolyte, since the dissolution products of the colloidal material or the dissociation products of water sometimes act as peptizing electrolytes. The amounts required for peptization are usually extremely small, often smaller than 1 meq/dm^3.

The surfaces of crystalline particles usually present active adsorption sites, since the valences of the lattice atoms which are exposed at the surface are not completely compensated, as they are in the interior of the crystal. These surfaces are called "broken-bond" surfaces. Two types of adsorption mechanisms of peptizing ions may be distinguished:

2a. Specific peptizing ions may be adsorbed on the surface by chemical bonds (chemisorption), whereas, particularly in the case of adsorption of organic ions, hydrogen bonds and van der Waals attraction forces may be responsible for the adsorption (physical adsorption). The creation of the double layer is sketched in Figure 7.

2b. Another adsorption mechanism which is very common in hydrophobic sols exists when the charging ions are identical with those constituting the crystallite. An illustrative example is the peptization of silver iodide sols by either silver ions or iodide ions. As sketched in Figure 7, at the surface of the cubic lattice of AgI, the growth of the crystal has been interrupted. Ag^+ and I^- ions (represented by dots and open circles, respectively) are adsorbed on the crystal surface and occupy the regular lattice positions, but they are not necessarily adsorbed in equivalent amounts. Depending on the relative availability of these ions in the solution, the surface can be positively or negatively charged by the slight excess of Ag^+ or I^- ions in the crystal surface.

As will be discussed later, the AgI particles behave as a reversible electrode for either Ag^+ or I^- ions. Therefore, the electric surface potential of the AgI particles is entirely determined by the concentration (or activity) of the Ag^+ or I^- ions in solution (Nernst potential). For this reason, the peptizing Ag^+ and I^- ions are also called *potential-determining ions* for the AgI sol.

Another example of peptization by potential-determining ions is the stabilization of aluminum hydroxide sols. Depending on the availability in solution of the potential-determining ions Al^{3+} or OH^- (or possibly more complex ions), either a positive or a negative sol is created. The situation in this system is somewhat more complicated than in the AgI sol. The availability of either peptizing ion depends on the pH of the solution. In acid solution, Al^{3+} ions are available, and the sol becomes positive. In alkaline solution, both OH^- and complex aluminate ions are present, and both may act as peptizing ions for the negative sol.

Not every type of sol is able to assume either a positive or a negative charge. For example, a silica sol, as such, only obtains a negative charge.

It may be mentioned that, because of the small amounts of peptizing electrolytes which are involved in the creation of the electric double layer, the qualitative and quantitative analysis of the composition or the structure of the inner and outer coating is extremely difficult, and in many sols exact information is still lacking on this point.

In oxides such as aluminum and ferric oxides which have a pH-dependent surface potential, it is customary to refer to H^+ and OH^- as the "potential-determining ions," although they may act as such only in an indirect way.

III. THE PREPARATION OF A STABLE HYDROPHOBIC SOL

Generally speaking, three steps are required in the preparation of a stable hydrophobic sol:

1. The creation of particles of colloidal size, either by aggregation of ions or molecules (*condensation method*) or by subdivision of larger particles (*dispersion method*).
2. The creation of a particle charge by the addition of a small amount of peptizing electrolyte (as discussed in the preceding section). Such an addition is not necessary when the particle obtains a charge from interior lattice imperfections.
3. Removal of any flocculating amount of salt in the system.

The following examples may illustrate the various techniques of sol preparation:

A. Condensation Method

A silver iodide sol is prepared by mixing fairly dilute solutions of a silver salt and an iodide. Silver iodide precipitates according to the following reaction equation:

$$AgNO_3 + KI \rightarrow KNO_3 + AgI\downarrow$$

Upon mixing, first a large number of AgI nuclei are formed which subsequently grow to particles of colloidal size. The final size of the crystallites depends on the course of the nucleation and growth processes. For a given system these processes are affected by the conditions of mixing and by the concentration of the reacting salt solutions (1). A positive or a negative particle charge is created by the potential-determining Ag^+ ions or I^- ions, depending on whether $AgNO_3$ or KI is present in excess. In the positive sol, nitrate anions act as counter-ions; in the negative sol, potassium ions act as such.

The excess of K^+ and NO_3^- ions in the sol flocculates the sol. In order to obtain a stable sol, part of these ions should be removed so that their final concentration is lower than that at which the sol flocculates.

B. Dispersion Method

A well-known dispersion method is the electric disintegration of metals by arcing two submersed electrodes of that metal (Bredig's method). In this process, locally melted metal is dispersed in the liquid. However, simultaneously a condensation process may occur in which vaporized metal associates to particles of colloidal size. In order for the sol to become stable, the water phase must contain a peptizing electrolyte or an electrolyte which forms peptizing ions by reacting with the metal.

Many sols are prepared by mechanical disintegration of a solid, either by dry-grinding of the material and subsequent suspension or by submitting a coarse suspension to high shear rates in mixers, blenders, or colloid mills of various designs. A modern technique for the preparation of fine dispersions is the disintegration of suspended particles with ultrasonic energy. Again, the creation of stable dispersions requires the addition of peptizing electrolytes, in most cases.

In clays a sufficiently small particle size has been created in nature by weathering and erosion processes. Therefore, lumps of dry clay are usually easily dispersed in water by means of simple mixers or blenders. Since an important part of the particle charge stems from lattice imperfections, it is often unnecessary to add peptizing electrolytes to obtain a stable suspension. The counter-ions, which balance the interior lattice charge, remain attached to the particles when the clay is dried. Therefore, when the dry clay is contacted with water, an electric double layer is spontaneously created. The preparation of clay suspensions is described in some detail in Appendix I.

C. Cleaning of the Sols

Sols prepared by various methods may contain a flocculating amount of salt, either as a reaction product or as an impurity. This salt must be removed in order to make a stable sol. Several different methods may be followed to free a sol or suspension from dissolved electrolytes:

The flocculated sol is allowed to settle under gravity or in a centrifuge, the supernatant liquid is decanted, and the residue is redispersed in fresh water. Upon repetition of this process, the sol becomes more and more peptized, and the larger gravitational fields of a supercentrifuge or an ultracentrifuge will be required for the separation.

The flocculated sol may also be filtered and washed to remove the flocculating electrolytes. An *ultrafilter* must be used which has fine enough pores to retain the peptized particles. Finally, the filter cake is redispersed in water.

Alternatively, the salt may be removed by *dialysis*. In this method the sol is separated from a fresh-water reservoir by a membrane which is permeable to the salt but not to the sol particles. Such a membrane (as well as an ultrafilter) is called *semipermeable.* Because of the concentration gradient across the membrane, the salt diffuses into the fresh water. The water in the reservoir is frequently replenished.

A more expedient method of dialysis is *electrodialysis*. In this procedure the flocculated sol is placed in the center compartment of three compartments which are separated by semipermeable membranes. Electrodes are placed in the two side compartments, and an electric field is applied across the sol in the middle compartment. The electric field accelerates the removal of the electrolyte from the sol. Fresh water is circulated through the side compartments in order to remove the products of electrolysis which are formed at the electrodes.

IV. FLOCCULATION OF SOLS BY ELECTROLYTES

The flocculating power of electrolytes for a given sol varies considerably for different types of electrolytes. A measure of the relative effectiveness of different electrolytes in this respect is obtained from a *flocculation series test.* In this test the minimum concentration of an electrolyte is determined which causes the flocculation of a certain sol in a given time. This minimum concentration is called the *flocculation value,* or in modern terminology, the *critical coagulation concentration* (c.c.c.) for a given sol and a given electrolyte. In its most simple form the test is carried out as follows:

In a series of test tubes, increasing amounts of an electrolyte are added to a certain amount of the sol. The contents of the tubes are mixed by shaking, and the tubes allowed to stand for an arbitrarily chosen time. At the end of this period, the contents are visually inspected to determine which sols are still stable and which are flocculated. The photograph in Figure 8 shows the appearance of the sols at the end of the test and demonstrates the sharpness of the test. In this way, the flocculation value is enclosed between an upper and a lower limit. If desired, this range may be narrowed down in a second series in which the electrolyte concentration is varied in smaller steps between the previously established limits.

The conditions of the test, that is, the sol concentration and the time of standing, are arbitrarily chosen. Therefore, the results of the flocculation series test have relative significance only, allowing a comparison of the flocculation values for different electrolytes. Nevertheless, as will be discussed later, the study of the mechanism of flocculation has shown that the test is a sensible and fundamental test, despite its arbitrariness.

In the classical literature, written in the days when the test tube had not yet become obsolete, as it seems to be in this age of electronic laboratory equipment, numerous flocculation experiments have been reported. Around 1900, certain regularities were observed in the flocculation values for a variety of hydrophobic sols and different electrolytes. One such regularity is formulated in the well-known *Schulze-Hardy rule,* which states that the flocculation value is predominantly determined by the valence—rather than

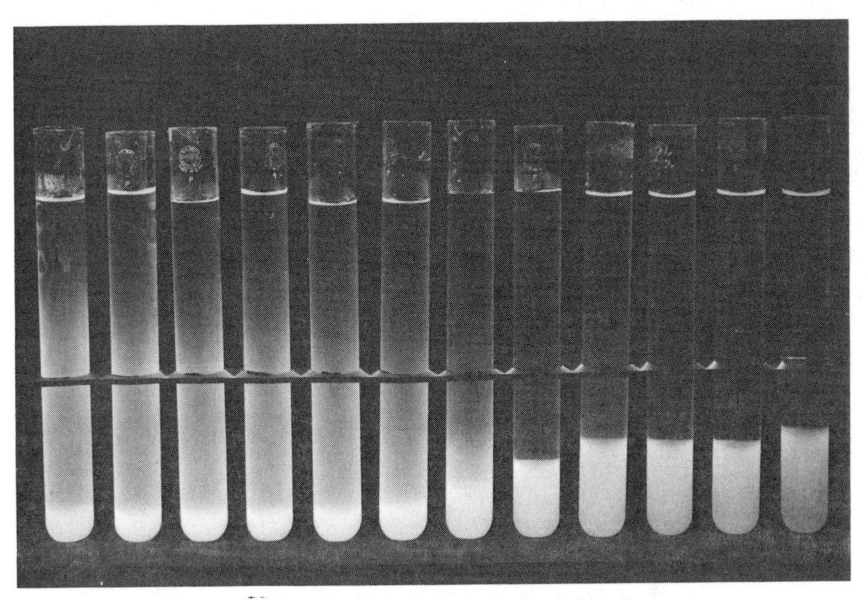

Figure 8. Flocculation series experiment.

the type—of those ions with a charge opposite to that of the particle. The effect of valence or type of the ions of the same charge is of secondary importance only. The higher the valence of the ions of opposite charge, the greater their flocculating power, and hence the lower their flocculating concentration. The following orders of magnitude of the flocculation values are observed: 25–150 mmol/dm^3 for monovalent ions, 0.5–2.0 mmol/dm^3 for divalent ions, and 0.01–0.1 mmol/dm^3 for trivalent ions.

In 1882, H. Schulze observed the effect of the valence of cations when studying the flocculation of the negative As_2S_3 sol. His findings were confirmed for other negative sols by Prost and by H. Picton and S. E. Linder. W. B. Hardy (2) was the first to discover the effect of the valence of anions in the flocculation of positive sols. In 1900, he formulated the rule which is now known as the Schulze-Hardy rule:

"The coagulative power of a salt is determined by the valency of one of its ions. This prepotent ion is either the negative or the positive ion, according to whether the colloidal particles move down or up the potential gradient. The coagulating ion is always of the opposite electrical sign to the particle."

The Schulze-Hardy rule is valid for a variety of sols and *indifferent* (or *inert*) electrolytes. These are electrolytes which do not engage in any sort of specific reaction with the sol. For example, they should not contain potential-determining ions for the sol or other ions which are specifically adsorbed on the sol particles, nor should they react chemically with the ions which constitute the electric double layer on the particle surface.

The flocculating power of cations for negative sols decreases slightly in the order Cs > Rb > NH_4 > K > Na > Li, and of anions for positive sols in the order F > Cl > Br > NO_3 > I > CNS. These sequences of the ion effects are frequently observed in physico-chemical phenomena. They are called *lyotropic series* or *Hofmeister series*.

In addition to flocculation by electrolytes, two other causes of flocculation of sols may be mentioned briefly. When two sols containing particles of opposite sign are mixed, so-called *mutual flocculation* occurs owing to the electrical attraction between the positive and negative sol particles. Furthermore, flocculation may be induced by adding watermiscible organic solvents to the sol, for example, acetone or alcohols.

V. REVERSAL OF PARTICLE CHARGE—IRREGULAR FLOCCULATION SERIES

So far, we have discussed how sols can be peptized or flocculated merely by making rather small adjustments in the composition of the liquid phase. It is also possible to reverse the sign of the charge of sol particles by appropriate additions to the fluid phase. Three different ways to achieve a reversal of charge may be distinguished:

1. As discussed before, in those sols for which certain anions and certain cations can act as potential-determining ions the sign of the particle charge depends on the relative availability of these ions. For example, a silver iodide sol is positively charged by Ag^+ ions and is negatively charged by I^- ions. In the presence of AgI, both silver and iodide ions will be present in the water phase owing to the slight solubility of silver iodide in water. The solubility of AgI can be expressed by the "solubility product," which is the product of the silver and iodide ion concentrations in solution. When the silver ion concentration is increased by the addition of a soluble silver salt, the iodide concentration decreases, keeping the solubility product constant. Alternatively, the silver ion concentration decreases upon the addition of a soluble iodide. Because of this interdependence of the concentrations of the two ions, it was stated that the sign of the particle charge depends on the *relative* availability of these ions. At a high ratio of silver to iodide ions the sol becomes positive, and at a high ratio of iodide to silver ions it becomes negative. At some intermediate ratio of the silver and iodide concentrations the sol is not charged and is therefore not stable. The sol is said to be at the *point of zero charge* (p.z.c.).

When a flocculation series experiment is run with a negative AgI sol and with $AgNO_3$ as the electrolyte, the following so-called *irregular series* is observed: The originally stable AgI sol becomes flocculated as soon as sufficient $AgNO_3$ is added for the ratio of Ag^+ and I^- ions to be such that the point of zero charge is obtained. Upon further addition of $AgNO_3$, a positive stable sol is created. At a still higher silver nitrate concentration, this salt acts as a flocculating electrolyte for the positive AgI sol, and flocculation occurs once more. In the series of test tubes, the sequence of peptized—flocculated—peptized—flocculated sol is clearly distinguishable. The same sequence is observed when increasing amounts of a soluble iodide are added to a positive AgI sol.

Another example of an irregular series is shown by an aluminum hydroxide sol when the pH of the sol is changed. Starting from a positive stable sol in slightly acid solution, the sol flocculates when the pH is increased by the addition of alkali. Upon further addition of alkali, a stable negative sol is created, and this sol finally flocculates with an excess of alkali. The reverse sequence occurs when acid is added to a stable negative sol in alkaline solution.

It has been found that the point of zero charge is usually not obtained at equal concentrations of the positive and negative peptizing ions but at a slight excess of one of the ions. For example, an aluminum hydroxide sol is not necessarily neutral at neutral pH. For this sol, the point of zero charge is known to be dependent on the crystal structure of the alumina particles. In silver iodide sols, the point of zero charge is not attained when the silver and iodide ion concentrations are both 10^{-8} mol/dm^3 (the solubility product is 10^{-16} mol^2/dm^6), but it is attained at a silver ion concentration of about 10^{-6} mol/dm^3 or at an iodide ion concentration of about 10^{-10} mol/dm^3. Apparently, the iodide ion, which is more polarizable, is more strongly held by the surface than the silver ion is.

2. A different type of charge reversal occurs when a negative sol is treated with a very dilute solution of a salt containing a trivalent cation such as Al^{3+} or a tetravalent cation such as Th^{4+}. When small increments of such salts are added to a negative sol, first flocculation occurs, but upon the further addition of the salt, stability of the sol is restored. In this condition, the sol particles have become positively charged. Finally, with an excess of the salt, the positive sol flocculates. Again, in a flocculation series experiment, the sequence peptized—flocculated—peptized—flocculated sol is observed.

The following explanation has been given for this type of reversal of charge: Owing to hydrolysis of the aluminum or thorium salts, polymeric hydroxides are formed in the solution. Since the hydrolyzed solution will be slightly acid, the polymeric hydroxides will be positively charged. These positive units are adsorbed by the negative particle, and mutual flocculation occurs. When an excess of the salt is added, an excess of positive hydroxide units is adsorbed by the sol particles, and a stable positive sol is obtained. Finally, this positive sol flocculates with a large excess of the aluminum salt solution.

3. A third method for the reversal of charge is to add ions to the sol which react chemically with the peptizing ions of the double layer. The added ions, which must be of opposite charge to the particle charge, first neutralize the peptizing charge, and the sol flocculates. If the neutral reaction products which remain adsorbed at the surface attract an additional amount of ions, either by adsorption or by complex ion formation, the surface becomes charged, and the sol regains stability with a reversed charge. Several examples of this mechanism of charge reversal will be discussed when we deal with clay suspensions, because this method is of great technological importance.

VI. COUNTER-ION EXCHANGE

The counter-ions of the double layer can be exchanged for other species of ions of the same sign if they are made available in the solution. In the presence of two species of ions, an equilibrium distribution is established which may be described by the relation between the ratio of the concentration of the two ion species in the bulk solution and this ratio in the counter-ion coating. It is customary to write this relation in the formalism of the mass law, for example,

$$\frac{[A]_i}{[B]_i} = \frac{K\,[A]_e}{[B]_e}$$

or

$$K = \frac{[A]_i[B]_e}{[B]_i[A]_e}$$

in which A and B are two species of monovalent ions of the same sign, and $[\ ]_i$ and $[\ ]_e$ denote their concentrations (or activities) in the counter-ion coating and in the equilibrium bulk solution, respectively. In this formulation the factor K would be called the "equilibrium constant." If there is no preference of the double layer for either ion species, the equilibrium constant would be unity. The deviation from unity would indicate a certain preference of the double layer for one of the ion species. Both situations have been observed in hydrophobic sols. In addition, in some sols the "equilibrium constant" appears to be not a constant at all but to vary with a varying ratio of the ions in the counter-ion coating.

When two ions of different valence are involved in the exchange reaction, there is, in general, a preference of the counter-ion coating for the ions of the higher valence.

VII. GELATION—A SPECIAL CASE OF FLOCCULATION

The addition of electrolytes to certain moderately concentrated sols results in gelation instead of flocculation. A *gel* (or *lyogel*) is a homogeneous-looking system, displaying some rigidity and elasticity. Gelation occurs in certain silica, alumina, and clay sols.

In the gel the particles are agglomerated to one single "floc" which extends throughout the available volume. When the particles are more or less spherical, they may be linked in "strings-of-beads" fashion and thus impart rigidity to the system. Platelike particles may associate to rigid "house-of-cards" structures, and rods or needles may build a scaffolding. The rigidity of the gel will depend on the number and the strength of the particle links in the continuous structure. From such a gel, the water may be removed by freezing of the gel and evaporation of the ice under vacuum ("freeze-drying"). During this process the volume of the system does not change, and the end product is a dry framework of the solid material which still retains some strength (*aerogel*).

The concentrations of different electrolytes which are required to cause gelation of the sols are again governed by the Schulze-Hardy rule, indicating that gelation is a special case of flocculation in which particle linking only proceeds in a different fashion.

These are some of the most important phenomena which occur in hydrophobic sols. It is noteworthy that quite a latitude for changing the degree of stability as well as the sign of the particle charge is offered simply by making small changes in the composition of the liquid phase. It will be demonstrated later how important this fact is in controlling the technological properties of dispersed systems.

The low level of concentration at which peptizers and flocculants are effective often makes their use economically attractive. However, on the other hand, high requirements of cleanliness must often be met in dealing with colloidal systems.

References

1. Zaiser, Ethel M., and LaMer, Victor K. (1948), The kinetics of the formation and growth of monodispersed sulfur hydrosols, *J. Colloid Sci.*, **3,** 571–598.
2. Hardy, W. B. (1900), A preliminary investigation of the conditions which determine the stability of irreversible hydrosols, *Proc. Roy. Soc.* (*London*), **66,** 110–125. *Z. Physik. Chem.* (1900), **33,** 385–400 (German Translation). In this paper references are made to Schulze, H. (1882), *J. Prakt. Chem.*, **25,** 431; Prost (1887), *Bull. de l'Acad. de Sci. de Belg.*, **14,** Ser. 3, 312; Picton, H., and Linder, S. E. (1895), *J. Chem. Soc.*, **67,** 63.

CHAPTER THREE

The Theory of the Stability of Hydrophobic Sols

In this chapter, the discussion of the repulsive and the attractive forces between sol particles is resumed, and the major results of the theory of the stability of hydrophobic sols is reviewed. This theory was developed by Verwey and Overbeek (1) and by Derjaguin and Landau (2), with important partial contributions by many others, for example, Langmuir (3). [See also Appendix V, reference B-5 (Volume I) for a full treatment of this theory.] Many empirical facts in colloid chemistry have been successfully explained by the modern theory of stability, in particular the general Schulze-Hardy rule for the flocculation of sols by electrolytes. These successes of the theory will be discussed in the next chapter.

In this chapter, the physical basis of the theory will be discussed; the reader who is interested in the mathematical handling of the problem is referred to a review of double-layer formulas in Appendix III.

Prior to a discussion of the repulsive forces between sol particles, which result from their electric charge, the configuration of the electric double layer will be analyzed.

I. CONFIGURATION OF THE ELECTRIC DOUBLE LAYER (4–6)

We have described the electric double layer as consisting of a surface charge and a compensating counter-ion charge, which is accumulated in the liquid in the neighborhood of the surface of the particles. The description "accumulation of counter-ions" requires elaboration.

The counter-ions are electrostatically attracted by the oppositely charged surface. At the same time, however, these ions have a tendency to diffuse away from the surface towards the bulk of the solution, where their concentration is lower. This situation is analogous to that in the earth's atmosphere, in which the gas molecules are subject to the competition between gravitation and diffusion. The action of the two competitive tendencies results in an equilibrium distribution of gas molecules in which their concentration gradually decreases with increasing distance from the earth's surface. Such a distribution is described as an "atmospheric" distribution. The same type of distribution is obtained in the double layer. The concentration of the counter-ions near the particle surface is high, and it decreases with increasing distance from the surface. This diffuse character of the counter-ion "atmosphere" was recognized by Gouy (4a, b) in 1910 and by Chapman (5) in 1913, who were the first to present a theoretical treatment of the counter-ion distribution. The counter-ion atmosphere is often referred to as the *diffuse* or *Gouy layer.*

More precisely formulated, the diffuse layer does not merely consist of an excess of ions of opposite sign; simultaneously, there is a deficiency of ions of the same sign in the neighborhood of the surface since these ions are electrostatically repelled by the particle. Therefore, one speaks of the adsorption of counter-ions and the *negative adsorption* of ions of the same sign or *co-ions.* In Figure 9, the distribution of positive and negative ions in the neighborhood of a negative surface is schematically represented. From electrostatic and diffusion theory (the Poisson-Boltzmann equation), the exact distribution of positive and negative ions as a function of the distance from the surface can be computed. Figure 10 shows the results of such a computation. The level indicated by *BD* is the concentration of both cations and anions at a large distance from the surface. The average local concentrations of the ions of opposite sign are given by curve *DA*, those of the ions of the same sign as the surface charge by curve *DC*. The total excess of counter-ions is represented by surface area *BAD*, the total defi-

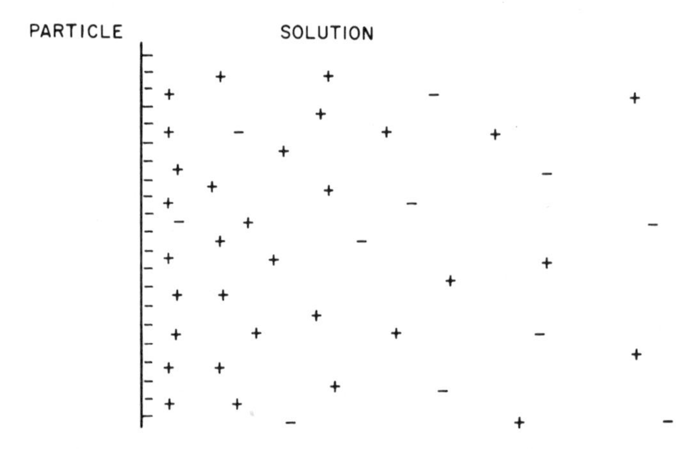

Figure 9. Diffuse electric double-layer model according to Gouy.

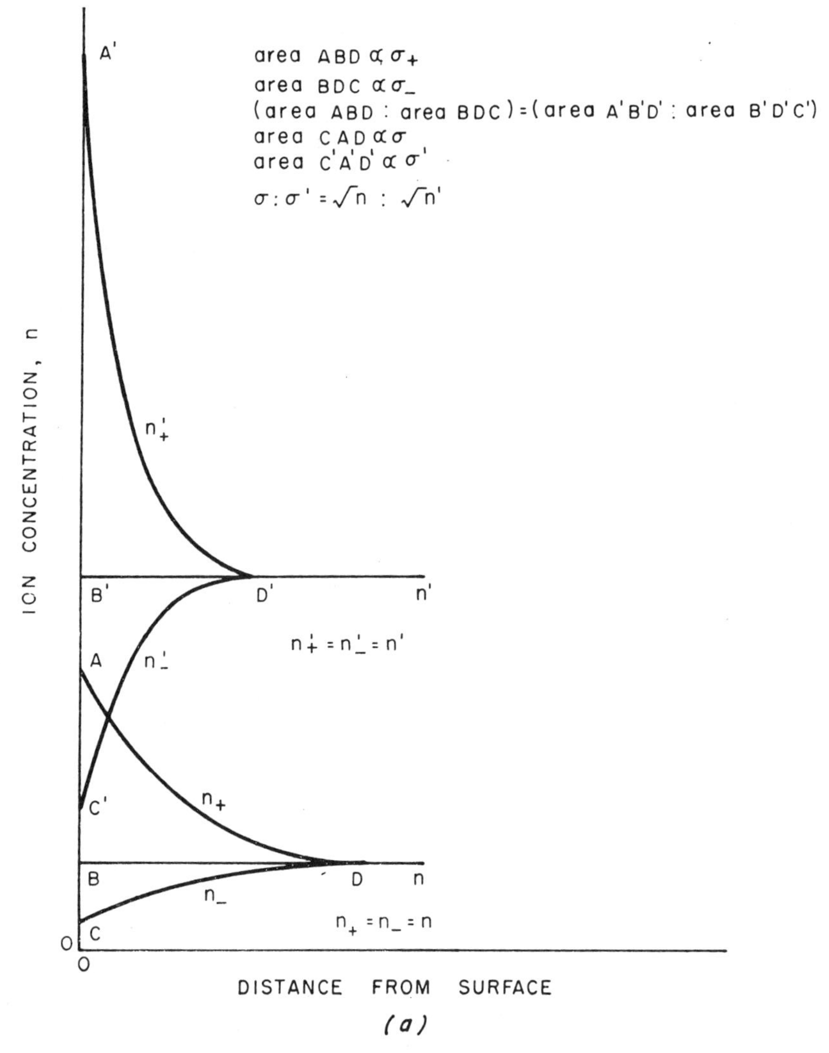

Figure 10. Charge distribution in the diffuse double layer of a negative particle surface at two electrolyte concentrations. (*a*) Constant surface potential. (*b*) Constant surface charge.

ciency of ions of the same sign by surface area *BCD*. The surface area *CAD* therefore represents the total net diffuse layer charge, which is equivalent to the surface charge.

In addition to the local concentrations of the ions, the average electric potential at any point with respect to that at a point far removed from the surface can be computed. In Figure 11, the electric potential is plotted as a

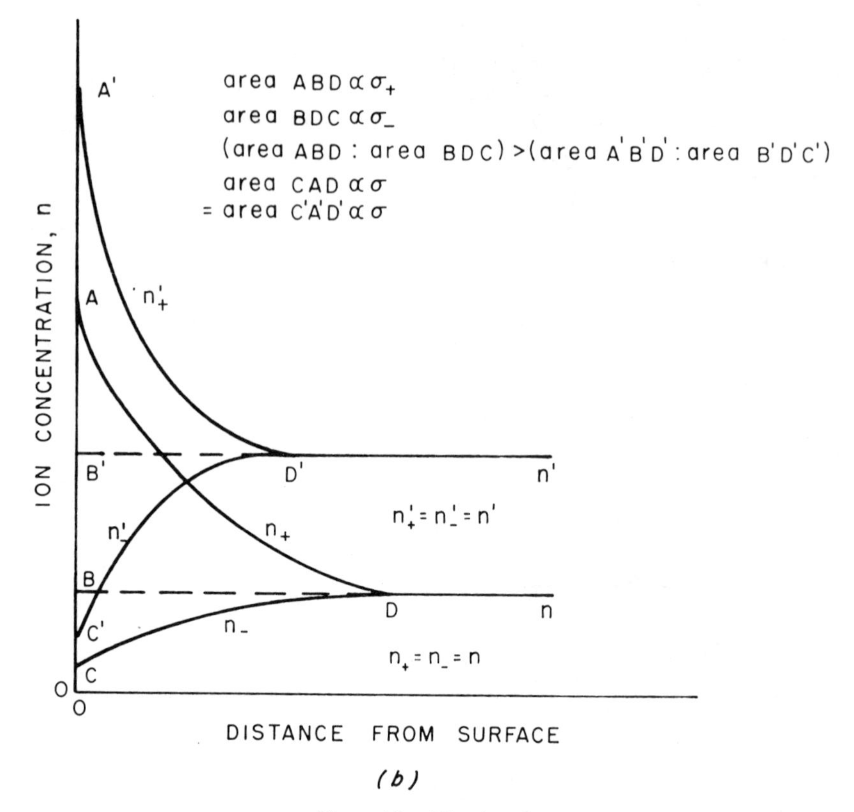

Figure 10. (*Continued*)

function of the distance from the surface. The potential has a maximum value at the surface, and it decreases roughly exponentially with the distance from the surface.

If the double layer is created by the adsorption of potential-determining ions, the electric potential at the particle surface is solely determined by the concentration (or activity) of these ions in solution, since the particle acts as a reversible electrode towards these ions. In this case the potential is given by the Nernst equation:

$$\Phi_0 = \frac{kT}{ve} \ln \frac{c}{c_0}$$

in which Φ_0 is the electric potential at the surface, k is the Boltzmann constant, T is the absolute temperature, e is the electronic charge, v is the valence of the potential-determining ion, c is the concentration of these ions in solution, and c_0 is their concentration at the point of zero charge when $\Phi_0 = 0$.

For example, in a silver iodide sol, the point of zero charge is at a silver ion concentration of about $c_0 = 10^{-6}$ mol/dm^3. A tenfold increase of the silver ion concentration to 10^{-5} mol/dm^3 results in the creation of a surface potential of about +57 mV at room temperature. For every tenfold increase of the silver ion concentration, the potential increases by 57 mV.

If the double layer results from interior crystal imperfections, the charge per unit surface area (the charge density) is a fixed quantity since it is determined by the imperfections in the interior of the crystal.

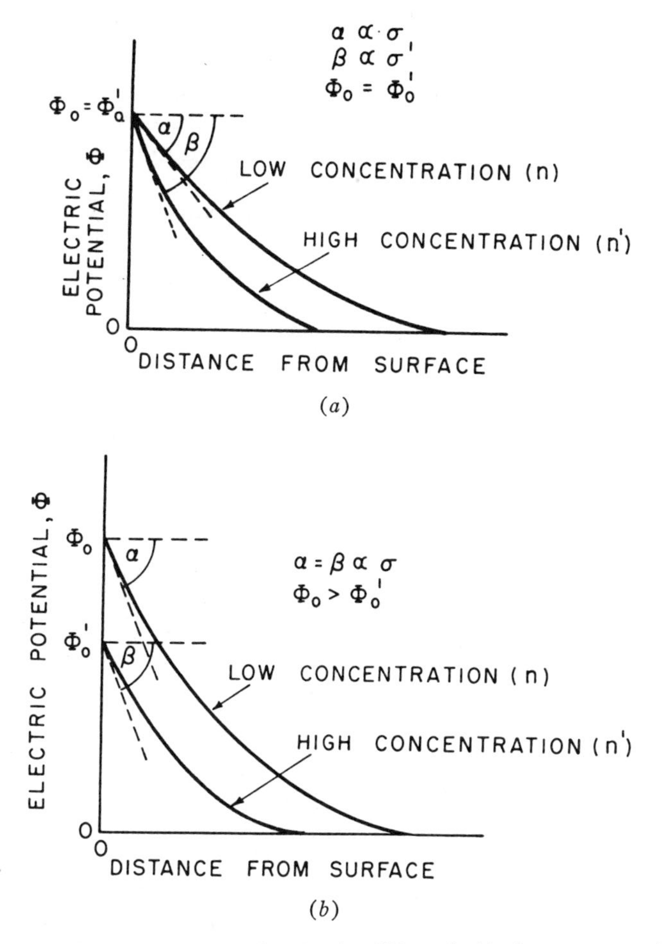

Figure 11. Electric-potential distribution in the diffuse double layer at two electrolyte concentrations. (*a*) Constant surface potential. (*b*) Constant surface charge.

II. EFFECT OF ELECTROLYTES ON THE CONFIGURATION OF THE ELECTRIC DOUBLE LAYER

The key question for the stability problem is, "What happens to the electric double layer when a flocculating electrolyte is added to the stable sol?"

The computation of the ion distribution in the diffuse double layer as a function of the electrolyte content of the bulk solution shows that *the diffuse counter-ion atmosphere is compressed* toward the surface when the bulk electrolyte concentration is increased. The effect of electrolyte addition on the surface potential and the surface charge of the particle depends on the type of double layer.

1. If the surface potential of the particles is determined by the concentration of potential-determining ions, the magnitude of this potential is not affected by the addition of an indifferent electrolyte so long as the concentration of the potential-determining ions, or rather their activity, is not affected by the presence of the electrolyte. For this type of double layer, it follows from the computations that the surface charge of the particles increases with increasing indifferent electrolyte concentration, and the surface potential remains constant.

The two sets of curves in Figures 10*a* and 11*a* illustrate these effects of electrolyte addition on the structure of the double layer of a constant surface potential. The diffuse charge is concentrated in a region closer to the surface, but the total net diffuse charge, given by surface area $C'A'D'$, which is equivalent to the new surface charge, is larger than the original charge (represented by the area CAD in Figure 10). Figure 11 illustrates that the surface potential remains the same as before, but owing to the compression of the double layer by electrolytes, the potential decays more rapidly with increasing distance from the surface.

2. If the surface charge of the particle is determined by interior crystal imperfections, the surface charge does not change with increasing electrolyte concentration. The diffuse double layer is still compressed, however, and in this case, the surface potential decreases with increasing electrolyte concentration. See Figures 10*b* and 11*b*.

A significant result of the computations of the effect of electrolytes on the spatial distribution of the electric potential and the counter-ion charge is that (at the high surface potentials prevailing in many practical systems) the degree of compression of the double layer is governed by the concentration and valence of the ions of opposite sign from that of the surface charge, whereas the effect of ions of the same sign is comparatively small. The higher the concentration and the higher the valence of the ions of opposite

Table 2

Approximate "Thickness" of the Electric Double Layer as a Function of Electrolyte Concentration at a Constant Surface Potential

Concentration of ions of opposite charge to that of the particle, mmol/dm³	"Thickness" of the double layer, Å	
	Monovalent ions	Divalent ions
0.01	1000	500
1.0	100	50
100.	10	5

sign, the more the double layer is compressed. This result will be important in the explanation of the Schulze-Hardy rule.

Some quantitative information about the computed extension of the diffuse layer in the solution is given in Table 2. A convenient way to express the changes in thickness of the diffuse layer, which is theoretically infinite, is to imagine the diffuse charge concentrated as a planar charge at such a distance from the surface that the capacity of the "condenser" thus formed is equal to the capacity of the completely diffuse double layer.

In dilute electrolyte solutions, the extension of the double layer thus defined is of the order of magnitude of the particle dimensions. In concentrated electrolyte solutions, the thickness of the double layer is considerably reduced.

In the next section we shall consider what happens when two particles approach each other.

III. THE BALANCE OF REPULSIVE AND ATTRACTIVE FORCES ON PARTICLE APPROACH

A. The Electric Double-Layer Repulsion (1–3)

When two particles approach each other in suspension owing to their Brownian motion, their diffuse counter-ion atmospheres begin to interfere. It can be shown that this interference leads to changes in the distribution of the ions in the double layer of both particles, which involves an increase in the free energy of the system. Work must therefore be performed to bring about those changes; in other words, there will be a repulsion between the particles. The amount of work required to bring the particles from infinite separation to a given distance between them can be calculated. This amount of work is the repulsive energy or the repulsive potential at the given distance. When the repulsive potential V_R is plotted as a function of distance, a so-called "potential curve" is obtained. The repulsive potential decreases roughly exponentially with increasing particle separation. In Figure 12, three potential curves are shown which are valid for the same particles but for different electrolyte concentrations indicated by "low," "intermediate," and "high." Due to the compression of the double layer at increasing electrolyte concentrations, the range of the repulsion is considerably reduced.

B. The van der Waals Attraction (7–32)

The phenomenon of flocculation demonstrates the existence of interparticle attractive forces. Obviously, in order to compete successfully with the double-layer repulsion which is still operative under flocculating conditions, the attraction must be of comparable range and magnitude. The general van der Waals attractive forces appear to satisfy this requirement. At first sight, it does not seem likely that the van der Waals forces would be large enough or far-reaching enough, since they are small and they decay rapidly with distance for a pair of atoms. However, the van der Waals attraction between atom pairs is additive; hence the total attraction between particles containing a very large number of atoms is equal to the sum of all the attractive forces between every atom of one particle and every atom of the other particle. The summation leads not only to a larger total force but also to a less rapid decay with increasing distance. For two atoms, the van der Waals attractive force is inversely proportional to the seventh power of the distance (or to the sixth power for the attractive energy), but for two large particles, the force is inversely proportional to the third power of the distance between the surfaces, and the attractive energy therefore to the second power of that distance, approximately.

In the lower part of Figure 12, the attractive energy V_A is plotted as a function of the distance. As pointed out before, the attraction remains practically the same when the electrolyte concentration of the medium is varied.

This approach to the evaluation of the van der Waals attraction between colloidal particles was proposed by Hamaker (8). The constant in the equation for the attractive energy as a function of particle distance is called the *Hamaker constant.* According to the classical treatment, van der Waals forces are attributed to charge fluctuations in atoms, resulting in an attraction between mutually induced dipoles in interacting atoms. The Hamaker constant can be evaluated from individual atomic polarizabilities and atomic densities in both the interacting particles and in the medium. Such

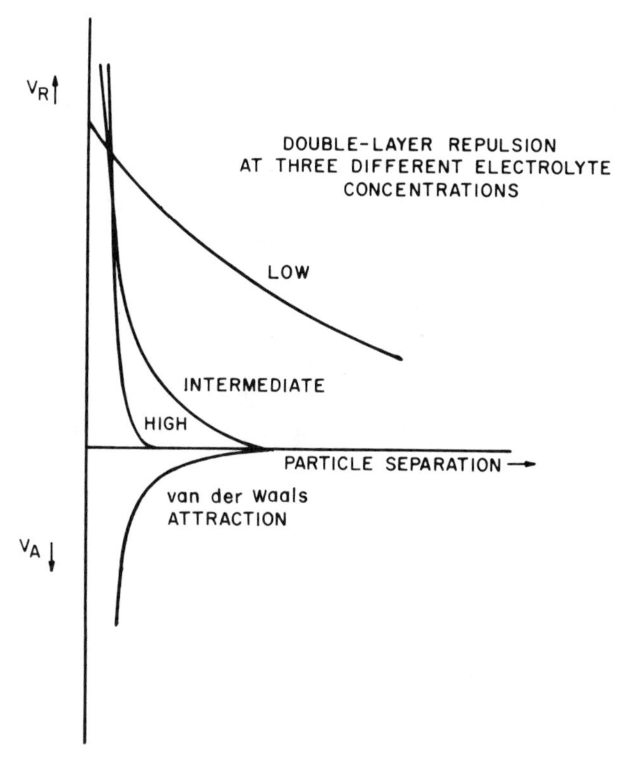

Figure 12. Repulsive and attractive energy as a function of particle separation at three electrolyte concentrations.

evaluations are in reasonable agreement with the results of more rigorous treatments, such as given by Lifshitz (24) and others.

Because of the importance in the stability problem of the operating range of the van der Waals forces between two particles consisting of many atoms, several attempts have been made in recent years to measure the van der Waals attraction between macroscopic objects at close separations. These experiments, requiring objects which have highly polished surfaces and are free of electrostatic charge, are extremely difficult to perform. Measurements have been made at distances covering only the larger distances of the range of interest in colloidal systems. Nevertheless, the measured forces in this range have given credence to the concept that the van der Waals forces between colloidal particles are of the right magnitude and range to compete with the double-layer repulsive forces. The reader who is interested in the discussion of these experiments, as well as in the theoretical treatment of van der Waals forces is referred to the literature references 7–32 at the end of this chapter. See Appendix V, reference B-12, for a recent review that appeared in a chapter by V. A. Parsegian, pp. 27–72 ("Long-range van der Waals interaction").

IV. THE SUMMATION OF REPULSION AND ATTRACTION

The summation of repulsive and attractive energy is carried out as follows: The net potential curve of particle interaction is constructed simply by adding the attractive and the repulsive potential at each particle distance, considering the attractive potential negative and the repulsive potential positive. Figure 13 shows the results of these additions for three electrolyte concentrations indicated as low (*a*), intermediate (*b*), and high (*c*).

In constructing these three net curves of interaction, an additional interaction force between the particles was taken into account. This interaction, not previously considered, is a repulsion of a very short range, to which there are two possible contributions. One, the Born repulsion,

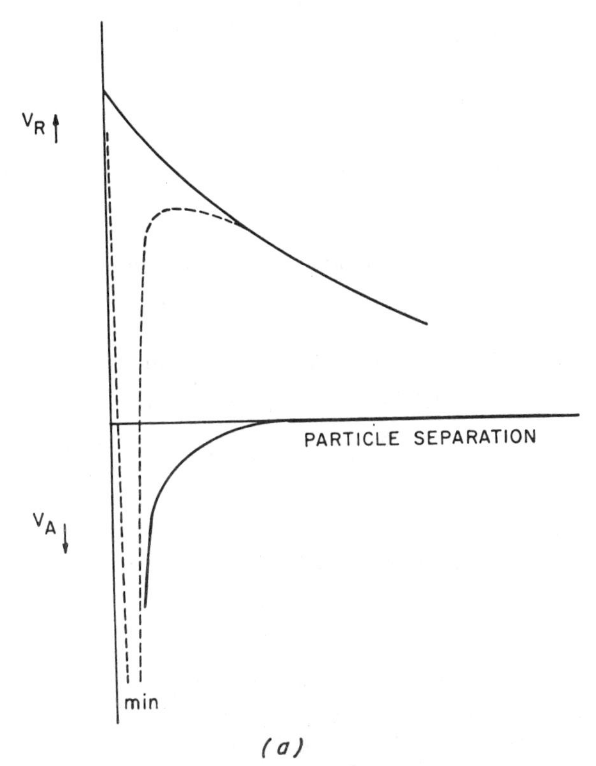

Figure 13. Net interaction energy as a function of particle separation. (*a*) Low electrolyte concentration. (*b*) Intermediate electrolyte concentration. (*c*) High electrolyte concentration.

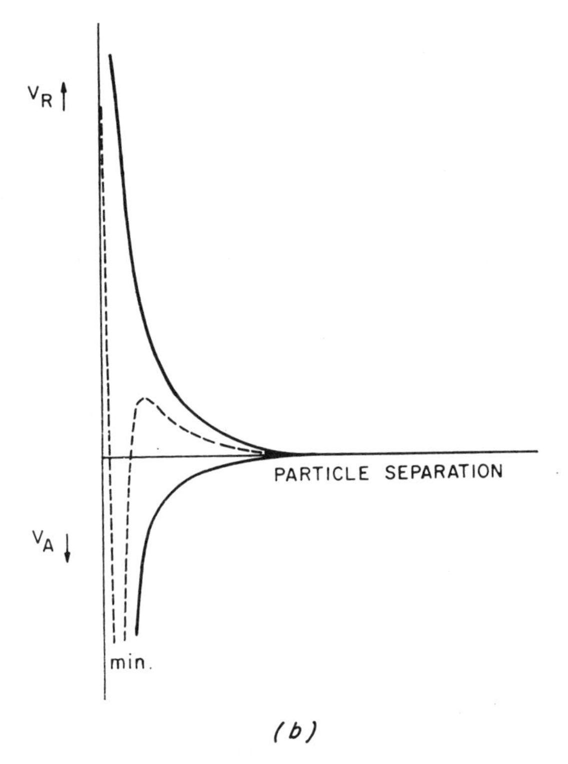

(b)

Figure 13. (*Continued*)

becomes effective as soon as extruding lattice points or regions come into contact; it resists the interpenetration of the crystal lattices. A second short-range repulsion is the result of specific adsorption forces between the crystal surface and the molecules of the liquid medium, that is, water. Owing to such adsorption forces, usually one or two monomolecular layers of water are held rather tightly by the particle surface. For the distance between the two particles to become less than the thickness of the adsorbed water layers on both particles, the adsorbed water must be desorbed. The work required for this desorption manifests itself as a short-range repulsion between the particles, which probably becomes appreciable at particle separations of the order of 10 A or less.

The increasing importance of the short-range repulsion for the net interaction is represented by the steep rise of the potential curves at very small values for the particle separation.

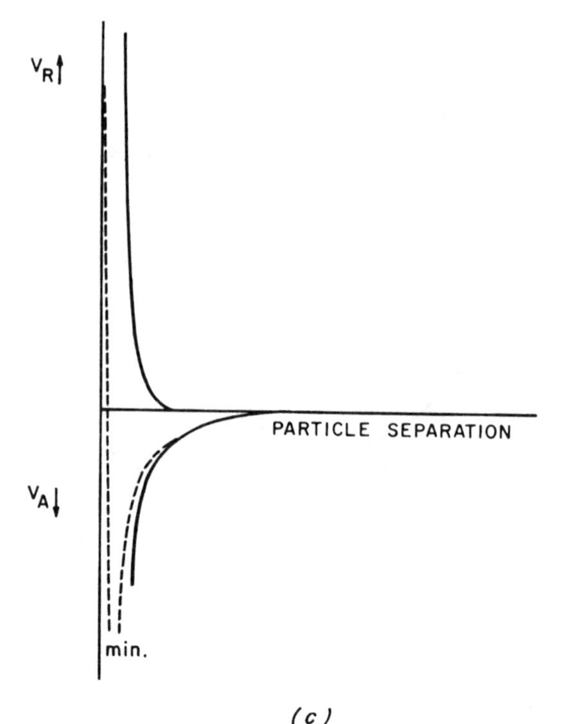

Figure 13. (*Continued*)

V. THE NET INTERACTION CURVE AND SOL STABILITY

The net interaction curves for a low and a medium salt concentration show a minimum with predominant attraction at close approach and a maximum with predominant repulsion at greater distances. No such maximum is displayed by the curve for a high electrolyte concentration.

The interpretation of stability and flocculation in terms of the shapes of the net potential curves of particle interaction is based on the following considerations:

At a high electrolyte concentration, the potential curve shows a deep minimum at close separation only, and attraction predominates at any particle distance except at very close approach. If two particles approach each other owing to their Brownian motion, they will agglomerate when they reach the position at which the deep attraction minimum occurs. The rate at which the agglomeration process occurs can be computed from the theory of diffusion (33–36).

In the classical kinetic treatment of this problem by von Smoluchowski (33), a fixed central particle is considered, and the number of collisions per second between the surrounding particles and the central particle is found from their average kinetic energy, the frictional forces of the medium on the particles, the number of particles per unit volume, and their collision radius. Hydrodynamic effects are considered in reference 36.

If repulsion dominates part of the way during the mutual approach of the particles (curves for medium and low electrolyte concentration), the diffusion will be counteracted by the extraneous repulsive force field, and the rate of agglomeration will decrease, as dealt with in the theory of Fuchs (34).

At a high electrolyte concentration at which the potential curve shows no repulsion at any distance (Figure 13*c*), particle agglomeration occurs at a maximum rate and the process is called *rapid coagulation.* It is determined almost solely by diffusion, although the still-prevailing attraction at great distances slightly accelerates the process. At intermediate electrolyte concentrations (Figure 13*b*), the coagulation process is retarded by the long-range repulsion. Under these conditions, *slow coagulation* takes place. At very low electrolyte concentrations (Figure 13*a*), the coagulation process is retarded to such an extent by the appreciable long-range repulsion that it may take weeks or months for coagulation to become perceptible. For all practical purposes, the sol is called "stable" under these conditions.

According to this analysis, the stability of a hydrophobic sol is not an absolute stability. The difference between a stable and an unstable sol is only gradual and is a matter of smaller or greater coagulation velocities. Stability is customarily expressed in terms of the *stability ratio, W,* which is the factor by which the coagulation velocity is decreased by the particle repulsion. The factor W can be evaluated from the total potential curve of interaction and the particle size. In common terminology, the sections of the potential curves above the distance coordinate which represent the long-range repulsion between the particles are called *energy barriers.* The particles which have passed the barrier in the collision process are said to have "jumped over the barrier." Particle association in the presence of an energy barrier is said to require "activation energy." When the position of the deep attraction minimum is reached, the rate of particle separation will be small since the particles would have to jump over a much higher energy barrier, given by the difference in the height of the maximum and the depth of the minimum. The particles are therefore said to be "trapped" in the minimum.

Figure 13 tells the story of stability and flocculation in a nutshell. From the way the net potential curves have been constructed, it is clear that the size of the energy barrier for a given sol is primarily governed by the degree of compression of the double layer by the added electrolytes. Since this compression effect is dominated by the concentration and valence of the ions of opposite sign, the theory is indeed able to explain the Schulze-Hardy rule as will be discussed in Chapter Four, Section I.

To ensure a reasonable stability of a sol, the height of the energy barrier should be of the order of $15kT$ (k = the Boltzmann constant; T = the absolute temperature) for a dilute sol. For a more concentrated sol, in which the rate of agglomeration in the absence of an energy barrier is already higher than that for dilute sols, the energy barrier should be as high as about $25kT$ to produce a reasonably stable sol. (Although the complete shape of the barrier should be taken into account, the maximum height appears to be the most important parameter.)

Occasionally, one encounters the general statement that a sol is flocculated by an electrolyte because of a neutralization of the charge. Although this mechanism would apply to those systems in which a chemical reaction between the stabilizing ions of the particle and the added ions can take place, in general the flocculation by inert electrolytes is governed by the compression of the double layers.

References

DOUBLE-LAYER AND STABILITY THEORY

1. Verwey, E. J. W., and Overbeek, J. Th. G. (1948), *Theory of the Stability of Lyophobic Colloids*, Elsevier, New York, Amsterdam, London, Brussels.
2. Derjaguin, B., and Landau, L. D. (1941), *Acta physicochim. U.R.S.S.*, **14,** 635; *J. Exp. Theor. Phys.* (*U.S.S.R.*), **11,** 802 (reprinted **15,** 662, 1945).
3. Langmuir, I. (1938), The role of attractive and repulsive forces in the formation of tactoids, thixotropic gels, protein crystals, and coacervates, *J. Chem. Phys.*, **6,** 873–896.

4a. Gouy, G. (1910), Sur la constitution de la charge électrique à la surface d'un électrolyte, *Ann. Phys.* (*Paris*), Série 4, **9,** 457–468.

4b. Gouy, G. (1917), Sur la fonction électrocapillaire, *Ann. Phys.* (*Paris*), Série 9, **7,** 129–184.

5. Chapman, D. L. (1913), A contribution to the theory of electrocapillarity, *Phil. Mag.*, **25,** No. 6, 475–481.
6. Sparnaay, M. J. (1972), *The Electrical Double Layer*, Pergamon, Oxford.

VAN DER WAALS FORCES IN COLLOIDAL SYSTEMS
AND BETWEEN MACROSCOPIC OBJECTS

7. de Boer, J. H. (1936), The influence of van der Waals forces and primary bonds on binding energy, strength and orientation with special reference to some artificial resins, *Trans. Faraday Soc.*, **32,** 10–38.
8. Hamaker, H. C. (1937), The London–van der Waals attraction between spherical particles, *Physica*, **4,** 1058–1072.

9. Casimir, H. B. G., and Polder, D. (1948), The influence of retardation on the London–van der Waals forces, *Phys. Rev.*, **73,** 360–372.
10. Overbeek, J. Th. G., and Sparnaay, M. J. (1952), Long-range attractive forces between macroscopic objects, *J. Colloid Sci.*, **7,** 343–345.
11. Overbeek, J. Th. G., and Sparnaay, M. J. (1954), London–van der Waals attraction between macroscopic objects, *Discussions Faraday Soc.*, **18,** (Coagulation and Flocculation), 12–24.
12. Sparnaay, M. J. (1958), Measurements of attractive forces between flat plates, *Physica,* **24,** 751–764.
13. Sparnaay, M. J. (1959), Additivity of London–van der Waals forces. An extension of London's oscillator model, *Physica,* **25,** 217–231.
14. Sparnaay, M. J. (1959), Van der Waals forces and fluctuation phenomena, *Physica,* **25,** 444–454.
15. Black, W., de Jongh, J. G. V., Overbeek, J. Th. G., and Sparnaay, M. J. (1960), Measurements of retarded van der Waals forces, *Trans. Faraday Soc.*, **56,** 1597–1608.
16. Derjaguin, B., and Abricossova, I. I. (1951), *J. Exp. Theor. Phys.* (*U.S.S.R.*), **21,** 945.
17. Derjaguin, B., and Abricossova, I. I. (1953), *Compt. rend. acad. sci. U.R.S.S.*, **90,** 1055.
18. Derjaguin, B., Titijevskaia, A. S., Abricossova, I. I., and Malkina, A. D. (1954), Investigations of the forces of interaction of surfaces in different media and their application to the problem of colloidal stability, *Discussions Faraday Soc.*, **18** (Coagulation and Flocculation), 24–41.
19. Howe, P. G., Benton, D. P., and Puddington, I. E. (1955), London–van der Waals attractive forces between glass surfaces, *Can. J. Chem.*, **33,** 1375–1383.
20. Kitchener, J. A., and Prosser, A. P. (1956), Direct measurement of the long-range van der Waals forces, *Nature,* **178,** 1339–1340.
21. Kitchener, J. A., and Prosser, A. P. (1956), Direct measurement of the long-range van der Waals forces, *Proc. Roy. Soc.* (*London*), Ser. A, **242,** 403–409.
22. Vold, M. J. (1954), Van der Waals attraction between anisometric particles, *J. Colloid Sci.*, **9,** 451–459.
23. Vold, M. J. (1961), The effect of adsorption on the van der Waals interaction of spherical colloidal particles, *J. Colloid Sci.*, **16,** 1–12.
24. Landau, L., and Lifshitz, E. M. (1960), *Electrodynamics of Continuous Media,* Addison-Wesley, Reading, Mass.
25. Chu, B. (1967), *Theory of Intermolecular Forces,* Interscience, New York.
26. van Silfhout, A. (1966), Dispersion forces between macroscopic objects, *Proc. Kon. Akad. Wetensch.*, **B69,** 501–541.
27. Tabor, D., and Winterton, R. H. S. (1969), The direct measurement of normal and retarded van der Waals forces, *Proc. Roy. Soc.*, **A312,** 435–450.
28. Israelachvili, J. N., and Tabor, D. (1972), The measurement of van der Waals dispersion forces in the range 1.5 to 130 nm, *Proc. Roy. Soc.*, **A331,** 19–38.
29. Israelachvili, J. N. (1972), The calculation of van der Waals dispersion forces between macroscopic bodies, *Proc. Roy. Soc.*, **A331,** 39–55.
30. Ninham, B. W., and Parsegian, V. A. (1970), Van der Waals interactions in multilayer systems, *J. Chem. Phys.*, **52,** 3398–3402.
31. Richmond, P., and Ninham, B. W. (1972), Calculation of van der Waals forces across mica plates using Lifshitz theory, *J. Colloid Interface Sci.*, **40,** 406–408.
32. Rouweler, G. C. J., and Overbeek, J. Th. G. (1971), Dispersion forces between fused silica objects at distances between 25 and 350 nm, *Trans. Faraday Soc.*, **67,** 2117–2121.

See also Appendix IV.

COAGULATION KINETICS

33. von Smoluchowski, M. (1916), *Physik. Z.*, **17**, 557, 585; *Z. Physik. Chem.*, **92**, 129 (1917).
34. Fuchs, N. (1934), *Z. Physik*, **89**, 736.
35. Hansen, R. S. (1975), Coagulation kinetics and bimolecular reaction kinetics, Appendix V, reference B-12, pp. 201–218.
36. Honig, E. P., Roebersen, G. J., and Wiersema, P. H. (1971), Effect of hydrodynamic interaction on the coagulation rate of hydrophobic colloids, *J. Colloid Interface Sci.*, **36**, 97–109.

CHAPTER FOUR

Successes of the Theory of Stability—Further Theories and Refinements

I. STABILITY, FLOCCULATION, AND THE SCHULZE-HARDY RULE

Since "stable" and "unstable" sols differ only in their rate of coagulation, the degree of stability or instability of a sol may be defined and measured in terms of coagulation rates. Therefore, a test of stability should be a dynamic test.

Experimentally, the coagulation process may be followed by ultramicroscopic counting of the number of particles at different times. From such observations, as well as from the kinetic theory of flocculation, it appears that the coagulation rate decreases during the process, since the number of kinetic units per unit volume and hence the chance of collision gradually decrease. It is customary to describe the coagulation process in terms of the *flocculation time,* the time in which the number of particles is halved.

In routine practice, stability and flocculation of a sol are studied by means of the simple flocculation series test described previously. In essence, this test is a dynamic test, although it is only a single-point test. In the flocculation series experiment, one measures the electrolyte concentration at which the sol flocculates almost completely in a specified time. In other words, one determines the electrolyte concentration at which the sol coagulates at a specified average rate or at which the sol reaches a certain degree of instability. Therefore, the flocculation series test is a more fundamental test than its arbitrariness might suggest.

The flocculation value, that is, the amount of electrolyte required to reduce the stability of a sol to a certain level, is a useful relative measure of the original stability of the sol. The higher the original stability, the greater the amount of electrolyte required for reduction of the stability to the given level which is fixed by the conditions of the test. For example, the effectiveness of a peptizer in improving the stability of a sol can be measured by the increase of the flocculation value, that is, the increased tolerance of the sol for electrolyte contamination.

Keeping the fundamental meaning of the flocculation test in mind, one can derive the Schulze-Hardy rule of flocculation from the stability theory (1). One of the most important supports for the double-layer theory of stability is the quantitative agreement between the theoretically predicted ratios of flocculation values for different kinds of electrolyte and the empirical Schulze-Hardy rule, although in many cases the agreement is only one of order of magnitude. There are two reasons for the lack of an exact agreement: On one hand, it is necessary to introduce several approximations in the handling of the theory, and on the other hand, the condition of the complete absence of specific interactions between the added electrolyte and the components of the double layer is often not fulfilled.

The prediction of flocculation behavior in accordance with the Schulze-Hardy rule is based on a calculation of the stability ratio W for a given sol with a variety of electrolyte types as a function of electrolyte concentration. For different electrolytes the respective flocculation values are then found by comparing the electrolyte concentrations that give identical values of W, that is, the same size* of the energy barrier.

The theory of stability can also explain why the flocculation test is such a sensitive test. Under the usual conditions of a flocculation series experiment, the coagulation velocity varies strongly with varying electrolyte concentration. Hence the flocculation value can be enclosed between very narrow limits of electrolyte concentration, despite the rather primitive character of the experiment.

II. LIMITS OF PARTICLE SIZE

It has been observed that small particles coagulate more rapidly than large particles; hence it is difficult to prepare stable sols of extremely small particles such as amicronic sols.

If small and large particles with identical double layers are compared, the repulsive-potential curves for the interaction of unit surface areas of the particles will be identical; however, the total particle repulsion will be smaller for the small particles. Evaluation of the diffusion of the large and small particles in the different force fields and application of the stability theory show that the low absolute value of the repulsive potential for small

* As mentioned, "equal size" does not necessarily mean equal height of the barriers; the complete shape is important (2).

particles is the major cause of the low stability of extremely finely dispersed sols. According to these considerations, the limit of particle size is of the order of 10^{-7} cm.

III. FLOCCULATION BY WATER-MISCIBLE ORGANIC SOLVENTS

The Gouy theory shows that the "thickness" of the double layer decreases when the dielectric constant of the medium is reduced. Such a reduction may be achieved by the addition of water-miscible solvents, such as alcohols or acetone, to the sol. Because of the compression of the double layer, the range of particle repulsion is reduced and the size of the energy barrier becomes smaller when the dielectric constant of the medium decreases. Therefore, the theory explains the well-known fact that most polar solvents enhance the flocculating power of an electrolyte markedly. If a sol in water is still stable at a certain level of contamination by an electrolyte, the addition of an organic solvent may induce flocculation.

Exceptions to these rules do occur, and they must be explained by considering secondary effects of the organic solvent on the double-layer structure, as will be discussed in Section VII.

IV. DIRECT EVIDENCE OF LONG-RANGE PARTICLE INTERACTION: SCHILLER LAYERS AND TACTOIDS

A. Schiller Layers

When certain stable sols containing plate- or rodlike particles are allowed to settle quietly for a long time, the sol becomes gradually more concentrated toward the bottom of the vessel. In the lower part of the sol, interference colors are observed. This phenomenon is shown by certain clay sols. The interference colors are attributed to the reflection of light at large numbers of horizontal parallel layers of oriented plate- or rodlike particles at distances which are of the order of the wavelength of light, that is, several thousand angstrom units. The particle distance can be computed from the interference colors. The parallel layers are called *Schiller layers* (from German, *Schiller,* meaning iridescence).

The parallel orientation of the plates or rods at large distances can be explained by the counteraction of the gravity forces by double-layer repulsion forces (3). This explanation is supported by the observation that the particle distance decreases with increasing electrolyte concentration in a manner which is predicted from the computed compression effect of electrolytes on the diffuse double layer. Therefore, the phenomenon of Schiller layers supplies direct evidence for the long-range character of the electric double-layer repulsion.

As mentioned previously, usually only gravitation and diffusion are considered in the settling of dispersed particles. For large particles, the gravity forces are predominant, and the suspension settles to a bottom sediment and a clear, particle-free supernatant liquid. For smaller particles, gravitation and diffusion may be of comparable magnitude, and an atmospheric-type distribution of particles is created. When the particles are extremely small, gravitation may be negligible with respect to diffusion, and the particles remain homogeneously dispersed like molecules in a true solution. A third force, particle repulsion, becomes important only when it is of comparable magnitude with gravitation and diffusion, as in the case of thin, light plates or rods with large interacting surfaces. Then, the phenomenon of Schiller layers may be observed.

B. Tactoid Formation

Another phenomenon of particle orientation is the formation of *tactoids.* These are spindle-shaped regions in a sol in which the particle concentration is higher than in the bulk of the sol. Occasionally, they separate as a bottom sediment. Tactoids are spontaneously formed in some sols of plate- or rod-like particles. In the tactoids, the particles are oriented parallel to each other at distances of the order of 100 A (4, 5). Tactoid formation is a special case of *coacervation,* that is, the separation of a sol in two liquid phases of different concentration.

The particle orientation in tactoids at the observed distances can be explained by considering double-layer repulsion and van der Waals attraction as the two opposing forces. According to the calculation of these forces, the double-layer repulsion decays more rapidly with distance than the van der Waals forces. Hence beyond the particle separations which have been considered previously in the discussion of the stability theory, there will be a distance where the van der Waals attraction once more becomes predominant, and a long-range minimum will occur in the interaction curve. Under favorable conditions, this *secondary minimum* (which was omitted in Figure 13) may become deep enough to "trap" the particles in this position, which is of the right order of separation to explain the tactoid formation. However, since some uncertainty still exists concerning the magnitude of the van der Waals forces at large separations, no definite proof has been given that this explanation is the right one. An alternative explanation has been proposed by Onsager (6) on the basis of entropy considerations.

The occurrence of equilibrium distances (7) of the order of 100 A between the plates in swollen clay flakes will be discussed in Chapter Ten (Section I-B).

V. COUNTER-ION EXCHANGE

A subject which is related to the double-layer theory itself rather than to the stability theory is the analysis of counter-ion exchange. The Gouy theory predicts that the ratio of the concentrations of two ion species of the same valence in the counter-ion atmosphere is the same as that in the bulk equilibrium solution—in other words, that the "equilibrium constant" is unity. Furthermore, the theory predicts that for two ions of different valence, the ion with the higher valence is predominantly accumulated near the oppositely charged surface. Therefore, the fraction of ions of highest valence is greater in the counter-ion atmosphere than in the bulk solution [compare Chapter Two (Section VI)].

These predictions have been quantitatively confirmed by experiments with sols in which specific interaction effects are absent, for example, for Na-H exchange and Ba-Na exchange in silver iodide sols (8).

Deviations from the predicted ion distribution must be attributed to specific interactions between the particles and the counter-ions or to specific ion effects which are related to ion size. In Appendix III, such effects are discussed in more detail.

VI. STERN'S MODEL OF THE DOUBLE LAYER AND OTHER REFINEMENTS

The Gouy model of the electric double layer, upon which the previous computations have been based, contains some unrealistic elements. For example, the ions are treated as point charges, and any specific effects related to ion size are neglected. Specific interactions between the surface and the counter-ions and the medium are not taken into consideration. Although many colloid chemical phenomena can be successfully interpreted in a general way on the basis of the Gouy model, deviations from the general conclusions of the theory are frequently encountered with specific colloidal systems. Evidently, the double-layer model must be refined in order to explain such deviations. Some of these refinements are the following:

1. One of the earliest attempts to create a more realistic double-layer model in which at least some of the idealizations of the Gouy model are eliminated was made by Stern (9). Stern considers that, unlike in the Gouy model, the distance of closest approach of a counter-ion to the charged surface is limited by the size of these ions. In his more or less schematic picture, the counter-ion charge is separated from the surface charge by a layer of thickness δ in which there is no charge. A "molecular condenser" is

formed by the surface charge and the charge in the plane of the centers of the closest counter-ions. In this *"Stern-layer,"* the electric potential drops linearly with the distance, from a value Φ_0 at the surface to a value Φ_δ, which is called the *Stern potential.* A low dielectric constant is assigned to the medium in the Stern layer (Section VII). Beyond the molecular condenser, the remainder of the counter-ion charge is distributed as in a diffuse Gouy atmosphere in which the electric potential decreases roughly exponentially with increasing distance. This model of the double layer is represented in Figure 14. The total counter-ion charge in this model is divided between the charge at atomic distance from the surface (σ_1) and the charge in the diffuse atmosphere (σ_2). The sum of these two charges is equal to the surface charge (σ). The actual distribution of the total charge over the Stern layer and the diffuse layer can be calculated from electrostatic theory combined with adsorption statistics. If necessary, specific adsorption forces between the surface and the counter-ions can be taken into account by assuming that there is a *specific adsorption potential energy* of the

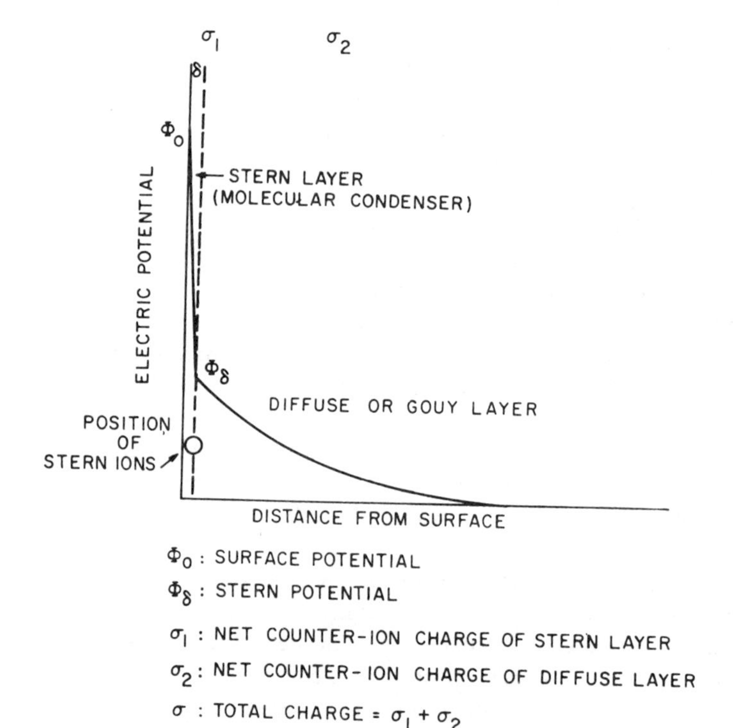

Figure 14. Stern's model of the potential distribution in the electric double layer.

counter-ions which must be added to the electrostatic attraction energy. With increasing adsorption energy, the ratio of Stern-layer charge to diffuse-layer charge increases, and the Stern potential, Φ_δ, decreases.

For the Stern model, the addition of electrolyte to the sol results not only in a compression of the diffuse part of the double layer but also in a shift of the counter-ions from the diffuse layer to the Stern layer and hence in a decrease of the Stern potential, Φ_δ. In Appendix III, some computations of the distribution of charge and potential in the Stern-Gouy double layer are presented.

The computation of the energy of interaction of two Stern-Gouy layers is somewhat more complicated than that for Gouy layers because of the changes occurring in the charge distribution and in Φ_δ. However, it has been shown by Mackor (10) that, for moderate interaction, the repulsive energy can be computed to a good approximation as though one were dealing with the interaction of two Gouy layers with a constant potential which is equal to the value of Φ_δ when the particles are at infinite separation. This result explains why the theory based on the simple Gouy model is still satisfactory in many cases. However, when specific adsorption of counter-ions occurs, the Stern model must be applied. There are indications that specific counter-ion adsorption often occurs in clay systems; therefore, Stern's model is of interest in clay colloid chemistry. Obviously, specific interactions between a surface and different species of counter-ions will be of considerable consequence for the counter-ion exchange equilibrium.

2. Another limitation of the Gouy theory is that secondary energy effects, such as counter-ion and solvent interactions, are not accounted for. Bolt and other authors (see Appendix III, References) corrected the original Gouy theory for such effects. By evaluating several correction terms, Bolt could show that the corrections, which compensate each other partly, are of little consequence in the computation of the repulsive energy between two particles carrying Gouy layers. However, Bolt points out that the corrections are of first-order significance for the treatment of the ion distribution in the ion exchange equilibrium.

3. A more sophisticated model than that of Stern was proposed by Grahame (11), who made a distinction between the plane through the centers of the counter ions at closest approach to the surface (*inner Helmholtz plane, ihp*) and the limit of the diffuse layer or the position of closest approach of the co-ions (*outer Helmholtz plane, ohp*). He has applied this model primarily to the interpretation of potential and charge data for the mercury-water interface.

4. In the sixties, considerable attention was given to the consequences of treating the Grahame model of the double layer, and particularly the inner regions, as an assembly of discrete charges surrounded by space depleted of

other counter ions, as in the Debye-Hückel theory. The mathematical treatment is rather specialized, and the interested reader is referred to some references to the work of Levine and co-authors (12–15).

VII. THE HYDRATION THEORY OF STABILITY AND ITS FALLACIES

The hydration theory of stability has been popular in colloid chemistry for a long time. In this theory, the repulsion between particles is attributed to the orientation of water dipoles in the electric field around the charged particle. In such a "solvation shell," the degree of orientation of the dipoles would gradually decrease with increasing distance from the charged surface. When these diffuse hydration shells begin to interfere when the particles approach each other, the mutual repulsion of the oppositely oriented water dipoles around the colliding particles would account for particle repulsion.

For hydrophilic colloids, the hydration repulsion was considered an additional stability factor over and above the double-layer repulsion. The high resistance of the hydrophilic colloids to added salt was attributed to the presence of the extra stability factor. Also, in certain hydrophobic sols, hydration repulsion was thought to contribute to the stability.

The argument against the hydration theory of stability is that the orientation effect of the charged particles on the water dipoles can be expected to be significant only up to a few water-molecule diameters away from the surface. The energies involved are not sufficiently large to affect the balance of the double-layer and van der Waals interaction energies at the large distances from the surface where the fate of the colliding particles is decided. Actually, many facts of colloid chemistry appeared to be incompatible with the hydration theory of stability, which is now obsolete. Nevertheless, hydration repulsion should not be entirely disregarded. As mentioned previously, it is important in the short-range interaction of particles. A thin layer of water a few molecules thick is usually more or less strongly adsorbed at the particles surface; hence in bringing the particles close together, the required work of desorption of the water is manifested as a short-range repulsion.

The presence of a thick hydration shell around sol particles has also been assumed to be the cause of gelation in hydrophilic sols and in certain hydrophobic sols, such as alumina, silica, or clay sols. The rigidity of these gels was thought to be the result of a rigidifying effect of the charged particles on the water phase. However, the interparticle distances in gels are often very large and beyond any sensible range of hydration forces emanating from the surfaces. The alternative explanation of gelation on the basis of skeleton formation by particle linking is perfectly sensible and appears to supply a good basis for an understanding of the mechanism of the creation as well as the destruction of gels by the addition of small amounts of chemicals to the liquid phase.

It should be mentioned that moderately long-range effects of charged surfaces on certain properties of the water phase, such as the density or the activation energy of viscous flow, are still claimed to exist (see Chapter Ten, references 1–15).

The properties of the water in the first few monolayers which are adsorbed at the surface may differ significantly from those of bulk water. It is generally accepted that the dielectric constant of the adsorbed water layers is much lower than that of free water. In the Stern model of the double layer, the lowering of the dielectric constant in this region must be taken into account in computing the capacity of the molecular condenser and the Stern potential.

When alcohols or other organic polar solvents are added to the sol, these compounds may be adsorbed on the surface in competition with water. Therefore, they may affect the capacity of the Stern layer and hence the Stern potential. In addition, the point of zero charge may be shifted, and possibly even the van der Waals forces may be affected. Therefore, the change of the composition of the medium may have several effects on the constitution of the double layer and the particle-interaction energy. An analysis of such effects in particular cases has explained why an alcohol does not always have the flocculating effect predicted by the Gouy theory as a consequence of the lowering of the dielectric constant of the bulk liquid.

VIII. THE “CRITICAL ZETA POTENTIAL”

For many years the concept of the zeta or electrokinetic potential has governed colloid chemical thinking. The zeta potential is the electric potential in the double layer at the interface between a particle which moves in an electric field and the surrounding liquid. The zeta potential is computed from the electrophoretic mobility of the sol particle. Its magnitude was considered a measure of the particle repulsion. Upon the addition of an electrolyte, the zeta potential usually decreases, and at the flocculation value of the electrolyte it was considered to have reached a critical value, below which the particle repulsion would no longer be strong enough to prevent flocculation. (At the time, the nature of the attractive forces causing particle agglomeration was not clear.)

It has been realized that the seat of the zeta potential is the *shearing plane* or *slipping plane* between the bulk liquid and an envelope of water which moves with the particle. Since the position of this shearing plane is not known, the zeta potential represents the electric potential at an unknown distance from the surface in the double layer. Therefore, the zeta potential is not equal to the surface potential, but it is to some extent, comparable with the Stern potential, although it is not necessarily identical with that potential. Like the Stern potential, the zeta potential may be expected to decrease with increasing electrolyte concentration because of a shift of counter-ions toward the Stern layer when the diffuse double layer is compressed. Therefore, it is not surprising that a relation exists between colloidal stability and the magnitude of the zeta potential. However, because of its ill-defined character, the zeta potential is not a useful quantitative criterion of stability. Moreover, the computation of the zeta potential from the observed electrophoretic mobility is subject to several corrections which are difficult to evaluate quantitatively. The recognition of the existence of such correction terms has still further reduced the usefulness of the zeta potential as a quantitative stability criterion and, as such, this parameter has lost its significance. However, this does not mean that electrophoresis and other electrokinetic phenomena no longer deserve attention. On the contrary, these phenomena are of considerable importance in technology as well as in nature, and they will be discussed further in Chapter Twelve. In systems showing reversal of charge, the electrophoretic mobility changes sign. The point at which zero mobility, hence a zero zeta potential, is observed is called the isoelectric point (i.e.p.). It should be noted that the surface charge is not necessarily zero at this point; in other words, the p.z.c. and the i.e.p. of colloidal systems are, in general, not identical. The p.z.c. is the more fundamental double-layer property.

IX. "ENTROPY" STABILIZATION

Certain suspensions and sols are colloidally stable although the particles do not carry an electric charge, simply because of the absence of ions in the system. This situation may be encountered in certain suspensions in hydrocarbons. These are sometimes stabilized by organic compounds which are adsorbed on the particle surface owing to some degree of polarizability of part of the organic molecule (16). For example, the benzene ring of an alkyl benzene will attach itself to the particle surface; the alkyl chain will be pointed toward the hydrocarbon phase in which it will move freely around the point of attachment, as around a ball joint.

When two particles with the adsorbed alkyl benzene molecules approach each other and reach a distance which is smaller than twice the length of the alkyl chains, the freedom of movement of these chains is curtailed by steric hindrance. Hence the entropy of the adsorbed molecules decreases. Since a system changes spontaneously in the direction of increased entropy, the particles will tend to separate again. Hence the entropy effect is manifested as a particle repulsion. The range of repulsion is determined by the length of the alkyl chains in this example, and if the chain length is large enough, the repulsion may exceed the van der Waals attraction at a distance of the order of twice the chain length. In this way, a potential curve with a long-range repulsion maximum can be obtained, and the sol will be stable. This mechanism of stabilization is particularly effective for small suspended particles, since the van der Waals attraction for small interacting particles is small. This theory of "entropy" stabilization has been proposed by Mackor (17, 18) who has worked out various examples.

The existence of electric double-layer repulsion in hydrocarbon systems cannot be ruled out, however. Certain compounds which are soluble in hydrocarbons ionize slightly in that medium, and the ions may act as peptizing ions for certain dispersed particles. Examples of such systems have been analyzed by van der Minne and Hermanie (19). Koelmans and Overbeek (20) have shown that double-layer stabilization, rather than entropy stabilization, must be relied upon in stabilizing hydrocarbon suspensions of large particles for which the van der Waals attraction is strong.

In hydrous systems, entropy stabilization may offer an explanation for the stabilizing effect of some nonionic polymers on certain suspensions.

In Chapter Eleven, which deals with clay-organic systems, some additional mechanisms of particle interaction will be mentioned; these are summarized in Section VII of that chapter.

After this discussion of the highlights of the rules and theories of colloid chemistry, we shall return to the treatment of clay systems.

The peculiar shape of the clay particles and their particular crystal structure are two unusual features which make clay suspensions behave in a somewhat more complicated manner than the common hydrophobic colloids. These two features will be discussed in the next chapter.

References

1. Verwey, E. J. W., and Overbeek, J. Th. G. (1948), *Theory of the Stability of Lyophobic Colloids,* Chapter XII, Elsevier, New York.
2. Sweeton, F. H., *Calculation of Suspension Peptization,* Oak Ridge National Laboratory, ORNL-2791, Office of Technical Services, Department of Commerce, Washington, D.C. (Extensive tables for the computation of coagulation velocities from the potential energy curves.)
3. Bergmann, P., Löw-Beer, P., and Zocher, H. (1938), The theory of iridescent layers, *Z. Physik. Chem.,* **A 181,** 301–314.
4. Bernal, J. D., and Fankuchen, I. (1937), Structure types of protein "crystals" from virus infected plants, *Nature* **139,** 923–924.

5. Bernal, J. D. (1941), X-ray and crystallographic studies of plant virus preparations, *J. Gen. Physiol.*, **25**, 111–165.
6. Onsager, L. (1949), The effect of shape on the interaction of colloidal particles, *Ann. N.Y. Acad. Sci.*, **51**, 627–659.
7. Norrish, K. (1954), The swelling of montmorillonite, *Discussions Faraday Soc.*, **18** (Coagulation and Flocculation), 120–134.
8. Van Os, G. A. J. (1943), Ionenuitwisseling en geleidingsvermogen van het zilverjodidesol, Thesis, Utrecht; Reference in Kruyt, H. R. (1952), *Irreversible Systems*, Vol. I of *Colloid Science*, Elsevier, Amsterdam, Houston, New York, London, 177–180.
9. Stern, O. (1924), Zur Theorie der Elektrolytischen Doppelschicht, *Z. Elektrochem.*, **30**, 508–516.
10. Mackor, E. L. (1951), The stability of the silver iodide sol in water-acetone mixtures, *Rec. Trav. Chim.*, **70**, 841–865.
11. Grahame, D. C. (1947), The electrical double layer and the theory of electrocapillarity, *Chem. Rev.*, **41**, 441–501.
12. Levine, S., Bell, G. M., and Calvert, D. (1962), The discreteness-of-charge effect in electric double layer theory, *Can. J. Chem.*, **40**, 518–538.
13. Levine, S., and Bell, G. M. (1962), An extension to the stability theory of lyophobic colloids, *J. Colloid Sci.*, **17**, 838–856.
14. Levine, S., Mingins, J., and Bell, G. M. (1965), The diffuse layer correction to the discrete-ion effect in electric double layer theory, *Can. J. Chem.*, **43**, 2834–2866. [See also a review paper in *J. Electroanal. Chem.*, **13**, 280 (1976)].
15. Levine, S. (1971), Adsorption isotherms in the electric double layer and the discreteness-of-charge effect, *J. Colloid Interface Sci.*, **37**, 619–634.

Appendix III contains additional references.

STABILITY IN NONAQUEOUS SYSTEMS

16. van der Waarden, M. (1950), Stabilization of carbon-black dispersions in hydrocarbons, *J. Colloid Sci.*, **5**, 317–325.
17. Mackor, E. L. (1951), A theoretical approach of the colloid chemical stability of dispersions in hydrocarbons, *J. Colloid Sci.*, **6**, 492–495.
18. Mackor, E. L., and van der Waals, J. H. (1952), The statistics of the adsorption of rod-shaped molecules in connection with the stability of certain colloidal dispersions, *J. Colloid Sci.*, **7**, 535–550.
19. van der Minne, J. L., and Hermanie, P. H. J. (1953), Electrophoresis measurements in benzene. Correlation with stability, II.—Results of electrophoresis, stability, and adsorption, *J. Colloid Sci.*, **8**, 38–52.
20. Koelmans, H., and Overbeek, J. Th. G. (1954), Stability and electrophoretic mobility in non-aqueous media, *Discussions Faraday Soc.*, **18** (Coagulation and Flocculation), 52–63.

CHAPTER FIVE

Clay Mineralogy

Colloid chemistry and clay* mineralogy have in common that the nomenclature and classification are somewhat confusing to the new-comer in the field. Mineral names were given before the crystal structures of the finely divided clays could be determined, and older classifications had to rely on chemical analysis. Only after some decades of intensive X-ray structure analysis did it become possible to devise a sensible classification of the clay minerals. In this text the recommendations of the AIPEA Nomenclature Committee have been followed (1). (See also Appendix V, reference P-6.)

The following brief survey is intended to acquaint the reader with the most important results of the structure studies on clay minerals since Pauling (2) clarified the general principles of the layer structure of clays in 1930. The review will go into as much detail as will be required to provide an understanding of the colloid chemical features of clay suspensions and to make the chemist aware of the complexity of the natural materials with which he is working.

The principal tools for structure analysis are X-ray and electron diffraction, complemented by infrared, nuclear magnetic resonance, and electron spin resonance techniques.

I. STRUCTURAL PRINCIPLES (2–25)

Clay minerals belong to the *phyllosilicates.* The principal building elements of the clay minerals are two-dimensional arrays of silicon-oxygen tetrahedra and two-dimensional arrays of aluminum- or magnesium-oxygen-hydroxyl octahedra. In most clay minerals, such sheets of tetrahedra and of octahedra are superimposed in different fashions. (See Table 3 for structural terms in different languages.)

* The term "clay" is applied here to the finely divided crystalline material which has the crystal structure discussed in this chapter. In soil science, the term "clay" is often used for the soil fraction with an equivalent Stokes diameter $< 2\ \mu m$. Therefore, the "clay fraction" of the soil also comprises other minerals such as quartz in particle sizes $< 2\ \mu m$.

Table 3

Structural Terms of Reference and Their Equivalents in Different Languages

English	French	German	Russian	Spanish	Italian
Plane	Plan	Ebene	ПЛОСКОСТЬ	Plano	Plano
Sheet	Couche	Schicht	СЕТКА	Capa	Foglietto
Layer	Feuillet	Schicht-paket	СЛОЙ	Estrato o Paquete (de capas)	Strato
Interlayer or Interlayer materials	Espace Interfol-taire	Zwischen-schicht	МЕЖСЛОЕВО ПРОМЕЖУТОК (Межслой)	Material interla-minar o Inter-stratificado	Interstrato
Unit structure	Unité structu-rale	Struktur einheit	ПАКЕТ	Unidad estructural	Unita strutturale

In the silicon-oxygen sheets, the silicon atoms are coordinated with four oxygen atoms. The oxygen atoms are located on the four corners of a regular tetrahedron with the silicon atom in the center (Figure 15*a*). In the sheet, three of the four oxygen atoms of each tetrahedron are shared by three neighboring tetrahedra. The fourth oxygen atom of each tetrahedron is pointed downward in the sketched arrangement in Figure 15*b*. Projections of this arrangement are shown in Figures 15*c* and *d*. These figures demonstrate the hexagonal symmetry of such a sheet, in which rings of six oxygen atoms appear. The holes in these rings are clearly visible in the three-dimensional model sketched in Figure 17*a*. The silicon-oxygen sheet is called the *tetrahedral sheet* or the *silica sheet*.

In the Al-, Mg-O-OH sheets, the Al or Mg atoms are coordinated with six oxygen atoms or OH groups which are located around the Al or Mg atom with their centers on the six corners of a regular octahedron (Figure 16*a*). The sharing of oxygen atoms by neighboring octahedrons results in a sheet such as that shown in perspective in Figure 16*b*. The oxygen atoms and hydroxyl groups lie in two parallel planes with Al or Mg atoms between these planes. The projection of the sheet (Figure 16*d*) shows that the oxygen atoms and hydroxyl groups form a hexagonal close packing. This sheet is called the *octahedral sheet* or the *alumina* or *magnesia sheet*, also called *gibbsite sheet* or *brucite sheet*, respectively. A three-dimensional model is shown in Figure 17*b*.

The analogous symmetry and the almost identical dimensions in the tetrahedral and the octahedral sheets allow the sharing of oxygen atoms between these sheets. The fourth oxygen atom protruding from the tetrahedral sheet is shared by the octahedral sheet. This sharing of atoms may occur between one silica and one alumina sheet, as is the case in the so-called 1:1 *layer minerals*. In the 2:1 *layer minerals*, one alumina or magnesia sheet shares oxygen atoms with two silica sheets, one on each side. The combination of an octahedral sheet and one or two tetrahedral sheets is called a *layer*. Most clay minerals consist of such layers, which are stacked parallel to each other. (See Table 3.)

Within each layer a certain unit repeats itself in a lateral direction. For

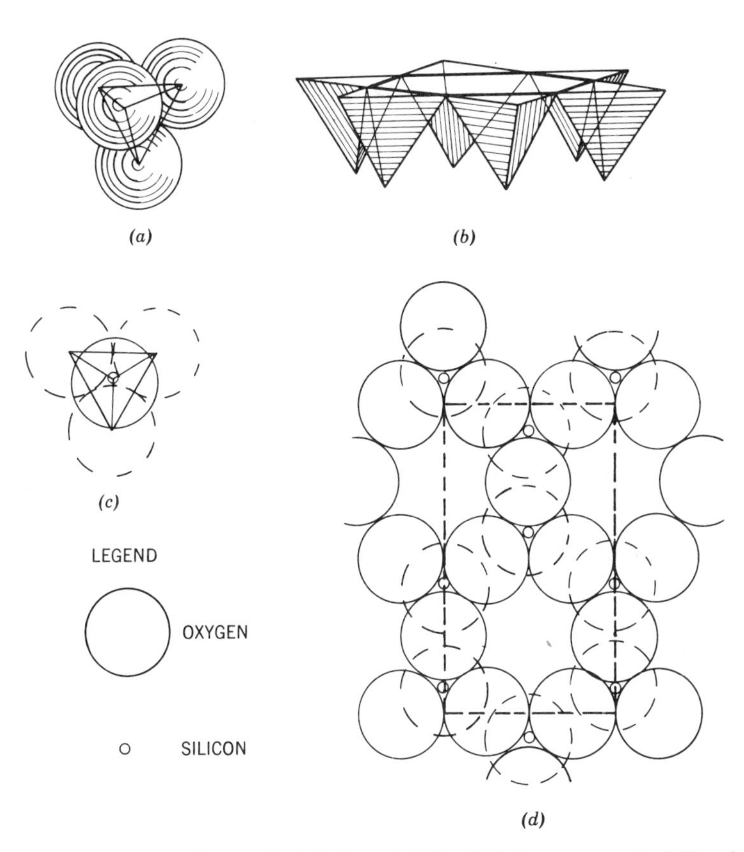

Figure 15. Structure of the tetrahedral sheet. (*a*) Tetrahedral arrangement of Si and O. (*b*) Perspective sketch of tetrahedral linking. (*c*) Projection of tetrahedron on plane of sheet. (*d*) Top view of tetrahedral sheet (dotted line: unit cell area). Large circles represent oxygen; small circles represent silicon.

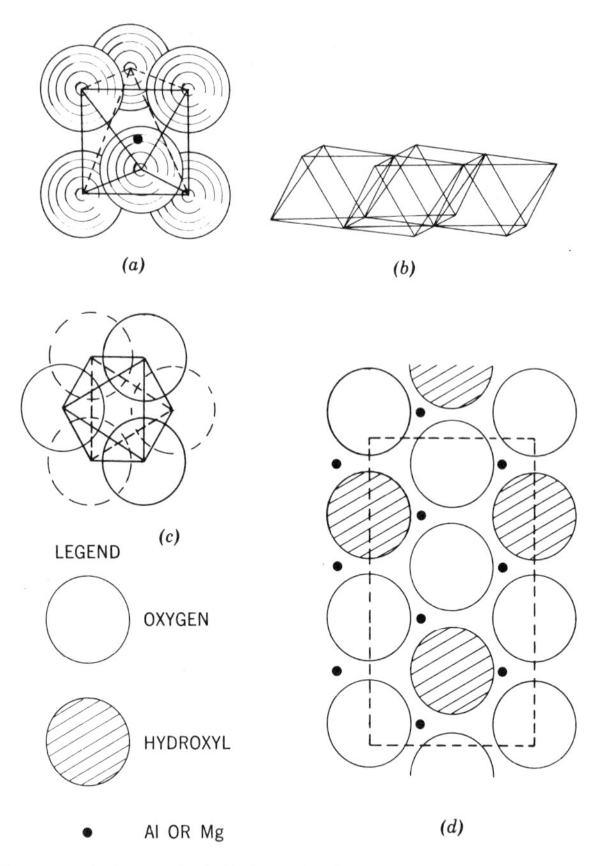

Figure 16. Structure of the octahedral sheet. (*a*) Octahedral arrangement of Al or Mg with O or OH. (*b*) Perspective sketch of octahedral linking. (*c*) Projection of octahedron on plane of sheet. (*d*) Top view of octahedral sheet (dotted line: unit cell area). Large circles represent oxygen; shaded circles represent hydroxyl; small black circles represent aluminum or magnesium.

easy reference, this unit will be called the *unit cell* in the following discussion. However, the crystallographically defined unit cell must extend from a certain plane in one layer to the corresponding plane in another parallel layer of the crystal. In this way, the features of the geometry of stacking of layers are included as well as the positions of any materials present between layers. The total assembly of a layer plus interlayer material is also referred to as a *unit structure.* (See Table 3.)

A schematic representation of the atom arrangements in a "unit cell" is shown in Figure 19 for a 1:1 layer clay and in Figure 18 for a 2:1 layer clay. The latter structure is usually referred to as the *Hofmann structure*

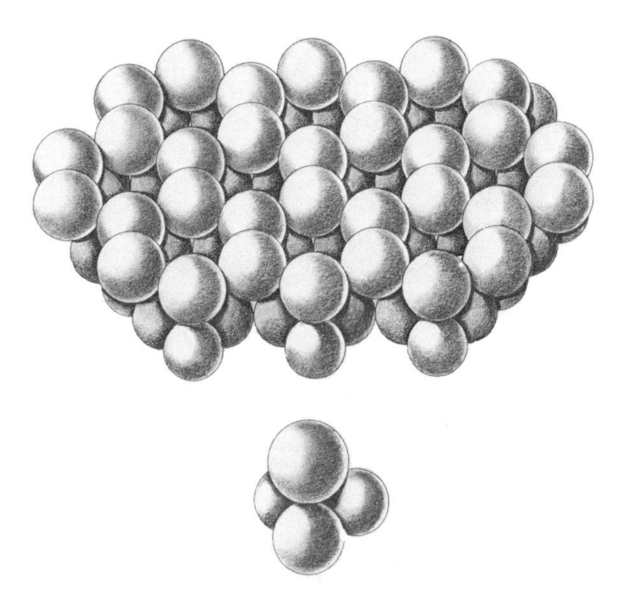

Figure 17*a*. Model of silicon-oxygen tetrahedron and of the tetrahedral sheet.

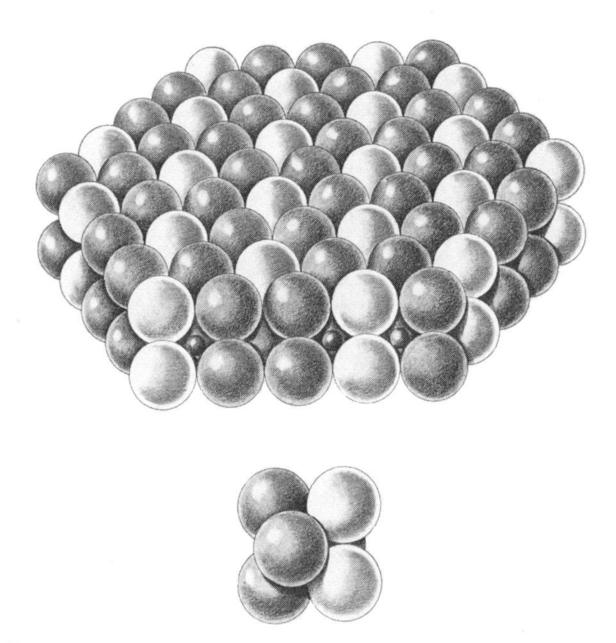

Figure 17*b*. Model of aluminum-oxygen octahedron and of the octahedral sheet. (In this sketch, hydroxyl groups have a lighter shading.)

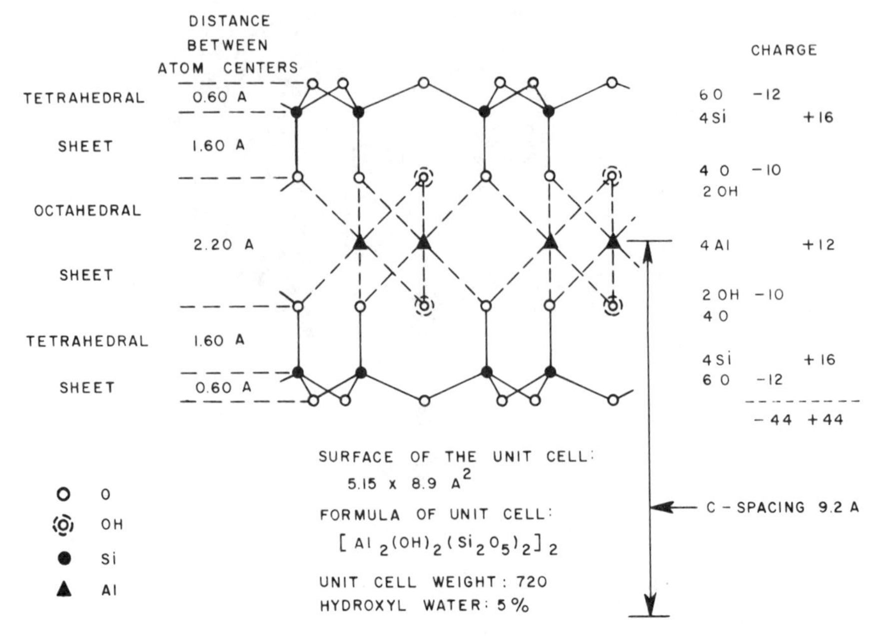

Figure 18. Atom arrangement in the unit cell of a 2:1 layer mineral (schematic).

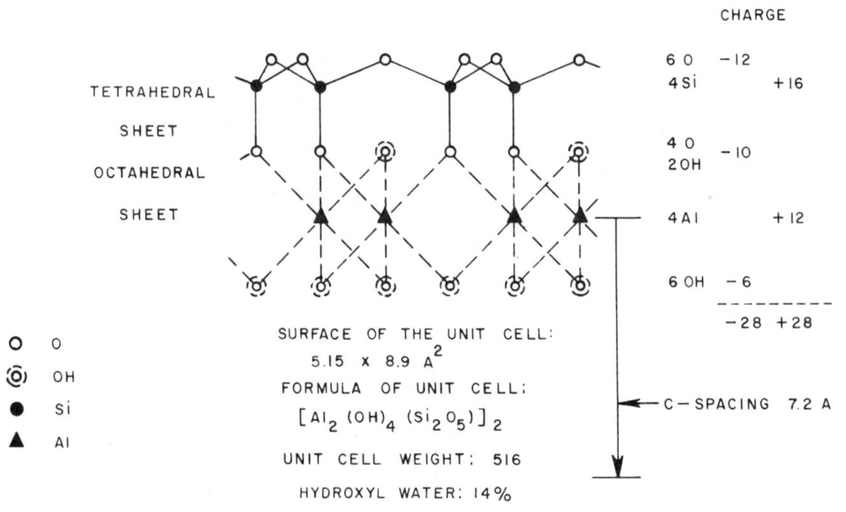

Figure 19. Atom arrangement in the unit cell of a 1:1 layer mineral (schematic).

(3).* In these figures, the atoms are not designed to scale in order to show the arrangement of the atom bonds. However, the scale models (Figures 17*a* and *b*) give an indication of the packing of the atoms in the crystal.

In Figures 18 and 19 the following information is given:

1. The unit-cell formula and the unit-cell weight.
2. The distances between the planes drawn through the centers of the atoms.
3. The dimensions of the unit cell ($a_0 \sim 5.15$ A; $b_0 \sim 8.9$ A) determined from the X-ray diffraction patterns.
4. The distance between a certain plane in the layer and the corresponding plane in the next layer. This distance is called the *001, basal* or *c spacing*. This repetition distance of the layers in the stack is very important. Its determination from X-ray diffraction patterns immediately enables one to distinguish between 1:1 layer clays with a *c* spacing of about 7.1–7.2 A and 2:1 layer clays with a *c* spacing of at least 9.2 A.

In the columns on the right of Figures 18 and 19, the positive and negative atom valences are tabulated. From the equality of the sums of positive and negative valences, it follows that in the depicted unit cells all valences are saturated. These two ideal electroneutral crystal structures are realized in nature.

The neutral 2:1 layer structure of Figure 18 represents the structure of the mineral *pyrophyllite*. Since in this mineral two of the three possible octahedral positions are occupied by trivalent Al, this structure is called *dioctahedral*. If the three octahedral positions are filled by three divalent Mg atoms, the electroneutral structure of the mineral *talc* is represented. This arrangement is called *trioctahedral*.

The minerals pyrophyllite and talc consist of stacks of the above described layers. The cohesive force between the layers is primarily electrostatic, augmented by van der Waals attraction (30). However, cleavage parallel to the layers is relatively easy, and this fact explains why these minerals occur in the form of flakes. Since these flakes are still rather thick and large, they apparently offer a reasonable mechanical resistance against the grinding forces to which they are subjected in nature. They do not disintegrate to a size which is typical for the clay minerals. Therefore, they are not classified as clay minerals proper, but as *mica-type minerals*.

The structure of these and some related micas are the prototypes of the structures of the 2:1 layer clay minerals of the *montmorillonite* and the *illite* groups. They are derived from the prototypes by introducing more or less random atom substitutions in the crystal structure. Calling these micas

* Two alternative propositions (5, 6) are mentioned in Chapter Eleven (Section VIII).

prototypes of the clay minerals, however, does not necessarily mean that they are the parent materials of the clays.

The clay minerals of the *kaolinite* group represent 1:1 layer structures as shown by Figure 19. The extent of atom substitutions in the kaolinite structure is relatively small.

These three principal groups of phyllosilicates will be discussed in more detail. In addition, two other groups of clay minerals, the *chlorites* and the *attapulgite*-type clays, which are built according to somewhat different structural principles, will be mentioned briefly. Outside the layer silicates, a large variety of silicate structures are known. (See Appendix V, reference B-30.) The *zeolites* (Appendix V, reference B-31) are silicates with open three-dimensional structures with "cages" of molecular dimensions, containing water and cations. They are used as "molecular sieves" and as catalysts. In soils of volcanic origin, "amorphous" clay materials exist; they occur as very small particles ("*allophanes*") or thin hollow threads ("*imogolite*") with low crystallographic structural order (31). However, in this text these other groups of silicates will not be discussed.

II. MONTMORILLONITES (EXPANDING 2:1 LAYER CLAYS) (3–8)

The structure of the minerals of the montmorillonite group of clay minerals is derived from that of the prototypes pyrophyllite and talc by substitution of certain atoms for other atoms. The following types of atom substitutions have been observed in representative minerals of this group.

In the tetrahedral sheet, tetravalent Si is sometimes partly replaced by trivalent Al. In the octahedral sheet, there may be replacement of trivalent Al by divalent Mg without complete filling of the third vacant octahedral position. Al atoms may also be replaced by Fe, Cr, Zn, Li, and other atoms. The small size of these atoms permits them to take the place of the small Si and Al atoms; therefore, the replacement is often referred to as *isomorphous substitution*. In many minerals an atom of lower positive valence replaces one of higher valence, resulting in a deficit of positive charge, or, in other words, an excess of negative charge. This excess of negative layer charge is compensated by the adsorption on the layer surfaces of cations which are too large to be accommodated in the interior of the crystal. This interpretation of the results of the analysis of the chemical composition of the clay minerals was first proposed by Marshall in 1935 (see Appendix II, References).

In the presence of water, the compensating cations on the layer surfaces may be easily exchanged by other cations when available in solution; hence they are called "exchangeable cations." The total amount of these cations may be determined analytically. This amount, expressed in milliequivalents per 100 g of dry clay, is called the *cation exchange capacity* (CEC) or the *base exchange capacity* (BEC) of the clay (56–76).

Since the exchangeable cations compensate the unbalanced charge in the interior of the layers due to isomorphous substitutions, the CEC is a measure of the degree of substitution. If the CEC and the chemical composition of the clay are known from a chemical analysis, it is possible to assign the substituting ions to the tetrahedral and the octahedral sheets, if the following arbitrary, but reasonable, assumptions are made:

1. There are 20 oxygen atoms and 4 OH groups (sometimes substituted by fluorine atoms) per unit cell.
2. All the Si present is assigned to the tetrahedral sheet.
3. The remainder of the tetrahedral positions is first filled by Al. Any additional Al is assigned to the octahedral sheet, together with Fe, Mg, and so on, except, of course, those cations which are in exchange position.

On the basis of such a computation, which is treated in detail in Appendix II, the formula of a clay may be derived. For example, the formula of a typical montmorillonite with a CEC of 70 meq per 100 g is as follows:

$$\underset{\text{(tetrahedral)}}{[(Si_{3.88}Al_{0.12})^{IV}}\ \underset{\text{(octahedral)}}{(Al_{1.64}Fe^{3}_{0.05}\,Mg_{0.36})^{VI}}\ O_{10}(OH)_2]_2 \atop \quad\quad\quad\quad\quad |\; M_{0.67}$$

in which M represents a monovalent exchangeable cation. The superscripts IV and VI refer to tetrahedral and octahedral coordination, respectively.

Care must be taken that mineral impurities are removed prior to the chemical analysis. Quartz, in particular, often occurs in a clay in such a fine particle size that it cannot be separated from the clay by centrifugation. In that case, the amount may be estimated from X-ray analysis, and the chemical analysis may be corrected accordingly.

In the stack of layers which form a montmorillonite particle, the exchangeable cations are located on each side of each layer in the stack; hence they are present not only on the external surfaces of the particle but also in between the layers. Their presence causes a slight increase of the basal spacing as compared with that of pyrophyllite from about 9.13 A to at least 9.6 A (for the dry clay) or slightly higher when the compensating cations are larger. The difference between the basal spacing of pyrophyllite and that of a montmorillonite is much less than the diameter of the compensating cations. Apparently, these cations are partly sunk in the holes of the tetrahedral sheet.

When montmorillonite clays are contacted with water or with water vapor, the water molecules penetrate between the layers (see References, Chapter Ten). This so-called *interlayer swelling,* or (*intra-*) *crystalline* swelling, of montmorillonites is evident from an increase of the basal spacing of the clays to definite values of the order of 12.5–20 A, depending on the type of clay and the type of cation. A more or less stable configuration of the hydrated clay is obtained, corresponding with the presence of one to four monomolecular layers of water between the layers. In a water-vapor-adsorption isotherm, a stepwise hydration with one, two, three, and four water layers could be observed in some clays by taking X-ray diffraction patterns of a flake of the clay at different water-vapor pressures. Cations, water, and other materials between layers are referred to as the *interlayer.* (See Table 3.)

Interlayer swelling leads to, at most, a doubling of the volume of the dry clay when four layers of water are adsorbed. The much larger degree of swelling which is observed for many montmorillonite clays is due to another mechanism ("osmotic" swelling), as will be discussed later.

The question immediately poses itself, Why do the prototype minerals pyrophyllite and talc not swell, as montmorillonite does? Two alternative explanations have been advanced:

According to one point of view, the interlayer cations in montmorillonites become hydrated, and the large hydration energy involved is able to overcome the attractive forces between the layers. Since in the prototype minerals interlayer cations are absent, there is no cation hydration energy available to separate the layers.

Alternatively, it has been proposed that the penetrating water does not hydrate the cations between the layers but becomes adsorbed on the oxygen surfaces by establishing hydrogen bonds. Certain geometric arrangements of the water molecules in the water layers have been proposed which would favor such a bonding.

In this explanation, the reason for the lack of interlayer swelling in the prototype minerals could be that the surface hydration energy is too small to overcome the van der Waals attraction between the layers, this attraction being stronger than for montmorillonites because of the smaller layer distance in the prototype minerals.

Actually, the problem of the mechanism of interlayer swelling is rather complicated. In addition to the van der Waals attraction and the hydration energy, the electrostatic energy between the charged layers and the cations between these layers must be considered in the evaluation of the balance of the forces which determine the separation of the layers. Direct evidence for either mechanism is difficult to obtain (Chapter Ten, Section I-A). In reality, in some clays and for certain species of exchange cations, ion hydration may dominate the interlayer swelling; in other clays and other species of cations, surface hydration may be more important.

The importance of the presence of the interlayer cations for the interlayer swelling process can be demonstrated by the following possibility to reduce the surface charge. By heating of dioctahedral montmorillonites in the lithium form, lithium ions diffuse through the layers and occupy vacant octahedral positions, thus reducing the layer charge. The mineral then looses its swelling ability with the charge reduction, and the layers "collapse" (77–84).

Montmorillonites also admit organic compounds of a polar or ionic character between the layers. The adsorption of the organic compounds leads to *organo-complexes* of montmorillonites. The basal spacings of these complexes depend on the size and the packing of the organic molecules. The

Table 4

Montmorillonoids[a]

Principal substitutions	Trioctahedral minerals	Dioctahedral minerals
Prototype (no substitutions)	Talc Mg_3Si_4[b]	Pyrophyllite Al_2Si_4
Practically all octahedral	Hectorite $(Mg_{3-x}Li_x)(Si_4)$	Montmorillonite[c] $(Al_{2-x}Mg_x)(Si_4)$
Predominantly octahedral	Saponite $(Mg_{3-x}Al_x)(Si_{4-y}Al_y)$ Sauconite $(Zn_{3-x}Al_x)(Si_{4-y}Al_y)$	Volchonskoite $(Al,Cr)_2(Si_{4-y}Al_y)$
Predominantly tetrahedral	Vermiculite $(Mg_{3-x}Fe_x)(Si_3Al)$	Nontronite $(Al,Fe)_2(Si_{4-y}Al_y)$

[a] The different species in the group of montmorillonites vary in type and degree of isomorphous lattice substitution. The names and formulas of some representatives of the group are listed. Since the term "montmorillonite" is also given to one of the members of the group, the proposal has been made that the whole group be designated *montmorillonoids*, or *smectites*. More recently it has been proposed to apply the term smectites only to the minerals in the lower substitution ranges to distinguish these from the higher charge vermiculites.

[b] To each formula the group $O_{10}(OH)_2$ should be added as well as the exchangeable cation.

[c] Montmorillonite is the principal clay mineral of bentonite rock which originates from volcanic ash. "Bentonite" is a rock name and not a mineral name, although it is often used as such in the chemical literature. For example, a "bentonite sol" means a sol of the montmorillonite clay separated from the bentonite rock material.

change in the basal spacing caused by the formation of such complexes is often used to detect the presence of montmorillonites in natural clay mixtures. Commonly, the complexing with ethylene glycol is used for this purpose. In Chapter Eleven, the interaction of clays and organic compounds will be discussed in detail.

III. ILLITES (NONEXPANDING 2:1 LAYER CLAYS) (9)

A different class of three-layer clays is formed by the illite clays. They are distinguished from the montmorillonites primarily by the absence of interlayer swelling with water or organic compounds. Substitutions occur predominantly in the tetrahedral sheet. The cations which compensate the net negative layer charge are usually potassium ions. Since the layers do not part upon the addition of water, the potassium ions between the layers are not available for exchange—they are *fixed.* Only the potassium ions on the external surfaces can be exchanged for other cations.

The minerals *muscovite* and *phlogopite* may be considered as the prototype mica minerals for the illites. In these two minerals, one of the four silicon atoms in the tetrahedral sheet is substituted by one aluminum atom, and the defect of one positive charge is compensated by one potassium ion which is located on the layer surface. There is no substitution in the octahedral sheet in these minerals. Muscovite is the dioctahedral mineral, with two aluminum atoms in the octahedral sheet, and phlogopite is the trioctahedral mineral, with three magnesium atoms in the octahedral sheet (29).

The basal spacing of the two prototype minerals as well as of the illites is about 10 A, which is the same as that of montmorillonites with potassium ions as exchange ions in the dry state. The illites are derived from the prototypes by variations in both the tetrahedral and octahedral substitutions. The total net negative layer charge which results from the substitutions, and therefore the amount of compensating potassium ions, is usually larger than for most montmorillonites—often one and one-half times larger.

The nonswelling character of the illite clays is attributed to a specific electrostatic linking effect of the layers by the potassium ions. These ions are of the right size to establish a favorable 12-coordination with opposite hexagonal oxygen rings of adjoining layers, being embedded in the space created by opposite holes.

Such a specific layer-linking force by potassium ions may be expected to be operative also in a montmorillonite in the potassium form. Nevertheless, a potassium montmorillonite still shows interlayer swelling. There may be two reasons for the different behavior of potassium montmorillonites and illites. In the first place, in illites there will be more such links per unit area because of the larger amount of potassium ions—possibly one and one-half times as many as in montmorillonites. Secondly, the links in montmorillonites may be weaker, since the negative lattice charge is concentrated more in the octahedral sheet and hence at a greater distance from the potassium ions than in illites, which have predominant tetrahedral substitution. It should be mentioned, however, that potassium montmorillonites do have a tendency to become nonexpandable after moderate heating, particularly after repeated drying and wetting (*collapse*). The potassium ions in a montmorillonite are said to become partially *fixed* by heating, since the loss of expansion involves a decrease of the exchange capacity. Vermiculites collapse immediately upon K exchange owing to their high charge density and preponderant tetrahedral substitution.

It is, however, possible to replace K ions in potassium micas with Na ions by treatment with a tetraphenylboron-NaCl solution. The resulting sodium-substituted clays do expand with water. In certain potassium micas, K can be replaced directly with alkylammonium ions (85–92).

Since only the external cations of illite clays are exchangeable, the cation exchange capacity of illites is smaller than that of montmorillonites, despite the higher degree of isomorphous lattice substitution in the illites. The CEC of illites is usually in the order of 20–40 meq per 100 g, whereas the total amount of compensating cations may be as high as 150 meq per 100 g.

The various species in the group which are all called illites, differ in the type and degree of isomorphous substitution.

IV. KAOLINITES (1:1 LAYER CLAYS) (10–12)

Almost perfect 1:1 layer clay structures are realized in the clay minerals of the kaolinite group. The main difference between various species is a difference in layer stacking geometry. Members of the kaolinite group are kaolinite, dickite, nacrite, and halloysite. In water these minerals are nonexpandable. However, halloysite does contain interlayer water, but upon heating it is irreversibly dehydrated. The dehydrated form is sometimes called metahalloysite. Preferred nomenclature is "halloysite (10A)" and "halloysite (7A)" for the hydrated and the dehydrated minerals, respectively. The cohesive energy in kaolinites is primarily electrostatic, augmented by van der Waals attraction and a certain degree of hydrogen bonding between the hydroxyl groups of one layer and the oxygen atoms of the adjoining layer (26, 27, 28). In the absence of interlayer cations, only surface hydration energy would be available to open up the crystallites, and this energy is apparently insufficient to overcome the rather large cohesive energy. Neither are those organic compounds intercalated which are commonly adsorbed between the layers of montmorillonites. Hence the kaolinites were generally considered to be nonexpandable until in the early sixties it was discovered that certain compounds were able to expand kaolinite crystals. These comprise such salts as potassium acetate which are intercalated as the total salt, hence the intercalation process is referred to as *intersalation.* Other intercalating compounds are strong hydrogen bonding compounds such as urea, hydrazine, dimethylsulfoxide, and so on. Their reactions with the kaolinites will be discussed in Chapter Eleven, Section VI.

The cation exchange capacity of the kaolinite minerals is quite low and usually amounts to 1–10 meq per 100 g. It has been suggested that the exchangeable ions are located on the broken edges of the kaolinite plates, where they would compensate charge deficiencies owing to broken bonds. However, as will be discussed later, it is not likely that the broken edges have cation exchange sites. It is more reasonable to assume that, as in montmorillonites and illites, the exchangeable cations are located on the flat-layer surfaces and compensate a net negative layer charge which is due to a small degree of isomorphous substitution. Since the basal spacing of kaolinites does not leave room for interlayer cations, all the charge-compensating cations must be adsorbed on the exterior surfaces of the stack of layers representing a particle.

Because of the extremely small degree of isomorphous substitution required to explain the cation exchange capacity, it is difficult to prove analytically the equivalence of exchange cation charge and lattice charge, particularly since small amounts of impurities would upset the computations. However, cases have been reported in the literature in which such an equivalence could be demonstrated (11). On the other hand, a recent detailed study (76) of cation and anion adsorption equilibria on a well-purified kaolinite has shown that, at least for the particular kaolinite studied, the theory of equivalence of cation exchange capacity and isomorphous substitution did not apply. The negative surface charge appeared to vary with electrolyte concentration in a manner to be expected for a surface of constant potential rather than of constant charge. This behavior was proposed to be the result of the modification of the exterior layer surfaces by strongly adsorbed alumina-silicate species.

V. CHLORITES (13, 14, 19, 21)

Chlorites are clay minerals which are structurally related to the 2:1 layer clays. In these minerals, the charge-compensating cations between montmorillonite-type layers are replaced by an octahedral magnesium hydroxide sheet—formerly called brucite sheet. Owing to some replacement of Mg by Al in the hydroxide sheet, this sheet carries a net positive charge. Since the cation exchange capacities of chlorites are very low, the positive charge of the hydroxide sheet practically compensates the net negative charge of the layers. The characteristic basal spacing of chlorites is 14.2 Å.

VI. ATTAPULGITE (PALYGORSKITE) (15, 16)

An entirely different principle of superposition of tetrahedral and octahedral elements of the unit cell is found in the mineral attapulgite or palygorskite. The unit cell has very narrow channels in which water molecules and exchangeable cations are located. These channels are brought about by the inversion of alternate pairs of silica tetrahedra. In an analogous mineral, sepiolite (16), alternate sets of four tetrahedra are inverted.

Attapulgite crystallizes in the form of long needles, as shown in the electron micrograph, Figure 4.

VII. MIXED-LAYER CLAYS

Natural clays sometimes consist of particles in which layers of different types of clay minerals are stacked together (*interstratification*). These clays are called *mixed-layer clays*. It is often possible to estimate the percentage of layers belonging to a certain clay group by quantitative X-ray diffraction analysis.

The mixed-layer clay can be distinguished from a mixture of clays, because the latter, but not the first, can be separated by physical means, such as fractionation in the ultracentrifuge.

VIII. THERMAL AND SPECTRAL ANALYSIS OF CLAYS (104–119)

A. Thermal Analysis

In addition to X-ray analysis, another useful tool for the characterization of clay minerals is *differential thermal analysis,* DTA. This analysis demonstrates water loss and phase changes upon heating of the clay by the accompanying thermal effects. The temperatures at which such changes occur are typical for various types of clay; therefore, the DTA is useful for identification purposes, particularly when studied in conjunction with X-ray diffraction data. Automatic equipment has been developed for the study of the changes in X-ray patterns which occur during heating of the clay sample. In addition, weight losses occurring during heating are being measured by means of thermal balances of varied design (*thermal gravimetric analysis*, TGA).

In the DTA apparatus, a small sample of the powdered clay is placed in a hole in a ceramic block. The block is heated at a certain constant rate up to about 900°C. The temperature of the block is measured by means of a thermocouple placed in the powder of an inert mineral which does not undergo any phase changes in the temperature range studied and which is mounted in the block in a completely identical manner as the clay sample. The actual temperature of the clay sample at any time is measured by determining the temperature difference between the sample and the inert mineral by means of a thermocouple—hence the term "differential" thermal analysis. The temperature difference between the two powders is plotted as a function of the temperature of the block. In these plots any changes in the clay sample which are accompanied by the absorption or the liberation of heat are shown as negative or positive peaks, respectively.

In Figure 20, an example of such a "DTA curve" is shown. For this clay, a calcium montmorillonite, a negative (or endothermic) peak is observed between 100° and 200°C. In this region, heat is absorbed by the clay, which is required for the desorption of the interlayer water. The shoulder on the peak, indicating an incompletely resolved double peak, shows that part of the water is more tightly adsorbed than another part. For sodium montmorillonite, only a single peak is observed in this region.

A second negative (endothermic) peak occurs at about 600°C. This peak indicates the loss of structural hydroxyl groups in the form of water

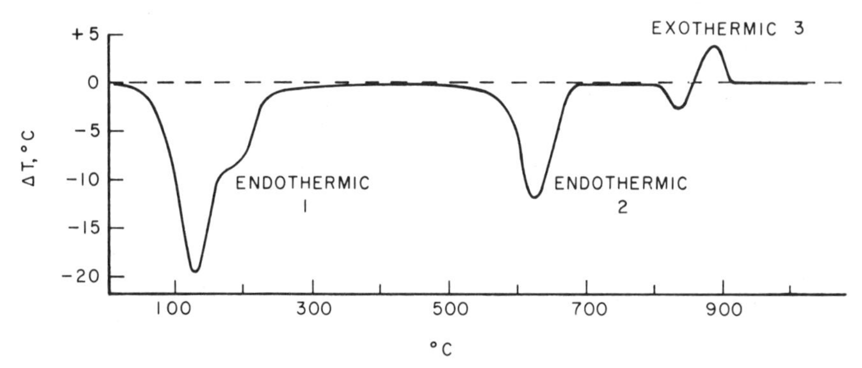

Figure 20. Differential thermal analysis (DTA) of a montmorillonite (schematic). Endothermic reaction 1 = loss of surface hydration water. Endothermic reaction 2 = loss of structural hydroxyl water. Exothermic reaction 3 = crystal phase change.

(*dehydroxylation*). At still higher temperatures, phase changes of the crystal take place, which are indicated by a positive (exothermic) peak around 900°C. A small negative peak, which occurs just prior to the exothermic change, is attributed to the loss of the last traces of hydroxyl water.

In Figure 21, results obtained with the *oscillating-heating X-ray diffraction technique* are shown. During heating of the sample, the position and height of a characteristic diffraction peak, for example, the peak indicating the *c* spacing, are measured by scanning at the corresponding diffraction angle. The type of information obtained in such a run is illustrated by Figure 21, which presents an automatic recording of the height and position of the diffraction peak corresponding to the basal spacing of sodium-, potassium-, and lithium-montmorillonite.

B. Spectral Analysis

Information on the bonding state of elements in the clay structure, as well as on bond directions, can be obtained from *infrared* (IR) *absorption spectroscopy*. Since IR spectroscopy also provides much insight in the state of bonding of adsorbed molecules, some of the principles of the method will be discussed briefly in Chapter Ten (Adsorbed Water), and Chapter Eleven (Adsorbed Organic Molecules).

Information on the location of paramagnetic ions (Fe^{3+}) in the clay structure can be obtained from *electron spin resonance* (ESR) *spectroscopy*, since the resonance frequency is affected by the ions surrounding the paramagnetic ion. *Nuclear magnetic resonance* (NMR) *spectroscopy* can provide important information on the location of protons, both in the structure and in the adsorbed phase. The analysis of IR, ESR, and NMR spectra is complicated by a number of extraneous influences as well as by the multitude of structural influences, and it is therefore the domain of the spectroscopist.

IX. GENESIS AND DIAGENESIS OF CLAY MINERALS (37–45)—SYNTHETIC CLAYS (46–55)

Clay minerals may be formed in nature either by alteration (diagenesis) of mica minerals or by synthesis from oxides (genesis). Diagenesis may occur in the solid state or it may involve subsequent dissolution and recrystallization processes. Some of these processes have been studied under laboratory conditions. The synthetic route, that is, the formation of clay minerals from mixtures of colloidal oxides, has been carried out in the laboratory and even on a technical scale. Synthetic clays, particularly synthetic hectorite, are commercially available. One advantage of synthetic clays is

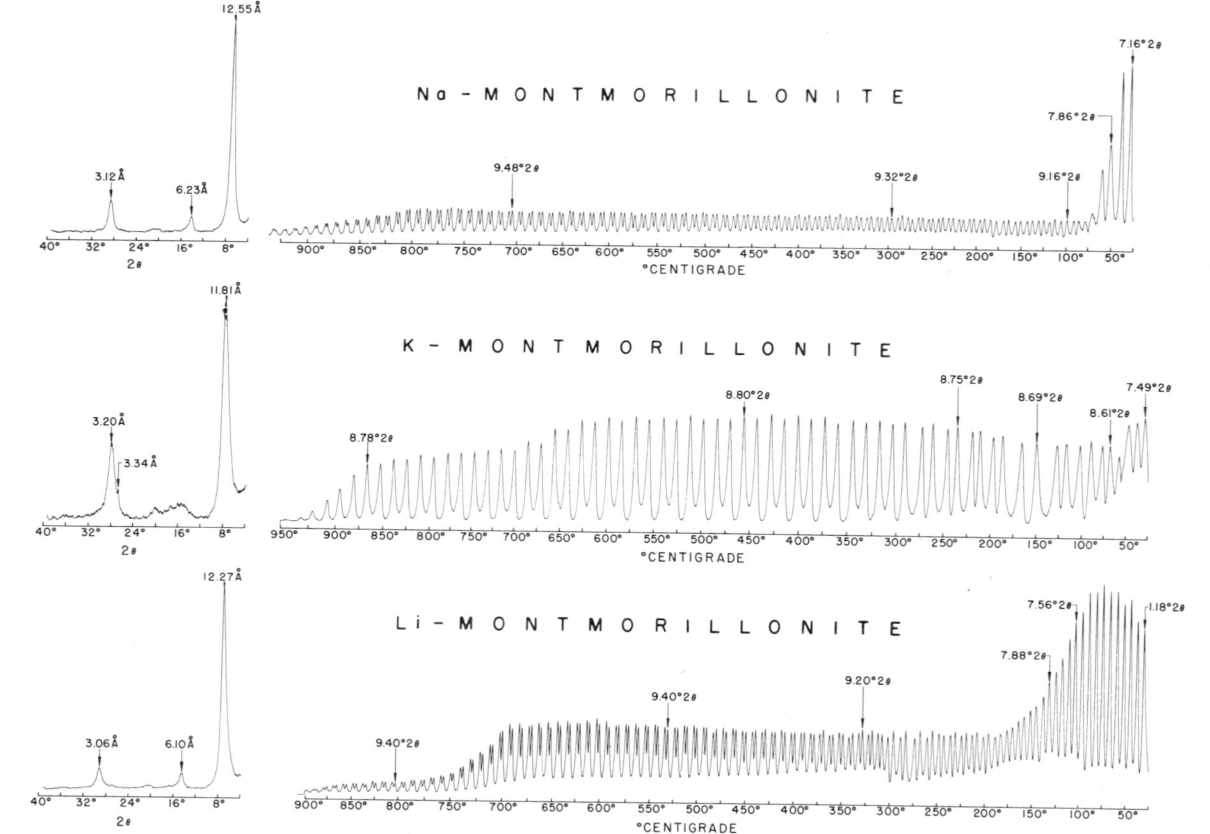

Figure 21. X-ray diffraction pattern and oscillating-heating X-ray powder diffraction diagrams of *c* spacing for sodium-, potassium-, and lithium-montmorillonite from room temperature to 900° at a rate of temperature rise of 5°C per minute. (The diagram on the left shows the first-, second-, and fourth-order peaks; on the right, the change in intensity and the shift in spacing is indicated for the first-order peak.) [Reprinted from R. A. Rowland, E. J. Weiss, and W. F. Bradley, "Dehydration of monoionic montmorillonites," *Clays, Clay Minerals*, **4**, 85–95 (1956).]

that they do not contain the abrasives present in natural clays, hence they are particularly suited for the formulation of lubricants.

The size of the clay particles is probably limited by the creation of tensions in the crystal. If they are formed from micas in the solid phase, the breakdown of the particles could be a result of the tensions created in the altered crystal. If they crystallize from solution by a nucleation and growth process, the growth of the crystallites will be limited by such tensions, but also the course of the nucleation process will affect the final dimensions of the particles.

There are several possible causes of structural tension. In the first place, the crystal imperfections due to isomorphous atom substitutions may introduce weak spots in the structure. Another cause of tensions which has been considered is the misfit between the oxygen-sharing superimposed octahedral and tetrahedral sheets. This misfit is clearly demonstrated by the projections of the tetrahedral and octahedral sheets in Figures 15*d* and 16*d*. In order to enable the sheets to share oxygen atoms, the atoms in the octahedral sheet cannot touch as they do in the tetrahedral sheet. It seems likely that the tensions due to this misfit would have greater consequences in the 1:1 layer minerals than in the symmetrically arranged 2:1 layer minerals. The occurrence of tubes, which are actually rolled up plates, in halloysite has been considered to be a result of the tensions created by the misfit of the two sheets (12). However, detailed X-ray studies of some layer silicates have revealed that the silicate sheet is able to adapt its dimensions to those of the octahedral sheet by a slight rotation of the tetrahedrons around an axis perpendicular to the sheet. By this rotation, the holes in the hexagonal oxygen rings of the ideal layer become smaller and the total area of the tetrahedral sheet is reduced (17, 18, 22, 23). (Such a reduced hole size agrees better with the size of the potassium ion occupying opposite holes in the illites; then, the potassium ions would only coordinate with six oxygen atoms, three on each side, instead of with six on each side.)

At the same time, the alternative proposal has been made that the rolling up of the halloysite plates may be due to lateral tensions caused by electrostatic attraction between separated cations on the layer surface and a charge center within the layer.

The mechanical strength of a clay particle will also depend on the average number of layers stacked in the particle and on whether or not water can penetrate between the layers.

X. CONCLUSIONS

The progress made in the analysis of clays, particularly by X-ray diffraction methods and by the electron-microscopic observation technique, has resulted in a sensible classification of clay minerals according to crystal structure. Major groups, such as 1:1 layer clays and swelling or nonswelling 2:1 layer clays, are distinguished. Depending on whether two octahedral positions are occupied by Al or whether three such positions are occupied by Mg, the subgroups of dioctahedral and trioctahedral clay minerals are separated. The great variety of species in each group results from different isomorphous substitution patterns, from different types of exchange cations, or from different ways in which layers are stacked in the crystal. Furthermore, in mixed-layer clays, different layer types are combined in a single stack.

It is the mineralogist's problem to unravel all these complications of the clay structure. The colloid chemist, who wants to interpret the colloid chemical behavior of clay systems, must be aware of these complications, and, therefore, the subject of clay mineralogy and morphology was discussed in some detail. The chemist must realize that he is working with a natural material and he should know what the analysis obtained from the mineralogist means in terms of homogeneity of the material. The results of X-ray analysis—and also of a chemical analysis—represent only a statistical-average composition of the unit cell of the clay. Still, the average composition and structure of one particle may be different from those of another particle. Within one particle, all stacked layers are not necessarily identical, and the same applies to the different unit cells in a layer (7).

Different particle size fractions obtained by selective centrifugation may show systematic variations of the average structure and composition with particle size. Actually, such variations may have been the very reason why the particle obtained that particular size.

Apart from the structural complications, one should be aware of the possible presence of organic and mineral impurities in the natural clay. When these impurities are of the same particle size as the clay particles, they are difficult to separate from the clay. For example, the presence of quartz particles in the colloidal size range is not rare in natural clays.

These words of warning do not mean that one has to be an optimist to work with clays, but a careful analysis of the minerals is necessary in interpreting their behavior.

References

NOMENCLATURE

1. Bailey, S. W., Brindley, G. W., Johns, W. D., Martin, R. T., and Ross, M. (1971), Summary of national and international recommendations on clay mineral nomenclature, *Clays, Clay Minerals*, **19**, 129–132.

X-RAY DIFFRACTION ANALYSIS OF CLAY STRUCTURE

2. Pauling, L. (1930), The structure of micas and related minerals, *Proc. Natl. Acad. Sci. U.S.*, **16,** 123–129.
3. Hofmann, U., Endell, K., and Wilm, D. (1933), Kristallstruktur und Quellung von Montmorillonit, *Z. Krist.*, **86,** 340–348.
4. Nagelschmidt, G. (1936), On the lattice shrinkage and structure of montmorillonite, *Z. Krist.*, **93A,** 481.
5. Edelman, C. H., and Favejee, J. C. L. (1940), On the crystal structure of montmorillonite and halloysite, *Z. Krist.*, **102,** 417–431.
6. McConnell, D. (1950), The crystal chemistry of montmorillonite, *Am. Mineralogist,* **35,** 166–172.
7. McAtee, J. L., Jr. (1958), Heterogeneity in montmorillonites, *Clays Clay Minerals,* **5,** 279–288.
8. Mathieson, A. McL., and Walker, G. F. (1952), The structure of vermiculite, *Clay Minerals Bull.*, **1,** 272–276.
9. Grim, R. E., Bray, R. M., and Bradley, W. F. (1937), The mica in argillaceous sediments, *Am. Mineralogist,* **22,** 813–829.
10. Gruner, J. W. (1932), The crystal structure of kaolinite, *Z. Krist.*, **83,** 75–88.
11. Robertson, H. S., Brindley, G. W., and Mackenzie, R. C. (1954), Mineralogy of kaolin clays from Pugu, Tanganyika, *Am. Mineralogist,* **39,** 118–139.
12. Bates, T. F., Hildebrand, F. A., and Swineford, A. (1950), Morphology and structure of endellite and halloysite, *Am. Mineralogist,* **35,** 463–484.
13. McMurchy, R. C. (1934), Structure of chlorites, *Z. Krist.*, **88,** 420–432.
14. Steinfink, H. (1958), The crystal structure of chlorite, I—A monoclinic polymorph; II—A triclinic polymorph, *Acta Cryst.*, **11,** 191–195 (I), 195–198 (II).
15. Bradley, W. F. (1940), The structural scheme of attapulgite, *Am. Mineralogist,* **25,** 405–410.
16. Preisinger, A. (1963), Sepiolite and related compounds, its stability and application, *Clays, Clay Minerals,* **10,** 365–371.
17. Radoslovich, E. W., and Norrish, K. (1962), The cell dimensions and symmetry of layer lattice silicates. I. Some structural considerations, *Am. Mineralogist,* **47,** 599–616.
18. Radoslovich, E. W. (1962), The cell dimensions and symmetry of layer lattice silicates. II. Regression relations, *Am. Mineralogist,* **47,** 617–636.
19. Eggleton, R. A., and Bailey, S. W. (1967), Structural aspects of dioctahedral chlorite, *Am. Mineralogist,* **52,** 673–689.
20. Bailey, S. W. (1969), Polytypism of trioctahedral 1:1 layer silicates, *Clays, Clay Minerals* **17,** 355–372.
21. Bailey, S. W. (1972), Determination of chlorite compositions by X-ray spacings and intensities, *Clays, Clay Minerals,* **20,** 381–388.
22. Leonard, R. A., and Weed, S. B. (1967), Influence of exchange ions on the b-dimension of dioctahedral vermiculite, *Clays, Clay Minerals* **15,** 149–161.
23. Low, P. F., Ravina, I., and White, J. (1970), Changes in b-dimension of Na-montmorillonite with interlayer swelling, *Nature,* **226,** 445–446.
24. Eirish, M. V., and Tret'jakova, L. I. (1970), The role of sorptive layers in the formation and change of the crystal structure of montmorillonite, *Clay Minerals,* **8,** 255–266.
25. Lahav, N., and Bresler, E. (1973), Exchangeable cation-structural parameter relationships in montmorillonite, *Clays, Clay Minerals,* **21,** 249–256.

COHESION IN LAYER SILICATES

26. Cruz, M., Jacobs, J., and Fripiat, J. J. (1972), The nature of the cohesion energy in kaolin minerals, *Int. Clay Conf., Madrid,* **1,** 59–70.
27. Giese, R. F., Jr. (1973), Interlayer bonding in kaolinite, dickite and nacrite, *Clays, Clay Minerals,* **21,** 145–149.
28. Wolfe, R. W., and Giese, R. F., Jr. (1974), The interlayer bonding in one-layer kaolin structures, *Clays, Clay Minerals,* **22,** 139–140.
29. Giese, R. F., Jr. (1974), Surface energy calculations for muscovite, *Nature, Phys. Sci.,* **248,** 580–581.
30. Giese, R. F., Jr. (1975), Interlayer bonding in talc and pyrophyllite, *Clays, Clay Minerals,* **23,** 165–166.

AMORPHOUS CLAYS

31. van Olphen, H. (1971), Amorphous clay materials, *Science,* **171,** 90–91.

SPECTRAL STUDIES

32. Serratosa, J. M., and Bradley, W. F. (1958), Determination of the orientation of OH bond axes in layer silicates by infrared absorption, *J. Phys. Chem.,* **62,** 1164–1167.
33. Stubičan, V., and Roy, Rustum (1961), Infrared spectra of layer-structure silicates, *J. Am. Ceramic Soc.,* **44,** 625–627.
34. Farmer, V. C., and Russell, J. D. (1967), Infrared absorption spectrometry in clay studies, *Clays, Clay Minerals,* **15,** 121–142.
35. Ishii, M., Nakahira, M., and Takeda, H. (1969), Far infrared absorption spectra of micas, *Proc. Intern. Clay Conf.,* Tokyo, **1,** 247–259.
36. Karickhoff, S. W., and Bailey, G. W. (1973), Optical absorption spectra of clay minerals, *Clays, Clay Minerals,* **21,** 59–70.

See also Appendix V, reference B-16.

CLAY FORMATION AND DIAGENESIS IN SEDIMENTARY ROCKS AND IN SOILS

37. Jenny, H. (1941), *Factors of Soil Formation,* McGraw, New York.
38. Barshad, I. (1959), Factors affecting clay formation, *Clays, Clay Minerals,* **6,** 110–132.
39. Jackson, M. L. (1959), Frequency distribution of clay minerals in major great soil groups as related to factors of soil formation, *Clays, Clay Minerals,* **6,** 133–143.
40. Weaver, C. E. (1958), A discussion on the origin of clay minerals in sedimentary rocks, *Clays, Clay Minerals,* **5,** 159–173.
41. Weaver, C. E. (1959), The clay petrology of sediments, *Clays, Clay Minerals,* **6,** 154–187.
42. Burst, J. F., Jr. (1959), Postdiagenetic clay mineral environment relationships in the Gulf Coast Eocene, *Clays, Clay Minerals,* **6,** 327–341.
43. Whitehouse, U. Grant, and McCarter, S. (1958), Diagenetic modification of clay minerals in artificial sea water, *Clays, Clay Minerals,* **5,** 81–119.
44. Correns, C. W. (1963), Experiments on the decomposition of silicates and discussion of chemical weathering, *Clays, Clay Minerals,* **10,** 443–459.
45. Millot, G. (1972), Data and tendencies of recent years in the field of genesis and synthesis of clays and clay minerals, *Proc. Intern. Clay Conf.,* Madrid, **1,** 205–212.

See also Appendix V, references B-7.

SYNTHESIS OF CLAY MINERALS

46. Roy, R. (1954), The application of phase equilibrium data to certain aspects of clay mineralogy, *Clays, Clay Minerals,* **2,** 124–140.
47. Caillère, S., Oberlin, A., and Hénin, S. (1954), Etude au microscope électronique de quelques silicates phylliteux obtenus par synthèse à basse température, *Clay Minerals Bull.,* **2,** 146–155.
48. Hénin, S. (1956), Synthesis of clay minerals at low temperatures, *Clays, Clay Minerals,* **4,** 54–60.
49. Yoder, H. S. (1959), Experimental studies on micas; a synthesis, *Clays, Clay Minerals,* **6,** 42–60.
50. Granquist, W. T., and Pollack, S. S. (1960), A study of the synthesis of hectorite, *Clays, Clay Minerals,* **8,** 150–169.
51. Fripiat, J. J., and Gastuche, M. C. (1963), L'état d'organisation des produits de départ et la synthèse des argiles, *Proc. Intern. Clay Conf.,* Stockholm, **1,** 53–66.
52. Neumann, B. S. (1965), Behaviour of a synthetic clay in pigment dispersions, *Rheologica Acta,* **4,** 250–255.
53. Granquist, W. T., and Pollack, S. S. (1967), Clay mineral synthesis-II: A randomly interstratified aluminian montmorillonoid, *Am. Mineralogist,* **52,** 212–226.
54. Warren, D. C., and McAtee, J. L., Jr. (1968), A morphological study of selected synthetic clays by electron microscopy, *Clays, Clay Minerals,* **16,** 271–274.
55. Granquist, W. T., Hoffman, G. W., and Boteler, R. C. (1972), Clay mineral synthesis-III: Rapid hydrothermal crystallization of an aluminian smectite, *Clays, Clay Minerals,* **20,** 323–330.

CATION EXCHANGE CAPACITY

56. Bower, C. A., and Truog, E. (1940), Base exchange capacity determinations of soils and other materials using colorimetric manganese method, *Ind. Eng. Chem., Anal Ed.,* **12,** 411–413.
57. Mackenzie, R. C. (1951), A micro-method for determination of cation exchange capacity of clay, *J. Colloid Sci.,* **6,** 219–222; *Clay Minerals Bull.,* **1,** 203–204 (1952).
58. van Olphen, H. (1951), A tentative method for the determination of the base exchange capacity of small samples of clay minerals, *Clay Minerals Bull.,* **1,** 169–170.
59. Perkins, A. T. (1952), Determination of cation exchange capacity of soils by use of "Versenate," *Soil Sci.,* **74,** 433–446.
60. Mortland, M. M., and Mellor, J. L. (1954), Conductometric titration of soils for cation exchange capacity, *Soil Sci. Soc. Am. Proc.,* **18,** 363–364.
61. Thomas, H. C., and Gaines, G. L. (1954), The thermodynamics of ion exchange in clay minerals. A preliminary report on the system montmorillonite-Cs-Sr, *Clays, Clay Minerals,* **2,** 398–403.
62. Frysinger, G. R., and Thomas, H. C. (1960), Adsorption studies on clay minerals, VII. Yttrium-Cesium and Cerium(III)—Cesium on montmorillonite, *J. Phys. Chem.,* **64,** 224–227.
63. Walker, G. F. (1959), Diffusion of exchangeable cations in vermiculite, *Nature,* **184,** 1392.
64. Schwertmann, U. (1962), Die selektive Kationensorption der Tonfraktion einiger Boden aus Sedimenten, *Z. Pflanzenernähr., Düng., Bodenkunde,* **97,** 9–25.
65. Hinckley, D. N. (1962), An X-ray spectrographic method for measuring base exchange capacity, *Am. Mineralogist,* **47,** 993–996.

66. Malcolm, R. L., and Kennedy, V. C. (1969), Rate of cation exchange on clay minerals as determined by specific ion electrode techniques, *Soil Sci. Soc. Am. Proc.*, **33**, 247–253.
67. Malcolm, R. L., Kennedy, V. C., and Jenne, E. A. (1969), Determination of cation exchange capacity with the potassium specific ion electrode, *Proc. Intern. Clay Conf.*, Tokyo, **1**, 573–583.
68. Bache, B. W. (1970), Barium isotope method for measuring cation-exchange capacity of soils and clays, *J. Sci. Fd. Agric.*, **21**, 169–171.
69. Brindley, G. W., and Thompson, T. D. (1970), Methylene blue absorption by montmorillonites. Determination of surface areas and exchange capacities with different initial cation saturations, *Israel J. Chem.*, **8**, 409–415.
70. Pham Ti Hang and Brindley, G. W. (1970), Methylene blue adsorption by clay minerals. Determination of surface areas and cation exchange capacities, *Clays, Clay Minerals*, **18**, 203–212.
71. Gast, R. G., and Klobe, W. D. (1971), Sodium-lithium exchange equilibria on vermiculite at 25°C and 50°C, *Clays, Clay Minerals*, **19**, 311–320.
72. Bundy, W. M., Murray, H. H., and Harrison, J. L. (1972), Influences of sample preparation on the surface properties of kaolinite, *Proc. Intern. Clay Conf.*, Madrid, *Kaolin Symposium*, 159–170.
73. van Bladel, R., Gaviria, G., and Laudelout, H. (1972), A comparison of the thermodynamic, double layer theory and empirical studies of the Na-Ca exchange equilibria in clay-water systems, *Proc. Intern. Clay Conf.*, Madrid, **2**, 15–30.
74. Banin, A. (1972), Factors affecting the kinetics and mechanism of ion exchange reactions between clays and other solid ion exchangers, *Proc. Intern. Clay Conf.*, Madrid, **2**, 31–42.
75. Busenberg, E., and Clemency, C. V. (1973), Determination of the cation exchange capacity of clays and soils using an ammonia electrode, *Clays, Clay Minerals*, **21**, 213–218.
76. Ferris, A. P., and Jepson, W. B. (1975), The exchange capacities of kaolinite and the preparation of homoionic clays, *J. Coll. Interface Sci.*, **51**, 245–259.

See also Appendix III, references 10–20.

PENETRATION OF THE CRYSTALS

77. Greene-Kelly, R. (1953), Irreversible dehydration in montmorillonite, *Clay Minerals Bull.*, **2**, 52–56.
78. White, J. L. (1956), Reactions of molten salts with layer-lattice silicates, *Clays, Clay Minerals*, **4**, 133–146.
79. White, J. L. (1958), Layer charge and interlamellar expansion in muscovite, *Clays, Clay Minerals*, **5**, 289–294.
80. Weiss, A., and Koch, G. (1961), Uber einen Zusammenhang zwischen dem Verlust des innerkristallinen Quellungsvermögen beim Erhitzen und dem Schichtaufbau bei glimmerartigen Schichtsilikaten, *Z. Naturforsch.*, **16**, 68–69.
81. Glaeser, R., and Mering, J. (1967), Effet de chauffage sur les montmorillonites saturées de cations de petit rayon, *C. R. Acad. Sci. Paris*, **265**, 833–835.
82. Brindley, G. W., and Ertem, G. (1971), Preparation and solvation properties of some variable charge montmorillonites, *Clays, Clay Minerals*, **19**, 399–404.
83. Calvet, R., and Prost, R. (1971), Cation migration into empty octahedral sites and surface properties of clays, *Clays, Clay Minerals*, **19**, 175–186.
84. Garcia Verduch, A., and Moya Corral, J. S. (1972), Reaction at low temperatures between kaolin and lithium carbonate, *Proc. Intern. Clay Conf.*, Madrid, **1**, 173–184.

POTASSIUM FIXATION AND EXCHANGE

85. Rich, C. I., and Lutz, J. A. (1965), Mineralogical changes associated with NH_4^+ and K-fixation in soil clays, *Soil Sci. Soc. Am. Proc.*, **29,** 167–169.
86. Weir, A. H. (1965), Potassium retention in montmorillonites, *Clay Minerals*, **6,** 17–22.
87. van Olphen, H. (1966), Collapse of potassium montmorillonite clays upon heating—"Potassium fixation," *Clays, Clay Minerals*, **14,** 393–406.
88. Thompson, T. D., Wentworth, S. A., and Brindley, G. W. (1967), Hydration states of an expanded phlogopite in relation to interlayer cations, *Clay Minerals*, **7,** 43–49.
89. Barshad, I., and Kishk, F. M. (1970), Factors affecting potassium fixation and cation exchange capacities of soil vermiculite clays, *Clays, Clay Minerals*, **18,** 127–138.
90. Sawhney, B. L. (1972), Selective sorption and fixation of cations by clay minerals: A review, *Clays, Clay Minerals*, **20,** 93–100.
91. Ross, G. J., and Rich, C. I. (1973), Effect of particle thickness on potassium exchange from phlogopite, *Clays, Clay Minerals*, **21,** 77–82.
92. Ross, G. J., and Rich, C. I. (1973), Effect of particle size on potassium sorption by potassium-depleted phlogopite, *Clays, Clay Minerals*, **21,** 83–88.

ELECTRON OPTICAL STUDIES

93. Brindley, G. W. (1961), Identification of clay minerals by single crystal electron diffraction, *Am. Mineralogist*, **46,** 1005–1016.
94. Mering, J., Mathieu-Sicaud, A., and Perrin-Bonnet, I. (1950), Observation au microscope électronique de montmorillonite saturée par differents cations, *Trans. Intern. Congr. Soil Sci., 4th Amsterdam*, Vol. III, 29–32.
95. Taggart, M. S., Milligan, W. O., and Studer, H. P. (1956), Electron-micrographic studies of clays, *Clays, Clay Minerals*, **4,** 31–64.
96. Cooper, W. D., Craik, D. J., and Parfitt, G. D. (1965), Analysis of particle size distribution in colloidal dispersions by electron microscopy, *Koll. Zeit.*, **205,** 108.
97. Mering, J., and Oberlin, A. (1967), Electron-optical study of smectites, *Clays, Clay Minerals*, **15,** 3–26.
98. Borst, R. L., and Keller, W. D. (1969), Scanning electron micrographs of API reference clay minerals and other selected samples, *Proc. Intern. Clay Conf.*, Tokyo, **1,** 871–901.
99. Bohor, B. F., and Hughes, R. E. (1971), Scanning electron microscopy of clays and clay minerals, *Clays, Clay Minerals*, **19,** 49–54.
100. Güven, N. (1974), Electron-optical investigations on montmorillonites.—I. Cheto, Camp Berteaux and Wyoming montmorillonites, *Clays, Clay Minerals*, **22,** 155–166.
101. Güven, N. (1974), Factors affecting selected area electron diffraction patterns of micas, *Clays, Clay Minerals*, **22,** 97–106.
102. Greene, R. S. B., Murphy, P. J., Posner, A. M., and Quirk, J. P. (1974), A preparative technique for electron microscopic examination of colloid particles, *Clays, Clay Minerals*, **22,** 155–188.
103. Jepson, W. B., and Rowse, J. B. (1975), The composition of kaolinite—An electron microscope microprobe study, *Clays, Clay Minerals*, **23,** 310–317.

See also Appendix V, references B-15 and B-19.

DIFFERENTIAL THERMAL ANALYSIS

104. Orcel, J., and Caillère, S. (1933), L'Analyse thermique différentielle des argiles à montmorillonite, *Compt. Rend.*, **197,** 774–777.
105. Grim, R. E., and Rowland, R. A. (1942), Differential thermal analysis of clay minerals and other hydrous materials, *Am. Mineralogist*, **27,** 746–761, 801–818.

106. Kerr, P. F., and Kulp, J. L. (1948), Multiple differential thermal analysis, *Am. Mineralogist,* **33,** 387–419.
107. Mackenzie, R. C. (1950), Differential thermal analysis of clay minerals, *Trans. Intern. Congr. Soil Sci., 4th, Amsterdam,* Vol. II, 55–59.
108. Rowland, R. A., and Lewis, D. R. (1951), Furnace atmosphere control in differential thermal analysis, *Am. Mineralogist,* **36,** 80–91.
109. Arens, P. L. (1951), A study of the differential thermal analysis of clays and clay minerals, Thesis, Wageningen, The Netherlands.
110. Bradley, W. F., and Grim, R. E. (1951), High temperature thermal effects of clay and related materials, *Am. Mineralogist,* **36,** 182–201.
111. Allison, E. B. (1955), Quantitative thermal analysis of clay minerals, *Clay Minerals Bull.,* **2,** 242–254.
112. Murray, P., and White, J. (1955), Kinetics of clay dehydration, *Clay Minerals Bull.,* **2,** 255–264.
113. Sewell, E. C. (1955), The consequences for differential thermal analysis of assuming a reaction of the first order, *Clay Minerals Bull.,* **2,** 233–241.
114. Mackenzie, R. C., Editor, *Differential Thermal Analysis,* Vol. 1–*Fundamental Aspects* (1970); Vol. 2—*Applications* (1972), Academic, New York.
See also Appendix V, reference B-14.

ELECTRON SPIN RESONANCE—NUCLEAR MAGNETIC RESONANCE

115. Adrian, F. J. (1968), Guidelines for interpreting ESR spectra of paramagnetic species adsorbed on surfaces, *J. Colloid Interface Sci.,* **26,** 317–354.
116. Angel, B. R., and Hall, P. L. (1972), Electron spin resonance studies of kaolins, *Proc. Intern. Clay Conf.,* Madrid, **1,** 71–86.
117. Clementz, D. M., Pinnavaia, T. J., and Mortland, M. M. (1973), Stereochemistry of hydrated copper(II) ions on the interlamellar surfaces of layer silicates. An electron spin resonance study, *J. Phys. Chem.,* **77,** 196–200.
118. Clementz, D. M., Mortland, M. M., and Pinnavaia, T. J. (1974), Properties of reduced charge montmorillonites: Hydrated Cu(II) ions as a spectroscopic probe, *Clays, Clay Minerals,* **22,** 49–58.
119. Ducros, P., and Dupont, M. (1962), Nuclear magnetic resonance study of the protons in clay, *Compt. Rend.,* **254,** 1409–1410.
See also Chapter Ten, references 58, 60, 61, and 68.

CHAPTER SIX

Particle Size and Shape, Surface Area, and Density of Charge

I. SIZE AND SHAPE OF CLAY PARTICLES (1–17)

The electron micrographs of clays (Appendix V, references B-15, B-19), as presented previously in Figures 2–5, demonstrate the variety of shapes and sizes of clay particles. As mentioned before, there is no guarantee that the electron micrographs of dry clay particles reflect the size and shape of the particles in the suspended condition. The same is true for microscopic slides prepared from suspensions of clay particles of microscopic size such as many kaolinites. However, with proper deposition and drying techniques useful information can be obtained in many situations. From such micrographs, particle size distributions can be rapidly obtained with computerized scanning equipment which eliminates tedious manual handling.

A. Direct Method—Ultramicroscopical Counting

When the particles of the suspension are of greater than amicronic size, their number can be counted directly in the ultramicroscope. From the weight concentration of the suspension and the number of particles in a given volume, the average weight of a single particle may be computed.

For qualitative observations, an ordinary microscope is easily adapted to dark-field observation by replacing the condenser system by a *dark-field condenser,* for example, a so-called *cardioid condenser.* Through such a condenser, light is admitted to the sample from the sides at such angles that it cannot enter the objective lens directly. Since in this arrangement, the volume of observation cannot be defined, quantitative particle counting should be carried out with a different arrangement, which is realized in the *Zsigmondy-Siedentopf ultramicroscope.* In this instrument, the solution is contained in a small rectangular cell (the *Biltz cuvette*) which is placed under the objective lens. A beam of light from an arc lamp is shaped into a thin, flat, horizontal beam by a system of lenses and a narrow precision-made slit. The thickness of this beam, which can be measured in the microscope by turning the slit 90 degrees, determines the depth of the volume in which the scattering particles are observed. A diaphragm placed in the eye-piece of the microscope determines the surface area of the illuminated volume element which is observed in the microscope. Owing to the Brownian motion, particles will continuously disappear from the illuminated volume, and others will enter. Therefore, when determinations are made of the number of particles in the volume element, the count varies at random. A large number of counts must be made at regular intervals to arrive at a reasonably accurate average number. The frequency of counting, the number of counts, and the resulting accuracy are evaluated from statistical theory. Before the measurement, the solution is diluted to such an extent that not more than about five, or at most six, particles appear in the volume element at any time, since the visual instantaneous counting of more particles appears to be impossible.

B. Indirect Methods

In a variety of indirect methods for the determination of particle size in dispersions, certain bulk physical properties of the suspensions are measured, and the average particle dimensions are computed on the basis of theoretical relations between particle dimensions and physical properties. Measurements of the following physical properties of suspensions have been used for the evaluation of particle size and shape: viscosity, sedimentation velocity, the decay of optical birefringence induced by an orientation of the particles in an electric field, or by shear, the angular dependence of scattered-light intensity, intensity of transmitted light, and low-angle scattering of X-rays.

In the theoretical treatment of the data, the shape of the particles is idealized. Plates are considered as thin cylindrical discs or as ellipsoids of revolution with a small *c* axis with respect to the *b* axis, whereas rods are treated as long cylinders or as ellipsoids of revolution with a large *c* axis as compared with the *b* axis.

It is always necessary to combine two methods of observation if one desires to compute both the *c* axis and the *b* axis of a particle idealized as an ellipsoid of revolution. For example, from viscosity measurements the ratio of the two axes may be computed (see Chapter Nine, Section III-C). Ultramicroscopical counting supplies the value of the average volume of one particle if its density is known. By combining the results of the two methods, one can compute the length of both the *b* axis and the *c* axis.

A serious difficulty in the evaluation of particle size is that natural samples always contain particles of a wide range of sizes, even after the sample is fractionated by sedimentation in a centrifuge, by which the size range is only narrowed. Therefore, with the various techniques, only an average value for the dimensions is obtained, and the way in which the method takes the average may differ. One method may yield a number average, another technique a weight average, and so on. Such differences must be accounted for when the results of two techniques are combined to obtain values for the average lengths of the b and c axes.

For example, when combining counting data with viscosity data, the resulting particle parameters are not necessarily exactly number averages: the viscosity depends on the total volume of all particles, hence the axis ratios of large particles have a relatively large effect if they differ significantly from the axis ratios of small particles. Fortunately, however, in natural clays the larger particles are usually thicker, and the variation in axis ratio is comparatively small. In narrow size fractions, the axis ratio may be considered constant, hence the combining of counting and viscosity data will yield averages which are very close to number averages.

Sedimentation rate measurements yield weight average values for the "equivalent spherical radius" by calculating the particle radius as if it were a sphere, applying Stokes' formula. The equivalent spherical radius is related to the a and b dimensions of ellipsoids of revolution by

$$r_{eq} = \left(\frac{bc}{F}\right)^{1/2}$$

in which F is a form factor which depends on the axis ratio. For large axis ratios, as commonly apply for clay particles, $F = 0.66$. (See Appendix V, reference B-1, pp. 70, 71.)

From birefringence decay curves, the rotational diffusion constant for the particles can be derived, from which the major axis or particle diameter may be calculated (4). The average obtained is close to but not exactly equal to a weight average.

Most indirect methods yield only information on average values for the particle parameters. One method with which a complete particle size distribution can be obtained is angular dependence of scattered light intensity at very low forward scattering angles. The interpretation of the scattering data is basically empirical and yields a weight distribution of particle diameters. Precision instrumentation is required since measurements should be made down to a few hundredths of a degree from the primary beam. (The more common method of measuring dissymmetry of scattering at 45 and 135 degrees is not applicable in the case of the large clay plates for which dissymmetry ratios as large as 5–9 are observed, for example, in Wyoming, bentonite sols.) From the derived size distribution curves, a weight average diameter, as well as a number average diameter, can be derived and compared with averages obtained from other methods, and in specific cases that have been studied, a reasonable agreement resulted. The two averages may differ significantly, for example, for sodium Wyoming bentonite sols, the weight average appeared to be twice the number average, even though the size fraction was rather narrow.

Finally, the large platelike particles in clay suspensions can interact at quite low concentrations. Therefore, in order to minimize the effects of particle interaction when obtaining information on particle dimensions, measurements should be taken at the lowest possible concentration range, and the results should preferably be extrapolated to zero concentration. As an example, the rate of sedimentation of sodium Wyoming bentonite particles can be determined from the rate at which the interface between suspension and supernatant moves, using concentrations as low as 0.01 g per 100 cm^3. This rate, which is nearly constant between 0.1 and 0.05 g per 100 cm^3, increases almost a factor 4 between 0.05 and 0.01 g per 100 cm^3, resulting in a calculated increase by a factor 2 of the equivalent spherical diameter of the particles, applying Stokes' formula. Apparently, only at very low concentrations is particle interaction absent, allowing free fall. In standard pipette analysis to determine particle size distributions in soils, much higher concentrations are prescribed, hence results of such tests should give low values.

In particle counting experiments it has been observed that the number of particles increases with time after preparation of the suspension until an equilibrium number is reached. This number is usually larger at smaller concentrations (14). Apparently, layer packages split into thinner packages. Combining counting and viscosity data, it was found that the average thickness of particles in Wyoming bentonite suspensions was about eight layers for the Ca clay, and about two to three layers for the sodium clay. The particles do not seem to split up completely into single layers. This result is in agreement with the estimated thickness of sodium Wyoming bentonite particles of about 20 A, as measured from the breadth of the shadows in the electron micrograph of Figure 2. A possible cause for such pair formation might be the presence of a permanent dipole perpendicular to the plates which could have resulted from nonrandom isomorphous substitutions. The existence of a permanent dipole on bentonite particles has been derived from the way in which particles orientate in an electric field as a function of time after applying an electric pulse. The electro-optical birefringence rise curve indicates that the response of the particle is not solely due to dipoles induced in the diffuse double layers on the particle surfaces.

II. DETERMINATION OF THE SURFACE AREA OF CLAYS

A. Computation from the Particle Dimensions

The total surface area per gram of clay may be derived from the average dimensions and the average weight of the clay particles measured according to the procedures described in the previous section.

B. Computation from the Crystallographic Cell Dimensions

For clays with platelike particles, the total surface area per gram of clay, including each side of each layer in the stack, may be computed from the unit-cell dimensions (known from X-ray analysis) and from the unit-cell weight (known from chemical analysis). In montmorillonites, this total layer surface area per gram of clay represents both the external and the internal surface area which is accessible to exchange ions and to water or other polar molecules. This total surface area may be as high as 800 m^2/g. The computation is carried out in detail in Appendix II.

The total external area of the flat part of the surface per gram of clay is found by dividing the total layer surface area by the number of layers which are stacked in the particles. This number may be derived from the thickness of the particles obtained by applying the methods described in the preceding paragraph.

The total edge surface area per gram of clay may be computed if the diameter of the particles is known and if their shape is idealized as flat cylinders.

Typical values for external surface areas of kaolinites are 10–30 m^2/g and for montmorillonites 40–100 m^2/g.

C. Direct Determination from Adsorption Data (18–30)

A direct method for the determination of the surface area of a powdered solid from the adsorption isotherm for a gas or vapor on the solid has been developed by Brunauer, Emmett, and Teller. The "BET equation" is based on an analysis of the process of multilayer adsorption of a vapor on a solid surface. The equation may be written in the following linear form:

$$\frac{x}{n(1-x)} = \frac{1}{cm} + \frac{(c-1)x}{cm}$$

in which x is the relative vapor pressure p/p_0; n is the weight of vapor adsorbed at a given vapor pressure; m is the weight of vapor adsorbed at monolayer coverage, at unit weight of the solid; c is a constant which is related to the energy of adsorption for the first monolayer. (From the second layer on, only the energy of condensation of the vapor is considered.) At relative vapor pressures from about 0.1 to 0.3, the plot of $x/n(1 - x)$ versus x is usually linear, and the values of m and c can then be obtained from the slope and intercept of the plot. Knowing the surface area covered per molecule of the vapor at monolayer coverage, as obtained from calibration with surfaces of known area, the total area per unit weight can be calculated from the monolayer coverage m.

When only monolayer adsorption takes place, a Langmuir isotherm is obtained which can be written as

$$\frac{x}{n} = \frac{1}{bm} + \frac{x}{m}$$

from which m and the constant b are found. From the value of m, the total surface area may again be calculated.

Commonly, nitrogen gas is used in the routine BET method. When applied to clays, the total external surface area per gram of clay, including the edge surface area, is determined. This is also true for montmorillonites since the nitrogen molecules do not penetrate between layers, unless the layers were previously parted by interlayer material, but not completely covered by such materials.

When the surface-area determination is based on water-vapor-adsorption isotherms, the internal surface area of montmorillonites is also measured. From the ratio of the "nitrogen surface area" and the "water-vapor-adsorption area" of a montmorillonite clay, the number of layers per particle in the dry state can be computed, with the small contribution of the edge surface area to the total area neglected. In nonswelling clays, the nitrogen and water-vapor areas should be identical, unless in the stacking of the individual plates, void spaces are enclosed which are accessible to water vapor but not to nitrogen gas.

Alternative methods to determine total interlayer and external surface area are based on the adsorption of polar organic compounds (e.g., glycols and glycerol: references 31–34) from the liquid phase, or of cationic compounds (e.g., methylene blue: reference 35) from hydrous solution. Monolayer coverage for water as well as for these organic compounds is more dependent on cation population on the clay surfaces than in the case of nitrogen. Hence one should expect larger variations from the average area per molecule and therefore a larger uncertainty in the calculated surface area. (See reference 25 for a survey of these methods.)

In the glycol and glycerol methods, monolayer coverage is achieved by removing excess liquid at specified conditions of heating and vapor pressure. In the methylene blue method, monolayer coverage is determined from the MB concentration at which the clay is completely flocculated.

This point differs usually from the concentration at which the adsorption of the MB cation is equivalent to the cation exchange capacity.

III. DENSITY OF CHARGE OF THE SURFACE

For the colloid chemical behavior of clays, the density of the charge on the surface is of primary importance. This parameter can be derived from the exchange capacity per gram and the surface area per gram.

Montmorillonites, in which the exchange cations are distributed over all internal and external layer surfaces, have a charge density on each surface of the order of 10 $\mu C/cm^2$, corresponding to an exchange capacity of 70–100 meq per 100 g (see the computations in Appendix II). For illites, the surface density of charge is about one and one-half times larger than that of montmorillonites, as derived from the analytically determined total number of cations on exterior and interior surfaces, although the exchange capacity of illites is lower than that of montmorillonites, since only those cations which are located on the exterior surfaces are exchangeable. The cation exchange capacity of kaolinites is rather low, but since the cations are exclusively located on the exterior surfaces, and the particles are rather thick, the density of charge of the small exterior surface area is still appreciable and is not much smaller than that of montmorillonites.

References

PARTICLE SIZE AND SHAPE, INDIRECT METHODS

1. Orr, Clyde, Jr., and Dallavalle, J. M. (1959), *Fine Particle Measurement—Size, Surface, and Pore Volume,* Macmillan, New York.
2. Kahn, A., and Lewis, D. R. (1954), The size of sodium montmorillonite particles in suspension from electro-optical birefringence studies, *J. Phys. Chem.,* **58,** 801–804.
3. Melrose, J. C. (1956), Light scattering evidence for the particle size of montmorillonite clay. Symposium on Chemistry in the Exploration and Production of Petroleum, Division of Petroleum, ACS, Dallas, Texas (preprinted).
4. Kahn, A. (1959), Studies on the size and shape of clay particles in aqueous suspensions, *Clays, Clay Minerals,* **6,** 220–236.
5. Hight, R., Jr., Higdon, W. T., and Schmidt, P. (1960), A small angle X-ray scattering study of sodium montmorillonite clay suspensions, *J. Chem. Phys.,* **33,** 1656–1661.
6. Shah, M. J. (1963), Electric birefringence of bentonite. Extension of saturation birefringence theory, *J. Phys. Chem.,* **67,** 2215–2218.
7. Shah, M. J., Thompson, D. C., and Hart, C. M. (1963), Reversal of electro-optical birefringence in bentonite suspensions, *J. Phys. Chem.,* **67,** 1170–1178.
8. Jennings, B. R., and Plummer, H. (1968), A study of the light scattered by hectorite solutions when subjected to electric fields, *J. Colloid Interface Sci.,* **27,** 377–387.
9. Jennings, B. R., Plummer, H., Closs, W. J., and Jerrard, H. G. (1969), Size and shape of Laponite (Type S) synthetic clay particles, *J. Colloid Interface Sci.,* **30,** 134–139.

10. Lahav, N. (1971), Particle geometry and optical density of clay suspensions, *Clays, Clay Minerals*, **19,** 283–288.
11. Jennings, B. R., Brown, B. L., and Plummer, H. (1970), Electrooptic study of an experimental synthetic clay in aqueous dispersion, *J. Colloid Interface Sci.*, **32,** 606–621.
12. Schweitzer, J., and Jennings, B. R. (1971), The association of montmorillonite studied by light scattering in electric fields, *J. Colloid Interface Sci.*, **37,** 443–457.
13. Jennings, B. R. (1973), Electro-optic scattering and birefringence for studying colloidal particles and their associations, *Particle Growth in Suspensions*, A. L. Smith, Editor, pp. 95–108, Academic, New York.
14. van Olphen, H. (1973), Unit layer association and dissociation in suspensions of montmorillonite clays, *Particle Growth in Suspensions*, A. L. Smith, Editor, pp. 83–94, Academic, New York.

SIZE FRACTIONATION

15. McEuen, R. B. (1964), Dielectrophoretic behavior of clay minerals I: Dielectrophoretic separation of clay mixtures, *Clays, Clay Minerals*, **12,** 549–556.
16. Müller-Vonmoos, M. (1971), Zur Korngrössenfraktionierung tonreicher Sedimente, *Schweiz. Mineralogische und Petrographische Mitteil*, **51,** 245–257.
17. Olivier, J. P., and Sennett, P. (1973), Particle size relationships in Georgia sedimentary kaolins II, *Clays, Clay Minerals*, **21,** 403–412.

SURFACE AREA FROM MONOLAYER ADSORPTION

18. Emmett, P. H., Brunauer, S., and Love, K. S. (1938), The measurement of surface areas of soils and soil colloids by the use of low temperature van der Waals adsorption isotherms, *Soil Sci.*, **45,** 57–65.
19. Brunauer, S., Emmett, P. H., and Teller, E. (1938), Adsorption of gases in multimolecular layers, *J. Am. Chem. Soc.*, **60,** 309–319.
20. Escard, J. (1950), Adsorption de l'azote à basse température par la montmorillonite. Influence de l'eau résiduelle et des cations échangeables, *Trans. Intern. Congr. Soil Sci., 4th Congr., Amsterdam*, Vol. III, 71–74.
21. Keenan, A. G., Mooney, R. W., and Wood, L. A. (1951), The relation between exchangeable ions and water adsorption on kaolinite, *J. Phys. & Colloid Chem.*, **55,** 1462–1474.
22. Mooney, R. W., Keenan, A. G., and Wood, L. A. (1952), Adsorption of water vapor by montmorillonite, I—Heat of desorption and application of the BET theory; II—Effect of exchangeable ions and lattice swelling as measured by X-ray diffraction, *J. Am. Chem. Soc.*, **74,** 1367–1371 (I), 1371–1374 (II).
23. Quirk, J. P. (1955), Significance of surface areas calculated from water-vapor-adsorption isotherms by use of the BET equation, *Soil Sci.*, **80,** 423–430.
24. Gregg, S. S., and Sing, K. S. W. (1967), *Adsorption, Surface Area and Porosity*, Academic, New York.
25. van Olphen, H. (1970), Determination of surface areas of clays—evaluation of methods, *Surface Area Determination*, Pure and Applied Chemistry Supplement, IUPAC, 255–271.
26. Everett, D. H., Parfitt, G. D., Sing, K. S. W., and Wilson, R. (1974), The SCI/IUPAC/NPL project on surface area standards, *J. Appl. Chem. Biotechnol.*, **24,** 199–219.

27. Knudson, M. L., Jr., and McAtee, J. L., Jr. (1973), The effect of cation exchange of tri(ethylenediamine)cobalt(III) for sodium on nitrogen adsorption by montmorillonite, *Clays, Clay Minerals,* **21,** 19–26.
28. Knudson, M. L., Jr., and McAtee, J. L., Jr. (1974), Interlamellar and multilayer nitrogen adsorption by homoionic montmorillonite, *Clays, Clay Minerals,* **22,** 59–66.
29. Fripiat, J. J., Cruz, M. I., Bohor, B. F., and Thomas, J., Jr. (1974), Interlamellar adsorption of carbon dioxide by smectites, *Clays, Clay Minerals,* **22,** 23–30.
30. Aylmore, L. A. G. (1974), Gas sorption in clay mineral systems, *Clays, Clay Minerals,* **22,** 175–184.
31. Dyal, R. S., and Hendricks, S. B. (1950), Total surface of clays in polar liquids as a characteristic index, *Soil Sci.,* **69,** 421–432.
32. Kinter, E. B., and Diamond, S. (1958), Gravimetric determination of monolayer glycerol complexes of clay minerals, *Clays, Clay Minerals,* **5,** 318–333.
33. Diamond, S., and Kinter, E. B. (1958), Surface areas of clay minerals as derived from measurements of glycerol retention, *Clays, Clay Minerals,* **5,** 334–347.
34. Carter, D. L., Heilman, M. D., and Gonzalez, C. L. (1965), Ethylene glycol monoethylether for determining surface area of silicate minerals, *Soil Sci.,* **100,** 356–360.
35. Pham Thi Hang and Brindley, G. W. (1970), Methylene Blue absorption by clay minerals. Determination of surface areas and cation exchange capacities, *Clays, Clay Minerals,* **18,** 203–212.

CHAPTER SEVEN

Electric Double-Layer Structure and Stability of Clay Suspensions

I. ELECTRIC DOUBLE-LAYER STRUCTURE (1–20)

A. The Double Layer on the Flat Layer Surfaces

In the preceding chapter it has been shown that the clay crystal carries a net negative charge as a result of isomorphous substitutions of certain electropositive elements by such elements of lower valence. The net negative layer charge is compensated by cations which are located on the layer surfaces. In the presence of water, these compensating cations have a tendency to diffuse away from the layer surface since their concentration will be smaller in the bulk solution. On the other hand, they are attracted electrostatically to the charged layers. The result of these opposing trends is the creation of an atmospheric distribution of the compensating cations in a diffuse electrical double layer on the exterior layer surfaces of a clay particle. The compensating cations between the layers of the stack are confined to the narrow space between opposite layer surfaces.

The compensating cations act as the counter-ions of the double layers, and, like all counter-ions, they are exchangeable for other cations. Since the exchange capacity of the double layer has a rather high value when expressed per unit weight of the clay, the exchange capacity is a rather outstanding property of clay. Therefore, it has become customary to refer to the counter-ions as the exchange cations of the clay.

The electric double layer on the layer surfaces has a constant charge which is solely determined by the type and degree of isomorphous substitutions. Therefore, the layer-surface charge density is independent of the presence of electrolytes in the suspension. Such a double layer is rather uncommon in hydrophobic colloids. The double layer in such systems is usually created by the adsorption of potential-determining ions; then the surface potential is a constant, and the charge varies with bulk electrolyte concentration.

There are indications that in many clays specific adsorption forces between the layers and the counter-ions exist. Hence a larger than normal fraction of the counter-ions will be located on the surface, and a smaller fraction will be in the diffuse layer. Therefore, the Stern-Gouy model would be more commonly applicable to the clay double layer than the Gouy model. An obvious example of specific layer-to-counter-ion interactions is an illite clay in which potassium ions provide a strong link between layer surfaces. To what extent cation-to-surface bonding energies play a part in various ion forms of montmorillonites and other clays is still a matter of controversy.

B. The Double Layer on the Edge Surfaces of Clay Plates

The flat layer surfaces are not the only surfaces of the platelike clay particles; they also expose an edge surface area. The atomic structure of the edge surfaces is entirely different from that of the flat-layer surfaces; therefore, the question should be raised whether possibly a different electrical double layer exists on the edge surfaces.

At the edges of the plates, the tetrahedral silica sheets and the octahedral alumina sheets are disrupted, and primary bonds are broken. This situation is analogous to that on the surface of silica and alumina particles in silica and alumina sols. On such surfaces an electric double layer is created by the adsorption of potential-determining ions. When speculating about the possible structure of an edge double layer, one might develop a picture on the basis of an analogy with the double layers on silica and alumina particles, keeping in mind that the crystal structure at the broken edges of the clay is not completely identical with that at the surface of either silica or alumina particles.

That part of the edge surface at which the octahedral sheet is broken may be compared with the surface of an alumina particle. We have seen earlier that such a surface carries a positive double layer in acid solution with Al ions acting as potential-determining ions and a negative double layer in alkaline solution with hydroxyl ions acting as potential-determining ions. The point of zero charge does not necessarily occur at neutral pH; it is actually known to vary with the crystal structure of the alumina particle (1). Hence there is a definite possibility that in a neutral clay suspension a positive double layer is created on the edge surfaces owing to the exposed alumina sheet. This double layer may become more positive with decreasing pH, and its sign may be reversed with increasing pH. Despite the fact that the net electrophoretic charge of the clay particles is always negative, the existence of a positive edge double layer may not be excluded a priori, since the negative double layer on the large flat surfaces may well predominate in the electrophoretic experiment.

The part of the edge surface at which the broken tetrahedral sheet is exposed may be compared with the surface of a silica particle. Although silica surfaces normally carry a negative double layer, their charge is known to become positive in the presence of very small amounts of aluminum ions in the suspension. Since such small concentrations of aluminum ions will occur in the equilibrium liquid of a clay suspension owing to the slight solubility of the clay, it is quite possible to have a positive double layer on the broken silica surface of the edge. Moreover, it is possible that the silica sheets are preferentially broken at the places where aluminum ions have substituted silicon, so that a surface is exposed which is comparable with an alumina rather than with a silica surface. Hence, under appropriate conditions the entire edge surface area may well carry a positive double layer.

Several observations support the concept of a positive edge charge. One interesting, relevant experiment in this respect has been performed by Thiessen (2). He mixed a kaolinite sol and a negative gold sol and prepared an electron micrograph of the mixture. It appeared that the small negative gold particles were exclusively adsorbed at the edge surfaces of the large kaolinite plates. An electron-microscopic picture of such a mixture is shown in Figure 22. It may be mentioned that Thiessen did not interpret his electron micrographs to be an indication of a positive edge charge on the kaolinite, but he considered his results as a proof for the generally higher adsorptive activity of crystal edges. However, it seems rather obvious that the preferred edge attachment of the gold particles is a result of mutual flocculation of the negative gold particles and the positive kaolinite edges.

Also supporting the concept of the positive edge charge is the fact that clays show a certain anion adsorption capacity under certain conditions, although this capacity is rather small. For example, a kaolinite shows a small anion adsorption capacity in a slightly acid environment but not in an alkaline suspension. It seems likely that a positive edge double layer is responsible for the adsorption of anions acting as counter-ions. Then, the simultaneously observed small cation exchange capacity may be attributed to a negative double layer on the flat surfaces which results either from isomorphous substitution in the kaolinite crystal structure, or from adsorbed alumina-silicate coatings on the face surfaces. This picture is more reasonable than that in which the cation exchange capacity of the kaolinites is attributed to a double layer with negative sign on the broken-bond edge surfaces, as has been assumed in the past.

Prompted by Thiessen's results with the kaolinite-gold sol mixture and by the above considerations of the edge surface structure, the author

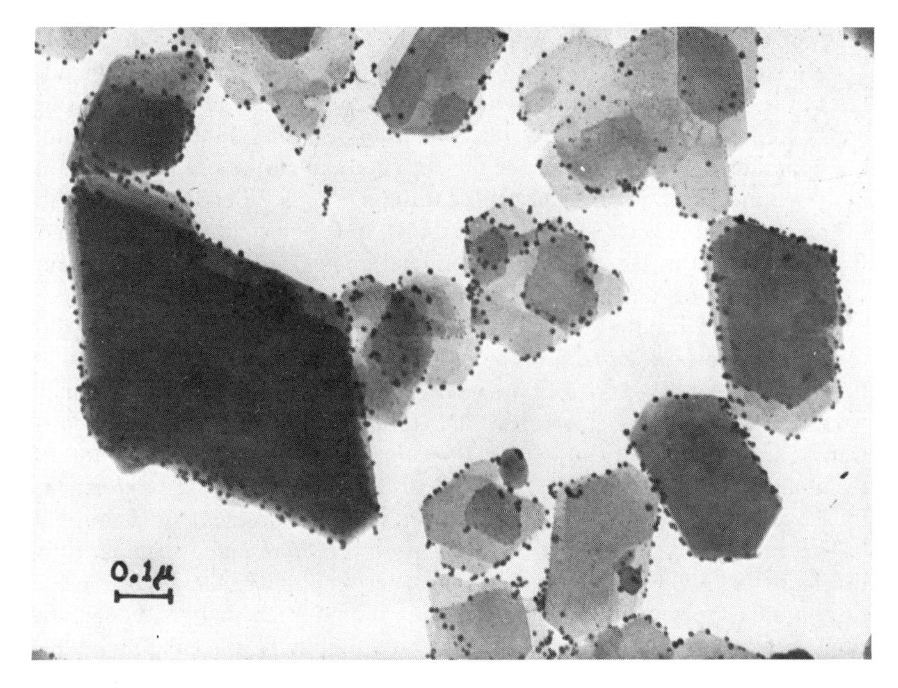

Figure 22. Electron micrograph of a mixture of a kaolinite and a negative gold sol. Micrograph by H. P. Studer. (After P. A. Thiessen.)

introduced the concept of a positive double layer in the interpretation of the complex stability behavior of montmorillonite suspensions (3, 4). Later, Schofield and Samson (5) adopted the same hypothesis in the interpretation of flocculation and anion adsorption in kaolinite sols.

It may be concluded that the double-layer structure of the clay particle is complicated by the fact that two crystallographically different surfaces are exposed by the platelike particles (6, 7), each carrying a different type electric double layer. Possibly, under certain conditions, these double layers are of opposite sign, or, in mythological terms, the clay particles have a Janus character.

II. FLOCCULATION AND GELATION

A. Modes of Particle Association

When a suspension of platelike clay particles flocculates, three different modes of particle association may occur: face-to-face (FF), edge-to-face (EF), and edge-to-edge (EE). The electrical interaction energy for the three types of association is governed by three different combinations of the two double layers. Moreover, the van der Waals interaction energy will be different for the three types of association, since a different geometry must be considered in the summation of the attraction between all the atom pairs of the approaching plates (8). Hence the net potential curves of interaction for the three modes of association will be different. In addition, the rate of diffusion in the three ways of mutual approach of the particles is not the same. Consequently, the three types of association will not necessarily occur simultaneously or to the same extent when a clay suspension is flocculated.

The physical results of the three types of association will be quite different. Face-to-face association merely leads to thicker and possibly larger flakes, whereas EF and EE association will lead to three-dimensional, voluminous card-house structures. The consequences of the various associations for the properties of the flocculated suspensions which are of technological interest may be expected to be quite different. Therefore, the intricacies of the clay flocculation problem must be considered in detail.

The various modes of particle association are represented schematically in Figure 23. A few notes about a special terminology in the description of the various types of particle association should be added. Although the three types of association are really three modes of flocculation in the colloid chemical sense, only the EE and the EF types of association lead to agglomerates which would deserve the name of flocs. The thicker particles which result from FF association cannot be properly called "flocs"; hence the suggestion has been made that a different term be coined for this type of association. The terms "oriented aggregation" or "parallel aggregation" or, briefly, "aggregation" have been proposed. The latter, unqualified term *aggregation* has been used most frequently, although it is a rather unspecific term. In the following discussions we shall quote the term to indicate its use for FF association of the plates.

At the same time, the proposal has been made that only the dissociation of EF and EE linked particles be described as "deflocculation," and that the splitting of FF associated "aggregates" into thinner flakes be described as *dispersion* of the clay. Again, the latter term is rather nonspecific, and it will also be quoted.*

In this terminology, the statement in the first paragraph may be reworded as follows: Flocculation and "aggregation" do not necessarily proceed simultaneously or to the same extent under certain conditions. For example,

* The term "aggregates" should not be confused with the same term which is commonly used in soil science to describe the macroscopic crumbs in a soil. When the soil scientist speaks of the "stability of aggregates," he does not mean the colloidal stability but the stability of the crumbs in a soil in the sense of their ability to resist mechanical disintegration.

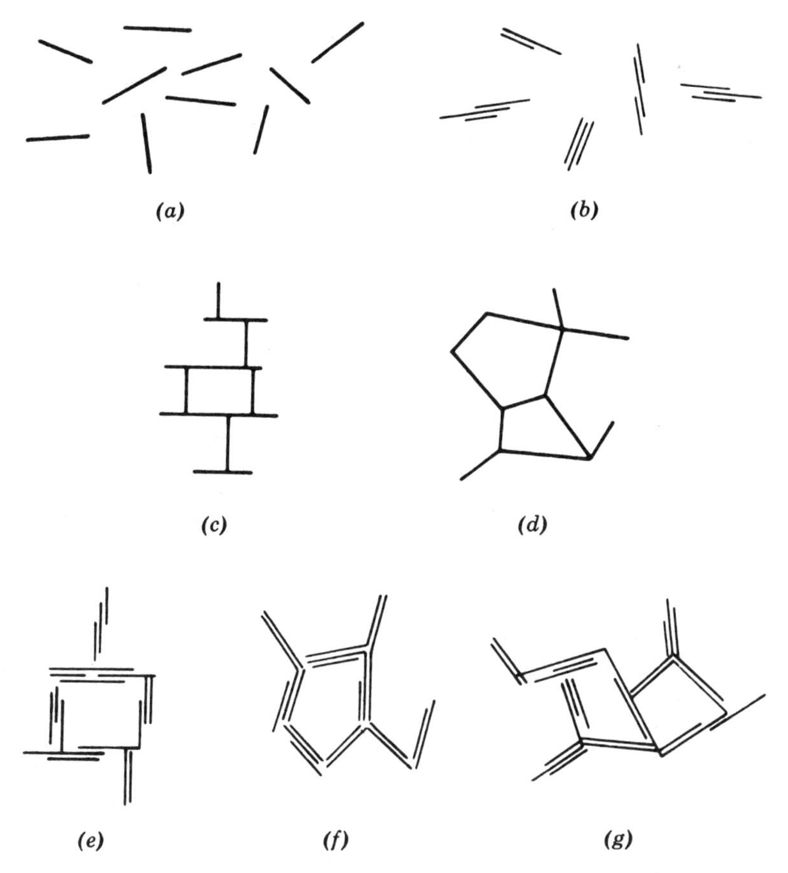

Figure 23. Modes of particle association in clay suspensions, and terminology. (*a*) "Dispersed" and "deflocculated." (*b*) "Aggregated" but "deflocculated" (face-to-face association, or parallel or oriented aggregation). (*c*) Edge-to-face flocculated but "dispersed." (*d*) Edge-to-edge flocculated but "dispersed." (*e*) Edge-to-face flocculated and "aggregated." (*f*) Edge-to-edge flocculated and "aggregated." (*g*) Edge-to-face and edge-to-edge flocculated and "aggregated."

the system may be well "dispersed" but not deflocculated, or it may be deflocculated but not too well "dispersed."

The term "aggregation" is used not only to describe the "aggregation" of several multilayer particles but also to describe the degree of layer stacking in a single particle, since one cannot distinguish between a particle of, say, 10 stacked layers and an aggregate of two particles consisting of 5 layers each. A calcium montmorillonite particle, which usually contains more layers per particle than a sodium montmorillonite particle, would be called a less well "dispersed" clay or a more "aggregated" clay.

B. Clay Flocculation and the Schulze-Hardy Rule

The flocculation value of a clay suspension can usually be determined without difficulty in a flocculation series by visual inspection of the contents of the tubes after several hours' standing. The flocculation values of a certain montmorillonite from a bentonite rock, which were determined in flocculation series tests are listed in Table 5.

In accordance with the Schulze-Hardy rule, $CaCl_2$ appears to be a more powerful flocculant for the negative montmorillonite sol than NaCl. However, for both the sodium and calcium montmorillonite sols, the ratio between the flocculation values for the monovalent and the divalent cation is of the order of 5, which is unusually small for a hydrophobic sol. This ratio is commonly of the order of 50–100. The smaller observed ratio for the clay can be explained by taking the simultaneously occurring exchange process into account. When the sodium clay is flocculated with $CaCl_2$, most of the sodium ions on the clay will be replaced by calcium ions. Thus calcium bentonite is formed, and this calcium bentonite will be flocculated by a small amount of $CaCl_2$. In quantitative terms, since the exchange capacity of the montmorillonite is of the order of 100 meq per 100 g the $\frac{1}{4}$ percent suspension can take 2.5 meq/dm^3 of calcium ions from the solution in forming the calcium clay. This calcium clay, according to the second line of the table, will be flocculated by a few tenths of 1 meq/dm^3 of $CaCl_2$, so that the flocculating concentration of the sodium clay will amount to slightly more than 2.5 meq/dm^3 of $CaCl_2$, which is indeed the correct order of the flocculation value, as shown by the first line in the table.

Consequently, the correct order of the ratio between the flocculation values for monovalent and divalent cations may be expected when the NaCl flocculation value for the sodium clay is compared with the $CaCl_2$ value of the calcium clay, and such ratio is indeed observed.

Table 5

Flocculation Values of a Montmorillonite Sol in the Sodium and in the Calcium Form

	Flocculation value, meq/dm^3	
Sol (conc. $\frac{1}{4}$ percent)	NaCl	$CaCl_2$
Na-montmorillonite	12–16	2.3–3.3
Ca-montmorillonite	1.0–1.3	0.17–0.23

Another way to eliminate the contribution of ion exchange is to determine the flocculation values of a sodium clay with salts of mono-, di-, and trivalent cations at different clay concentrations and to extrapolate the values for each of the salts to zero clay concentration. The ratio of the extrapolated flocculation values has been found to be of the normal order of magnitude (9).

Also in accordance with the Schulze-Hardy rule, it is found that the flocculation values for salts with anions of higher valence and which do not react with the clay, such as sulfates, are of the same order as those for chlorides.

It is important to note that the applicability of the Schulze-Hardy rule in the flocculation of clay suspensions indicates that the flocculation process is dominated by a negative double layer. Apparently, if the edge surfaces do carry a positive double layer, this double layer does not appear to upset the Schulze-Hardy rule for the negative clay particle. Yet, more detailed observations on the effects of salts on the properties of clay suspensions indicate that, under certain conditions, the presence of a positive double layer on the edge surfaces must be assumed to explain the flocculation phenomena. In the following paragraphs we shall discuss the physical changes which take place during flocculation of clay suspensions, and we shall relate these changes with the different possible modes of particle association and the existence of a positive double layer on the edge surfaces.

C. Particle Association and Flow Properties (21–41)

The flow behavior (or "rheological" behavior) of clay suspensions is a very sensitive and rather direct criterion for particle interaction. Hence the study of rheological suspension properties is most suitable for the analysis of changes in particle-interaction forces and modes of particle association which occur in the flocculation processes. At the same time, the rheological properties of clay suspensions are very important in technology, as will be discussed in detail in Chapter Nine, Section III.

At this point, the principal relations between flow behavior and particle interaction may be summarized briefly as follows:

The viscosity of dilute clay suspensions increases when conglomerates are formed by EE and EF association; the viscosity decreases when the particles become thicker by FF association.

In concentrated clay suspensions, EE and EF association leads to the formation of continuous, linked, card-house structures, which extend throughout the total available volume; thus, a gel is obtained. Such gels appear to behave as "Bingham systems," which are characterized by a "Bingham yield stress" (see Chapter Nine, Section III). The Bingham yield stress is a measure for the number and the strength of the links in the card house. If FF association occurs simultaneously, the number of units building the card-house structures is reduced. Hence the number of links in the card house is reduced, and the Bingham yield stress will decrease.

Keeping these relations in mind, we can analyze the particle-association processes occurring in a clay suspension at various concentrations of flocculating electrolyte from the changes in flow behavior with electrolyte concentration.

An example of a system in which a variety of pronounced changes in the flow properties occurs when increasing amounts of electrolyte are added is the sodium montmorillonite suspension. Figure 24*a* shows the relative viscosity (viscosity of suspension divided by viscosity of medium) of a dilute sodium montmorillonite suspension as a function of the concentration of NaCl in the suspension. In Figure 24*b*, the Bingham yield stress of a more concentrated system is plotted as a function of the salt concentration. The addition of a few milliequivalents per liter of NaCl causes a drop of the relative viscosity of the dilute suspension and a pronounced decrease of the Bingham yield stress of the concentrated system. Upon further addition of NaCl, both the viscosity and the yield stress increase, slowly at first, and rather sharply when the flocculating concentration of NaCl for the clay is approached. At a very high concentration of NaCl, the yield stress of the concentrated system tends to decrease again. This effect is not shown in the figure.

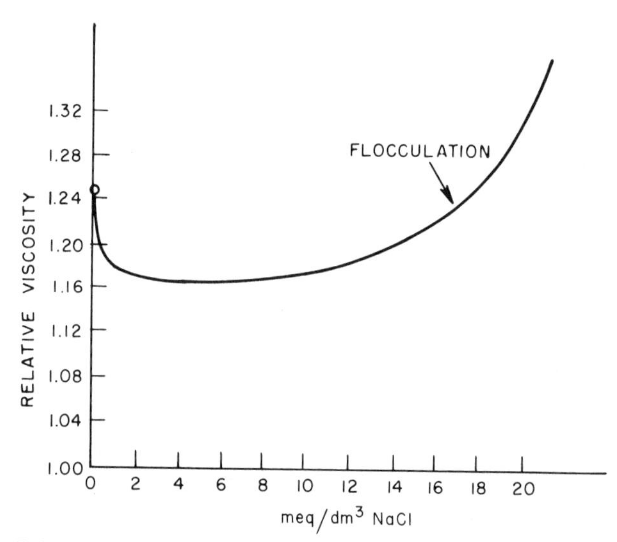

Figure 24*a*. Relative viscosity of a 0.23 percent sodium montmorillonite sol as a function of the amount of NaCl added.

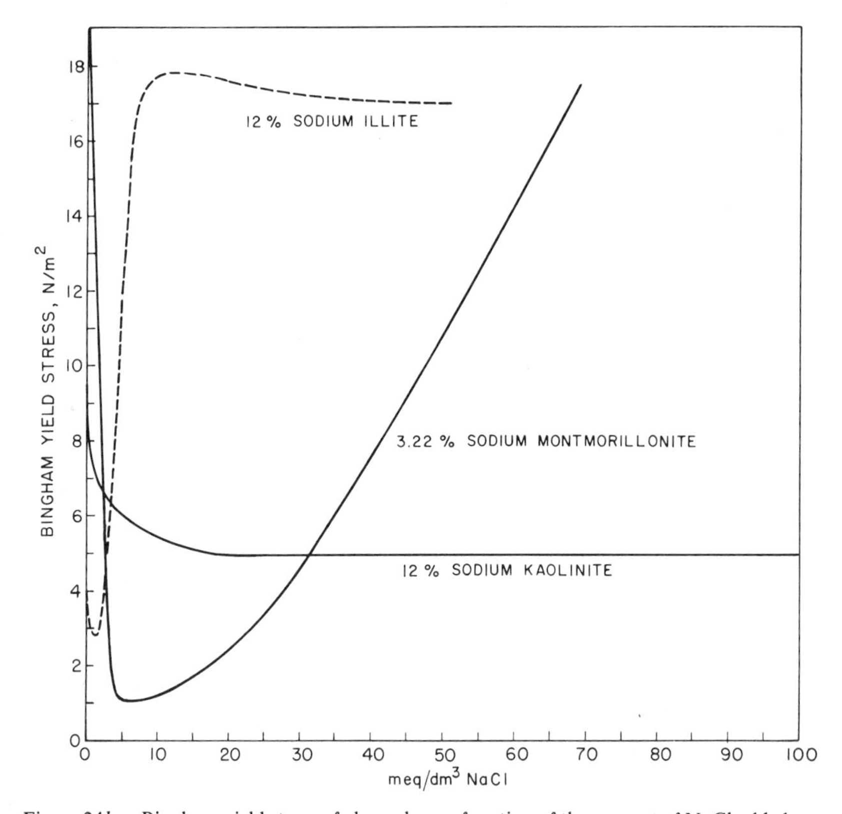

Figure 24*b*. Bingham yield stress of clay gels as a function of the amount of NaCl added.

On the basis of several detailed studies, and with the assumption that the edge double layer is indeed positive, these curves have been analyzed in terms of particle-interaction forces and modes of particle association which predominate in various regions of NaCl concentration. The results of these studies will be summarized here, but for the benefit of the interested reader, a detailed argumentation is presented in the next section.

1. In the electrolyte-free system, both double layers on the clay particle are sufficiently well developed to prevent particle association by van der Waals attraction. However, owing to the opposite charge of the edge and the face double layers, EF association takes place (internal mutual flocculation), causing a relatively high viscosity and yield stress in dilute and concentrated systems, respectively. The particles are associated in a double-T fashion, in which the EF attraction outweighs the FF repulsion.

2. In the presence of a few milliequivalents of NaCl, both double layers

are compressed, and their effective charge will be reduced. Consequently, both the EF attraction and the FF repulsion will diminish. Apparently, under these conditions, the attraction has become less than the repulsion in the double-T arrangement, and the particles become disengaged. Hence the card-house structure breaks down, and the yield stress of the concentrated suspension is dramatically reduced. In dilute solutions, the conglomerates are dispersed, and the viscosity decreases.

3. When the amount of NaCl in the suspension is increased, both double layers are further compressed. The van der Waals attraction between edges and faces enhances what remained of the opposite-charge attraction, and once more the subtle balance between EF attraction and FF repulsion becomes favorable for the formation of the card-house structure. Simultaneously, EE association by van der Waals attraction may contribute to the resulting increase of the viscosity and yield stress.

4. At very high salt concentrations, the yield stress decreases somewhat. This observation may be explained by the simultaneous occurrence of FF association by which the number of particles in the card house is reduced, and therefore its strength is decreased.

The sodium montmorillonite system is rather exceptional in the great variations in flow behavior with changing salt concentration. For other clays the various features of the salt-effect curves are often less pronounced, or even absent, because of the domination of one type of association in a wide salt-concentration region. For example, a sodium kaolinite system does not show the spectacular increase of viscosity and yield stress after the pure gel is broken down by the addition of a small amount of salt. Indications are that in this system FF association dominates as soon as the EF links have been weakened by the addition of a small amount of salt. In other clays—for example, in calcium montmorillonite—the initial sharp decrease of the viscosity and yield stress does not occur.

Apparently, different clays react differently on electrolyte addition because of differences in their initial double-layer structures. In practice, it will be difficult to predict the behavior of a particular clay in the presence of salts. However, the flow properties are easily determined. Then, an interpretation of the rheological behavior in terms of modes of particle association will be helpful in determining how to change the system to obtain the desired flow properties for a particular technological application (see Chapter Nine). The peculiar behavior of the clay suspensions at the very low electrolyte concentrations or in pure water is usually of little practical interest, since seldom are systems of such purity encountered in technology.

D. Argumentation

1. Why should the development of a yield stress in the pure sodium montmorillonite system be considered a result of positive-edge-to-negative-face linking? Are there any alternatives which do not require the postulation of opposite charges on faces and edges?

Consider the alternative possibility that the edge surfaces carry a weak negative double layer or none at all because of a low concentration or absence of potential-determining anions for these surfaces. Then the van der Waals attraction between edges or between edges and faces will predominate, and a card-house gel structure is created. Since, upon the addition of a small amount of salt, the van der Waals attraction would not diminish, an increase of the double-layer repulsion must be assumed to cause the observed breakdown of the structure. Hence it would have to be assumed that the added chloride anions charge the edge surfaces. However, there is no special reason to assume that the chloride anions would do so, since they do not act as potential-determining ions for sols with particles which have a surface structure analogous to that of the edge surfaces of the clay particles.

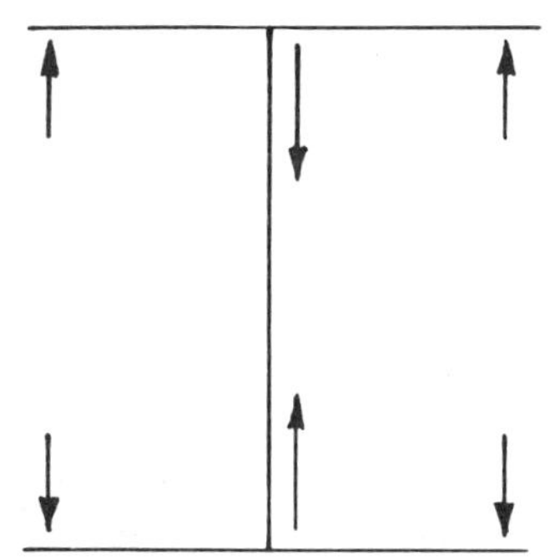

Another alternative could be that a ribbonlike network is created by the association of particle faces in such a way that only part of their surfaces become attached. However, such a ribbon formation would become stronger with increasing salt concentration, since the face double layer would become more compressed.

One is therefore left with the simple explanation of an association of negative faces and positive edges, leading to a voluminous cubic card-house structure.

The cubic card house may be visualized as a system of parallel plates which are held together by cross-linking particles perpendicular to the parallel plates. The cross-linking force in the double-T-shaped units of the card house is supplied by the electrostatic attraction between the oppositely charged edges and faces. This cross-linking force has to be higher than the repulsive force between the negatively charged parallel plates in the double-T unit.

2. The breakdown of the cubic card-house structure upon the addition of small amounts of salt may be seen as a reduction of the cross-linking force. This reduction may be the result of a decrease of the effective attracting charges on edge and face surfaces owing to a shift of counter-ions toward the surfaces. In view of the very small concentrations of salt involved, it seems remarkable that the compression of the double layers would be significant enough to reduce the attraction to the extent that the card house breaks down. However, it should be kept in mind that in the double-T structure a balance of repulsive and attractive forces exists which is possibly only slightly in favor of the attraction. Hence a small reduction of the attraction may be enough to make the repulsion predominant. (Alternatively, the reduction of the attraction may not be a result of double-layer compression only; it may be a result of a definite chemisorption of the anions at the positive edges, by which the positive charge would be neutralized. On the other hand, chemisorption of anions such as chloride or sulfate does not seem likely, and there is actually no indication of such a neutralization effect in alumina sols or in positive quartz sols, which are likely models for the edges of the clay plates.)

3. Upon the addition of more salt to the system, the double layers on both edge and face surfaces are compressed further. Several possibilities for particle association are now opened:

Edge-to-edge association by van der Waals attraction may no longer be prevented by the double-layer repulsion between the two positive-edge double layers. Although such association probably occurs in the salt-flocculation process, it cannot be a dominating event, since salt flocculation is governed by the valence of the cations. Therefore, according to the Schulze-Hardy rule, salt flocculation is governed by the changes in the negative-face double layer.

The compression of the face double layer in the presence of flocculating amounts of salt may have the following two consequences: (*1*) The degree of FF association may be increased by van der Waals attraction. (*2*) The FF repulsion may be reduced only to such an extent that the cross-linking EF forces (electrostatic as well as van der Waals) become once more larger than the repulsion between the parallel plates in the double-T unit.

The yield stress will rise if (*2*) predominates, but not if (*1*) prevails. The latter is apparently the case in certain kaolinite suspensions, whereas in sodium montmorillonite, (*2*) is the dominating event. At very high concentrations of salt in a sodium montmorillonite system, FF association becomes more and more important, and consequently the yield stress begins to decrease.

In conclusion, it appears that the very striking and technologically important variations in the rheological properties of clay-water systems are determined by a rather delicate balance of the three potential curves of interaction for EF, EE, and FF association.

E. Further Experimental Support

1. The Structure of the Pure Gel. If the card-house structure in the pure gels is idealized as an EF associated cubic network of clay plates, the weight of clay required to create a continuous card house in a certain volume can be computed from the dimensions of the particles. This amount of clay may be expected to be the minimum amount which is necessary to give the system a measurable yield stress. For a certain particle-size fraction of sodium montmorillonite, the average particle size was determined. For this material, a minimum concentration of 2 percent was computed to be required to fill the available volume with a continuous cubic card house. A homogeneous gel was prepared by concentrating the dilute suspension in the ultracentrifuge. It appeared that the minimum clay concentration at which a measurable yield stress was displayed did indeed amount to 2 percent. After conversion of the sodium clay into a calcium clay, the particles were found to be between three and four times thicker than those in the sodium clay suspension. Accordingly, the required minimum concentration of calcium montmorillonite for the development of a gel was found to be three to four times greater than that for the sodium clay (28, 29). The reason that the gelating concentrations of the sodium montmorillonites are so low is because the particles are extremely thin. One should realize that a ribbon obtained by lining up side by side all the particles contained in 0.01 g of sodium montmorillonite would reach almost across the state of Texas. Therefore, it is not surprising that far below the gelating concentration, particle interaction affects the viscosity and other physical properties of the suspensions.

2. Criterion for Face-to-Face Association ("Aggregation"). A simple test developed by H. C. H. Darley (10) the so-called "clay-volume test"—has practical merit when estimates are made of the degree to which FF association or "aggregation" occurs under certain conditions. In this test, the particles, as they exist in a certain environment, are instantaneously flocculated by the addition of a high dosage of flocculating electrolyte. The flocculated system is then centrifuged under standard conditions, and the volume of the flocculated mass is measured. Since it may be

reasonably assumed that in the rapid coagulation the particles associate edge to face and edge to edge before they have a chance to associate face to face, the sediment volume will be a relative measure of the original number of plates in the suspension. The thinner the plates, the larger their number, and the larger the sediment volume.

The test is carried out, for example, by shaking 1 ml of the suspension with 100 ml of a 1*M* NaCl solution in an ASTM oil-centrifuge tube. The flocculated suspension is centrifuged for five minutes at 2000 rpm in an "International No. 2" centrifuge. The sediment volume is read in the calibrated centrifuge tubes.

On the basis of this type of test, it was concluded that in the sodium montmorillonite sol very little, if any, "aggregation" takes place up to NaCl concentrations which are well beyond the flocculation value, and that a large excess of salt is required to induce "aggregation." This observation agrees with the fact that at such high salt concentrations the yield stress of the flocculated suspensions begins to decrease.

F. Particle Association in Dilute Sols and Spontaneous Swelling of Montmorillonites

1. Particle Association in Dilute Sols. Since the development of a Bingham yield stress in electrolyte-free suspensions at moderately high clay concentrations demonstrates the existence of particle association in this system, a clay suspension may not be called colloidally stable in the absence of electrolyte. This conclusion seems paradoxical, since a dilute electrolyte-free clay sol usually has a perfectly stable appearance. However, this appearance does not prove that there is no particle association at all in the dilute sols; it only means that any agglomerates present in the sol are so light that they do not settle rapidly or give the sol a flocculated appearance. Actually, the relatively high viscosity of the sol in the absence of electrolyte compared with that when a few milliequivalents per liter of salt are present suggests that in the salt-free sol, agglomerates exist indeed. Owing to the EF association tendency of the particles, such agglomerates may be T-shaped duplets or double-T shaped triplets, and such agglomerates indeed would be very light and therefore would not settle fast. Further growth of the agglomerates would probably be prevented by hydrodynamic convection forces in the sol, which would have a considerable dispersive effect on the very large plates in the agglomerates. Moreover, the repulsion between the face surfaces will assist in the dispersion by hydrodynamic forces. Therefore, particle association in the salt-free sols is likely to be restricted to duplets or triplets which are EF associated, and thus the sol retains its colloidally stable appearance.

2. Spontaneous Swelling of Montmorillonites. A dry montmorillonite powder swells spontaneously when contacted with water. The dry clay usually imbibes water and becomes a gel, and it can be stirred up with more water to yield a suspension or sol. The swelling of a clay is particularly spectacular in the case of montmorillonites. The montmorillonite clay first takes up one to four monolayers of water between the layers. This interlayer or intracrystalline swelling causes, at most, a doubling of the volume of the dry clay. However, the swelling process continues, and an amount of water is imbibed which is many times the volume of the original clay. The additional swelling is a result of the double-layer repulsion between the surfaces of the individual particles, which pushes them apart. Formalistically speaking, the swelling may be called *osmotic swelling* since the water tends to equalize the high concentration of ions between two particles, which are so close together that their double layers overlap, and the low concentration of ions far away from the particle surfaces in the bulk solution. Under suitable confining conditions, a fluid pressure is created which is called the *osmotic* or the *swelling pressure* of the clay. This pressure is a direct measure of the balance of the forces between the particle faces (see also Chapter Ten, Section I-B).

Sometimes, the question is debated whether spontaneous swelling of the clay finally leads to complete disintegration of the gel to a sol, or whether the swelling is limited and stops as soon as a certain gel volume is established. The reason that there is doubt about this question is because it is difficult to perform an unambiguous experiment in which all external forces are excluded, such as hydrodynamic convection forces which tend to disperse the particles, or gravity forces which usually operate against the further dispersion of the gel.

In principle, sol formation will be spontaneous only if at any particle distance at any configuration of the particles in space, repulsion predominates. In that case, the sol formation properly may be called spontaneous. However, it seems that most clays—even certain sodium montmorillonites—are not spontaneously dispersed. Usually, at certain particle separations and at certain particle configurations, attractive forces cancel the repulsion, for example, van der Waals forces and the important edge-to-face cross-linking forces. Entropy effects are probably relatively small in these systems.

G. Deflocculation of Clay Suspensions

In technology it is often necessary to use rather concentrated clay suspensions which are nevertheless fluid enough to be poured or pumped. As long as the clay suspension is in the flocculated condition, and EF or EE links exist, one deals with a rather stiff suspension which cannot be handled properly. Obviously, this stiff suspension could be turned into a more fluid system by breaking the particle links, thus deflocculating the system. An elegant way to achieve deflocculation would be to reverse the positive-edge charge and to create a well-developed negative-edge double layer. Then, the positive-edge-to-negative-face attraction would be eliminated, and a strong EE as well as EF repulsion would be created, resulting in a breakdown of the gel structure.

In the following chapter it will be shown that charge reversal of the edges of the clay plates is the basic mechanism underlying many empirical procedures for the thinning of clay suspensions by chemical treatment.

An alternative method of deflocculation would be to reverse the negative-face charge into a positive one. However, this method appears to be economically less attractive, and it is also less effective (see Chapter Eleven).

References

1. van Schuylenborgh, J. (1951), The electrokinetic behavior of freshly prepared γ-AlOOH, α- and γ-$Al(OH)_3$, *Rec. Trav. Chim.*, **70,** 985–988.

2a. Thiessen, P. A. (1942), Wechselseitige Adsorption von Kolloiden, *Z. Elektrochem.*, **48,** 675–681.

2b. Thiessen, P. A. (1947), Kennzeichnung submikroskopischer Grenzflächenbereiche verschiedener Wirksamkeit, *Z. Anorg. Chem.*, **253,** 161–169.

3. van Olphen, H. (1950), Stabilization of montmorillonite sols by chemical treatment, *Rec. Trav. Chim.*, **69,** 1308–1312 (I), 1313–1322 (II).

4. van Olphen, H. (1951), Rheological phenomena of clay sols in connection with the charge distribution on the micelles, *Discussions Faraday Soc.*, **11,** (The size and shape factor in colloidal systems), 82–84.

5. Schofield, R. K., and Samson, H. R. (1954), Flocculation of kaolinite due to the attraction of oppositely charged crystal faces, *Discussions Faraday Soc.*, **18** (Coagulation and Flocculation), 135–145, 220.

6. Hofmann, U., Weiss, A., Koch, G., Mehler, A., and Scholz, A. (1956), Intracrystalline swelling, cation exchange, and anion exchange of minerals of the montmorillonite group and kaolinite, *Clays, Clay Minerals*, **4,** 273–287.

7. Fripiat, J. J. (1957), Propriétés de surface des alumino-silicates, *Bull. Groupe Franç. Argiles.*, **9,** 23–45.

8. Vold, M. J. (1954), van der Waals attraction between anisometric particles, *J. Colloid Sci.*, **9,** 451–459.

9. Kahn, A. (1958), The flocculation of sodium montmorillonite by electrolytes, *J. Colloid Sci.*, **13,** 51–60.

10. Darley, H. C. H. (1957), A test for degree of dispersion in drilling muds, *J. Petr. Technol., Trans. A.I.M.E.*, **210,** 93–96.

11. Cashen, G. H. (1959), Electric charges of kaolin, *Trans. Faraday Soc.*, **55,** 477–486.

12. Weiss, A., and Russow, J. (1963), Uber die Lage der austauschbaren Kationen bei Kaolinit, *Proc. Intern. Clay Conf.*, Stockholm, **1,** 203–213.

13. van Olphen, H. (1964), Internal mutual flocculation in clay suspensions, *J. Colloid Sci.*, **19,** 313–322.

14. Edwards, D. G., Posner, A. M., and Quirk, J. P., (1965), Discreteness of charge on clay surfaces, *Nature*, **206**, 168.
15. Lagaly, G., and Weiss, A. (1969), Determination of the layer charge in mica-type layer silicates, *Proc. Intern. Clay Conf.*, Tokyo, **1**, 61–80.
16. Friend, J. P., and Hunter, R. J. (1970), Vermiculite as a model system in the testing of double layer theory, *Clays, Clay Minerals*, **18**, 275–284.
17. Fordham, A. W. (1973), The location of Iron-55, Strontium-85 and Iodide-125 sorbed by kaolinite and dickite particles, *Clays, Clay Minerals*, **21**, 175–184.
18. Stul, M. S., and Mortier, W. J. (1974), The heterogeneity of charge density in montmorillonites, *Clays, Clay Minerals*, **22**, 391–396.
19. van Olphen, H. (1975), Theories of the stability of lyophobic colloidal systems, see Appendix V, reference *B-12*, 5–15.
20. Swartzen-Allen, S. L., and Matijevic, E. (1974), Surface and colloid chemistry of clays, *Chem. Rev.*, **74**, 385–400.

RHEOLOGICAL PROPERTIES OF CLAY SUSPENSIONS

21. Philippoff, W. (1942, 1944), *Viskosität der Kolloide*, Theodor Steinkopf, Dresden und Leipzig (1942), Edwards Brothers, Inc., Ann Arbor, Mich. (1944).
22. Hauser, E. A., and Reed, C. E. (1937), Studies in thixotropy, II—The thixotropic behavior and structure of bentonite, *J. Phys. Chem.*, **41**, 910–934.
23. Goodeve, C. F. (1939), A general theory of thixotropy and viscosity, *Trans. Faraday Soc.*, **35**, 342–358.
24. Norton, F. H., Johnson, A. L., and Lawrence, W. G. (1944), Flow properties of kaolinite-water suspensions, *J. Am. Ceram. Soc.*, **27**, 149.
25. Fahn, R., Weiss, A., and Hofmann, U. (1933), Uber die Thixotropie bei Tonen, *Ber. Deut. Keram. Ges.*, **30**, 21–25.
26. Hofmann, U. (1952), Neue Erkenntnisse auf dem Gebiete der Thixotropie, insbesondere bei tonhaltigen Gelen, *Kolloid-Z.*, **125**, 86–99.
27. Weiss, A., and Frank, R. (1961), Uber den Bau der Gerüste in thixotropen Gelen, *Z. Naturforsch.*, **16**, 141–142.
28. van Olphen, H. (1956), Forces between suspended bentonite particles, *Clays, Clay Minerals*, **4**, 204–224.
29. van Olphen, H. (1959), Forces between suspended bentonite particles, Part II—Calcium bentonite, *Clays, Clay Minerals*, **6**, 196–206.
30. M'Ewen, M. B., and Pratt, M. I. (1957), The gelation of montmorillonite, Part I—The formation of a structural framework in sols of Wyoming bentonite, *Trans. Faraday Soc.*, **53**, 535–547.
31. M'Ewen, M. B., and Mould, D. L. (1957), The gelation of montmorillonite, Part II—The nature of interparticle forces in sols of Wyoming bentonite, *Trans. Faraday Soc.*, **53**, 548–564; Discussion, *Trans. Faraday Soc.*, **54**, 144–145, 1958.
32. Rehbinder, P. (1954), Coagulation and thixotropic structure. *Discussions Faraday Soc.*, **18** (Coagulation and Flocculation), 151–160.
33. Ventriglia, U. (1954), La plasticité des argiles, *Clay Minerals Bull.*, **2**, 176–178.
34. Street, N. (1956), The rheology of kaolinite suspensions, *Australian J. Chem.*, **9**, 467–479.
35. Foster, W. R., Savins, J. G., and Waite, J. M. (1955), Lattice expansion and rheological behavior relationships in water-montmorillonite systems, *Clays, Clay Minerals*, **3**, 296–316.

36. Wood, W. H., Granquist, W. T., and Krieger, I. M. (1956), Viscosity studies on dilute clay mineral suspensions, *Clays, Clay Minerals,* **4,** 240–250.
37. Granquist, W. T. (1959), Flow properties of dilute montmorillonite dispersions, *Clays, Clay Minerals,* **6,** 207–219.
38. Mitchell, J. K. (1960), Fundamental aspects of thixotropy in soils, *J. Soil Mech. Found. Div., Am. Soc. Civil Engrs.,* **SM 3,** 19–52.
39. Watanabe, T. (1960), Studies on thixotropy, II—On the thixotropic gel structure of bentonite, *Bull. Chem. Soc. Japan,* **33,** 523–527.
40. Christensen, R. W., and Kim, J. S. (1969), Rheological model studies in clay, *Clays, Clay Minerals,* **17,** 83–94.
41. Franklin, A. G., and Krizek, R. J. (1969), Complex viscosity of a kaolin clay, *Clays, Clay Minerals,* **17,** 101–110.

See also Chapter Nine, References, under "Rheology."

CHAPTER EIGHT

Peptization of Clay Suspensions

I. PEPTIZATION (DEFLOCCULATION) BY SPECIAL INORGANIC SALTS

In clay technology, particularly in drilling-mud technology, a variety of chemicals are known which are able to liquefy stiff, flocculated clay suspensions. These chemicals include inorganic salts, alkalies, and organic compounds. The inorganic salts and alkalies will be discussed in the sections of this chapter, but the interaction of clays and organic compounds will be dealt with in Chapter Eleven.

It is most remarkable how the addition of a few tenths of one percent of these chemicals is often sufficient to turn a stiff concentrated clay suspension into a rather free-flowing liquid. Since the same chemicals are able to disperse flocs in a diluted flocculated clay suspension, it is obvious that they act as deflocculants or peptizers.

When a small amount of peptizer is added to a pure clay gel, the yield stress decreases drastically. This effect of the peptizer is comparable with that of salt. Upon further addition of the peptizer, however, the yield stress remains low; this is contrary to the rise observed with further additions of salt. Only at very high concentrations of the peptizer does the yield stress begin to increase.

If salt is added to the system after the addition of small amounts of peptizer, the yield stress still remains low, unless a rather large amount of salt is added.

The observed decreases and increases of the yield stress correlate with observations of deflocculation and flocculation in dilute clay suspensions. The result of deflocculation-flocculation experiments with peptizers and salts can be presented in a *flocculation diagram,* in which the salt-flocculation value of the clay is plotted versus the amount of peptizer present in suspension. Curve 1 in Figure 25*a* shows the type of results usually obtained. The rather low flocculation value of the suspension without a peptizer is read on the vertical coordinate at zero peptizer concentration. The

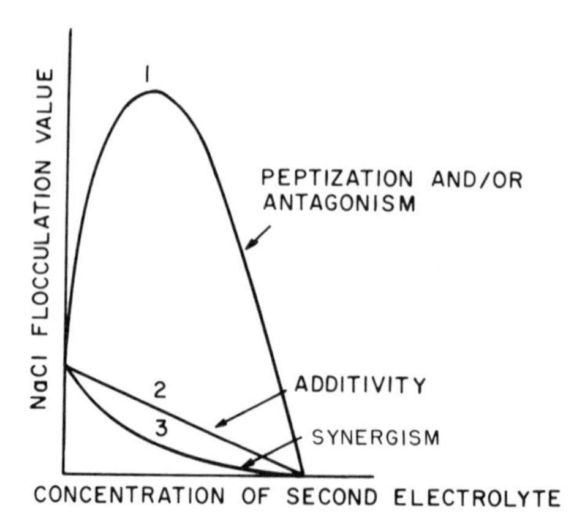

Figure 25*a*. Effects of a mixture of two electrolytes on the stability of a sol.

flocculation value of the treated suspension increases rather rapidly with increasing peptizer concentration, until an optimum effect is obtained. Upon the further addition to peptizer, the flocculation action of the peptizing salt, which contains sodium ions, makes itself felt, and the salt flocculation value begins to decrease. This effect is called *overtreatment* in technology. Finally, at a rather high concentration of the peptizing salt, the clay suspension flocculates with the peptizer alone. However, the flocculation value of the peptizing salt is much higher than that of salt (Figure 25 *b*). This behavior of a sol in the presence of two electrolytes is called *antagonism,* or, as in this special case, *peptization antagonism.*

Formerly, it has been suggested that the peptizing effect of the special peptizing chemicals is governed by ion exchange. Therefore, it should be stressed that the upper peptization curve, shown in the example of Figure

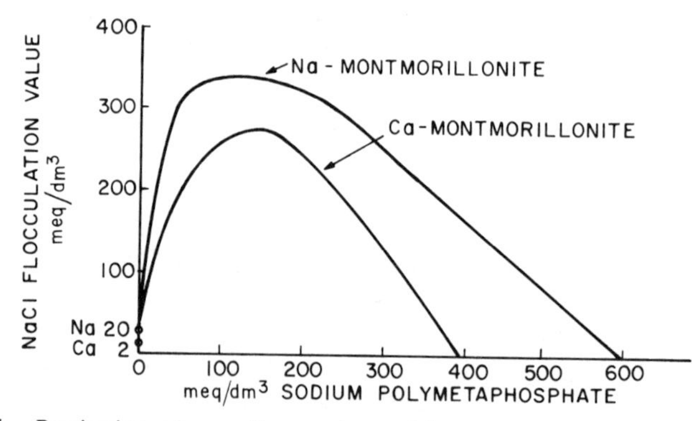

Figure 25*b*. Peptization of a sodium and a calcium montmorillonite sol with sodium polymetaphosphate.

25*b*, is valid for the *sodium* chloride flocculation value of a *sodium*-clay suspension as determined in the presence of *sodium* polymetaphosphate. In this case, only sodium ions are present in the system, and any contribution of an ion exchange reaction between the counter-ions of the clay and the cation of the added peptizer is excluded.

In Figure 25*a*, two more curves are drawn which represent other possible effects when a suspension is flocculated in the presence of two different electrolytes. If there are no special interactions between the suspended particles and either electrolyte, for example NaCl and KCl, the electrolytes being of the same type, the flocculation values are usually *additive*. This situation is sketched in curve 2 of Figure 25*a*. Sometimes, the presence of a second electrolyte makes the sol more susceptible for flocculation by the first electrolyte. In this case one speaks of *synergism*; curve 3 depicts this type of flocculation.

II. THE MECHANISM OF PEPTIZATION (1, 2, 3)

We have already indicated the possibility of stabilizing a clay suspension by the reversal of the positive edge charge into a negative one. In this way, both EF and EE association of the clay plates would be eliminated, or, at least, higher electrolyte concentrations would be required to establish such associations by van der Waals attraction.

There are indeed reasons to assume that the well-known peptizing agents for clay suspensions do reverse the edge double-layer charge. Since the anions of the peptizing agents are known to be adsorbed by clays, it is likely that edge-charge reversal is the result of anion adsorption at the edges. The adsorption capacity of the clay for the anions is usually rather small and is sometimes hardly measurable. Such a low adsorption capacity would be expected indeed if the comparatively small edge surface area were the site of adsorption. A reversal of the positive edge charge is also in agreement with the observation that the electrophoretic mobility of the clay particles increases upon the addition of small amounts of peptizing chemicals.

These observations, as well as other evidence to be mentioned shortly, support the concept that the reversal of the edge charge by anion adsorption is the basic mechanism of chemical clay peptization. With this picture, it becomes at once clear why such small amounts of peptizing chemicals have such a spectacular effect on the flow behavior of flocculated clay suspensions. In the first place, only small amounts of peptizer are needed to build a negative double layer on the small edge surfaces, and second, the breakdown of the EE and EF card house of clay plates must have a drastic effect indeed on the rigidity of the originally flocculated suspension. Actually, these features make the use of peptizing chemicals economically attractive.

When we consider the alternative possibility that the observed anion adsorption takes place on the flat layer surfaces, the following arguments may be raised against this assumption: (*a*) The adsorption capacity of the clay for anion adsorption would be expected to be much higher if the large layer surface areas were the sites of adsorption. (*b*) In the case of treatment of montmorillonite clays with peptizing salts, one would expect to observe an increase in the *c* spacing of the clay in the dried condition in accordance with the usually large size of the adsorbed anions, but no such increased *c* spacing is observed.

If the anions are indeed adsorbed at the edges of the clay plates, the next question is why they should be specifically adsorbed at these sites. There are good reasons for this. In the first place, the anions would be attracted by the positive edge surface. However, the electrostatic attraction of the anions alone would lead to the neutralization of the edge charge only. In order to create a stabilizing negative edge charge, the anions must be adsorbed in excess. When the known types of peptizing anions are considered, they appear to have in common a specific reactivity with aluminum with which they form either complex anions or insoluble salts. Hence these anions will be chemisorbed at the edges by reacting with the exposed aluminum. When chemisorption leads to the formation of complex anions, as in the case of polymetaphosphates, the creation of a negative edge charge is readily understood. When an insoluble, neutral aluminum salt is the reaction product, it seems likely that additional anions will be preferentially adsorbed on the aluminum salt which is attached to the edge surface and will act as potential-determining ions for the modified edge surface. Alternatively, the spatial distribution of the positive charge sites on the edge surfaces may be such that the two or more valences of the peptizing anions cannot be satisfied simultaneously. Therefore, perhaps one valence may be neutralized by the attachment of the anion at one positive charge site, whereas the other negative groups of the anion may build up a negative charge. An anion with a rather high valence, such as the polymetaphosphate anion, which may have a valence of 30 or more, may become attached with a fraction of its ionic groups to a number of charge sites, depending on the geometrical fit between the ionized groups of the anion and the location of the edge charges, as sketched below:

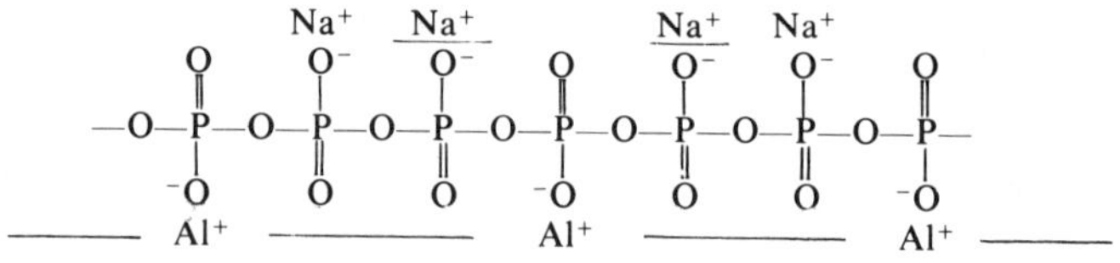

```
            Na+   Na+          Na+   Na+
     O      O-    O-     O     O-    O-     O
     ‖      |     |      ‖     |     |      ‖
 —O—P—O—P—O—P—O—P—O—P—O—P—O—P—
     |      ‖     ‖      |     ‖     ‖      |
    -O      O     O     -O     O     O     -O
 ——— Al+ ————————— Al+ ————————— Al+ ———
```

Edge surface of particle

As is so often the case in colloid chemistry, an exact analysis of the changes in the structure of the double layer upon the addition of salts which react with the constituents of the double layer is extremely difficult because of the small quantities involved. Frequently, such changes must be inferred on the basis of indirect evidence.

In analyses of the mechanism of peptization of clay suspensions, it has been very helpful to compare the response to peptizers shown by clays with that of simpler sols which may be considered as models for the clay sols insofar as the edge surface structure is concerned. It has been mentioned before that positive alumina sols, as well as silica sols which have been converted to positive sols by the addition of traces of aluminum salts, may be considered as models. The responses of these model sols to peptizers are remarkably analogous to those of clay sols, as is demonstrated by the flocculation diagrams presented in Figure 26 for sodium polymetaphosphate and sodium chloride. In the region of peptizer concentrations where a considerable increase in salt flocculation values is observed, it can be shown electrophoretically that the charge of the orginally positive alumina or quartz sol particles is reversed. A difference between the flocculation diagrams for the model sols and for clay suspensions is that in the model sols flocculation occurs upon the addition of very small quantities of peptizer. At this stage, the sol particles appear to be neutralized—evidently by anion adsorption. In clay sols, no visible flocculation is observed at the low peptizer concentrations at which the edge charge may be supposed to be neu-

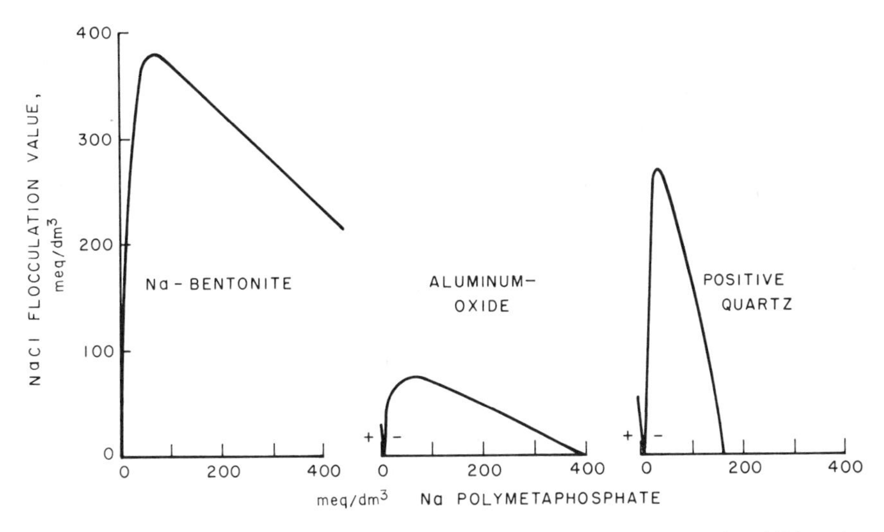

Figure 26. Response of sodium montmorillonite sol and of model sols to the addition of a peptizer.

tralized before the charge is reversed. Apparently, neutralization of the edge charge eliminates the electrostatic EF attraction, but any association of the unprotected edges does not lead to visible flocculation, possibly since the still strong FF repulsion gives the plates little chance to associate edge to edge.

These analogies between the responses of clay suspensions and of model sols may be considered a valuable support for the concept that clay peptization is staged at the edges, where chemisorption of the anions of the peptizers with the exposed octahedral cation occurs. At the same time, a possible explanation is presented of the fact that different clays respond quite differently toward the same peptizing salt, as is well known in technology. Clays often differ in the type of substituting ions in the octahedral sheet; therefore, in different clays different cations will be exposed at the broken edges. Instead of, or in addition to aluminum, magnesium,* ferric, and other ions may be exposed. For such sols the corresponding hydroxide sols should be considered as model sols instead of alumina sols. It has been shown that different hydroxide sols respond in a different way to peptizing salts, the effectivity of the various peptizers is different, and the sequence of their peptizing power varies with the type of hydroxide sol. For example, in one particular case studied, the sequence of peptizing power of several peptizing salts was determined for both alumina and ferric hydroxide sols. It was found that the difference in sequence observed for the two hydroxides was the same as that for two montmorillonites, with a large and with a small amount of ferric ions in octahedral position relative to the amount of aluminum ions.

In practice, it is difficult to predict which chemical will be the most efficient peptizer for the clay suspension on hand, but the best peptizer can be easily chosen on the basis of simple empirical tests. An interesting practical result of the above analysis of the peptization mechanism is that it supplies a guide for the screening of new chemicals appearing on the market from the point of view of clay peptization potentialities. There will be a good chance that salts which contain anions which are able to react or complex with aluminum and other octahedral cations will improve the stability of clay suspensions.

In conclusion, from all the evidence, there seems to be little doubt that edge charging by the anions of the peptizing salts is the basic mechanism of the stabilization phenomenon. The relatively small amounts of peptizer required with respect to the amount of clay present, and the small adsorp-

* Since reversal of charge phenomena have also been observed with magnesium (4), a positive edge charge is also possible for minerals which expose Mg at the edges.

tion capacity of the clay for the anions of the peptizers, point to their adsorption on the clay crystal edges. The spectacular effect of the peptizing salts on the flow properties of clay suspensions can be readily explained by the breakdown of the linked card-house structures involving the edge surfaces. The observed increase of the electrophoretic mobility indicates a charging effect. Finally, the analogous behavior of model sols toward the clay peptizers strongly supports this analysis of the peptization mechanism.

Although this mechanism will be the principal one, there are still two other factors which possibly contribute to the increased salt tolerance of the chemically peptized suspensions. These factors will be discussed in the next sections.

III. ACTIVITY REDUCTION AND ION EXCHANGE IN CLAY PEPTIZATION

In addition to the edge-charging effect due to the adsorption of an excess of the anions of the clay peptizers, the following two factors may be expected to contribute to the increase in salt tolerance of treated clay suspension: cation activity reduction and conversion of the clays into the sodium form by ion exchange with the added sodium salts.

A. Cation Activity Reduction

The flocculating power of an electrolyte on a sol is determined by the "activity" rather than the actual concentration of the ions of opposite sign. In the Debye-Hückel theory of electrolyte solutions, it is considered that in the neighborhood of a cation there will be, on the average, a higher concentration of anions than the average bulk concentration, as in the Gouy atmosphere around a positive sol particle. This anion "cloud" curtails the flocculating power of the cations on a negative sol. The theory predicts that the active concentration or "activity" of the cations is smaller than their actual concentration, and more so when the valency of the anions increases.

Since most peptizing salts for clay suspensions contain anions with a valency greater than 1, the reduction of the "activity" and therefore of the flocculating power of the sodium ions will be a definite contribution to the increased salt tolerance of the clay suspension in the presence of the peptizing salts. This effect is sometimes called *ion antagonism.* Particularly with such polyvalent anions as occur in sodium polymetaphosphate, which may have a valency of the order of 20–30, this effect may be important.

For salts with two- or three-valent anions the magnitude of the effect may be roughly computed from the theory. The results of the calculations show that the ion activity reduction will certainly account for part of the observed increase of the salt tolerance, but it is certainly not possible to explain the flocculation diagrams exclusively on the basis of activity reduction. The remaining effect must be due to the charging effect of the anion on the edges.

In the absence of the contribution of the activity reduction to the increased salt tolerance, only the charging effect remains. In such cases the total effectiveness of the peptizer is less than with the usual peptizers with polyvalent anions. For example, a clay suspension is peptized with salts containing monovalent fluoride anions, since these ions will effect a charge reversal of the edges by the formation of complex AlF_6^{3-} anions. The total effect of a fluoride is, however, much smaller than that of sodium polymetaphosphate because of the much smaller activity reduction of the alkali ions by the monovalent fluoride anion.

B. Ion Exchange

Historically, the first concepts about the mechanism of peptization of clays attributed the improved stability of the clay suspensions to a conversion of rather unstable, naturally occurring calcium clays into more stable sodium clays. The calcium ions were thought to be replaced by the sodium ions of the added peptizer, and the liberated calcium ions were considered to be made harmless by complex formation or by precipitation with the anions of the peptizing salts.

It is true that sodium clays are generally more stable than calcium clays, as shown, for example, in Table 5 and in Figure 25*b* (the points on the vertical coordinate). However, the gain in stability due to this conversion alone would be comparatively small. From Figure 25*b* one reads that the flocculation value increases from about 2 to about 20 meq/dm^3 NaCl when the calcium clay is converted to the sodium clay, which is a small gain compared with the increase of the stability due to the presence of the polymetaphosphate anion, by which the flocculation value is raised to about 300 meq/dm^3. Therefore, although the conversion to the sodium clay may be a contributing factor, it is relatively unimportant. Moreover, with highly concentrated clay suspensions, the amount of peptizing salt added is usually far below the amount which would be needed for complete conversion of the clay to the sodium form. Nevertheless, a complete, or nearly complete, conversion is sometimes practiced in technology, for example, by adding rather large amounts of sodium carbonate or by mixing the clay with organic resin exchangers in the sodium form; however, the last mentioned procedure is not economically attractive, nor is it very effective.

IV. PEPTIZATION BY ALKALI

It has been mentioned previously that the charge of a positive alumina sol is reversed by the addition of alkali. By analogy, in a clay sol the positive edge charge may be expected to be reversed upon the addition of small amounts of alkali; consequently, alkalies will probably have some peptizing effect on clay suspensions. With certain clay suspensions, such an effect has been observed indeed—for example, in kaolinite sols. The peptizing effect is much less pronounced than with peptizing salts, probably for the same reasons that fluorides are only moderately effective. In kaolinites, the reserval of the edge charge by alkalies may be demonstrated by the decrease of chloride ion adsorption with increasing pH (3).

In suspensions of sodium montmorillonite clays, the addition to alkali usually results in a slight sensization toward the flocculation by salt, instead of a peptization. Still, a beneficial effect of alkali resulting from edge charge reversal is observed in these systems. The addition of a few milliequivalents per liter of NaOH to a pure sodium montmorillonite gel causes a sharp reduction of its yield stress. So far, the effect of NaOH is comparable with that of NaCl. Upon further addition of NaOH the yield stress of the system remains low until the flocculating concentration of NaOH is reached, which is about the same as that of NaCl. With NaCl, however, we have seen that the yield stress rises immediately beyond a concentration of a few meq/dm^3 of NaCl in the system.

When natural raw clays are treated with alkali, more spectacular beneficial effects on the stability are often observed than predicted from the effect on sodium clays. This is particularly true when the raw clay shows an acid reaction. In the acid form most clays are less stable than in the sodium form; therefore, the conversion of the hydrogen clay into a sodium clay will result in an increased stability. In addition, organic acids are often present in the raw clay. These will be neutralized by the added alkali, and the organic salts which are thus formed frequently act as a peptizing chemical (see Chapter Eleven, Section II–A).

In the systems which were discussed in this chapter, the major effect of the peptizing chemicals has been interpreted in terms of deflocculation or peptization in the sense of the disengagement of EF and EE links in the gel card house. In these systems, changes in the degree of "aggregation" (or FF association) are probably of secondary importance.

In principle, a still better liquefying effect could be expected if the deflocculation in the above sense were accompanied by "aggregation," since less particles would be available to interact to build a rigid network. Such a combined effect is achieved in certain chemical-treatment procedures for clay suspensions involving organic chemicals. These systems will be discussed in Chapter Eleven (Section II–A–2).

Another special treatment procedure which is necessary in suspensions with a very high salt content (e.g., suspensions in sea water) involves the addition of so-called protective colloids, which are macromolecules of an organic nature.

Before the special features of the interaction between clays and organic compounds are discussed, however, it seems appropriate to deal with the technological applications of stability control in hydrophobic sols in general and of clay suspensions in particular.

References

1. van Olphen, H. (1950), Stabilization of montmorillonite sols by chemical treatment, *Rec. Trav. Chim.*, **69,** 1308–1312 (I), 1313–1322 (II).
2. van Olphen, H. (1951), Rheological phenomena of clay sols in connection with the charge distribution on the micelles, *Discussions Faraday Soc.*, **11** (The size and shape factor in colloidal systems), 82–84.
3. Schofield, R. K., and Samson, H. R. (1954), Flocculation of kaolinite due to the attraction of oppositely charged crystal faces, *Discussions Faraday Soc.*, **18** (Coagulation and Flocculation), 135–145, 220.
4. Troelstra, S. A. (1941), *Uitvlokking en omlading,* Thesis, Utrecht, p. 118. See also Troelstra, S. A., and Kruyt, H. R. (1943), *Kolloid-Beih.*, **54,** 225.
5. Goldsztaub, S., Hénin, S., and Wey, R. (1954), Sur l'adsorption d'ions phosphoriques par les argiles, *Clay Minerals Bull.*, **2,** 162–165.

CHAPTER NINE

Technological Applications of Stability Control: Sedimentation, Filtration, and Flow Behavior

I. SEDIMENTATION AND STABILITY

A. Principles

The rate of sedimentation of suspended particles increases markedly upon flocculation because of the large size of the aggregates. The effect is more pronounced in sols than in suspensions since the larger particles of a stable suspension settle rather rapidly.

Often of greater importance in technology is the difference between the structure of sediments of flocculated and of stable suspensions. Sediments of flocculated suspensions are usually much more voluminous than those of stable suspensions of the same concentration. At first sight, it seems somewhat paradoxical that the repelling particles of a stable suspension should obtain a higher degree of compaction in the sediment than the attracting particles of a flocculated suspension. However, when the individual particles of a stable suspension reach the lower part of the vessel, they are able to slide and roll past each other because of their mutual repulsion, and therefore they reach the lowest position in the vessel. In this way, a rather closely packed and dense sediment is obtained, although the mutual repulsion between the particles may keep them from coming into actual contact (illustrated in Figure 27*a*). In the flocculated suspension, on the other hand, the haphazardly formed voluminous flocs settle as such and pile up at the bottom of the vessel to form a voluminous sediment with large void spaces in and between the agglomerates (see Figure 27*b*). When the particles are nearly spherical, the flocs and the sediment may have a string-of-beads character, whereas rods or plates will assume a scaffolding or house-of-cards type sediment structure.

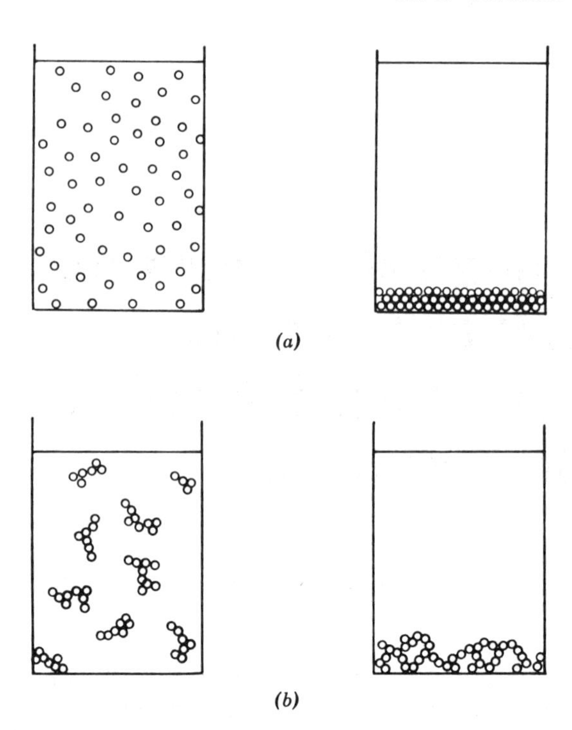

Figure 27. Sedimentation in a peptized and in a flocculated suspension. (*a*) Peptized suspension; dense, close-packed sediment. (*b*) Flocculated suspension; loose, voluminous sediment.

It should be stressed that such differences in sediment volume will not occur when the particles are so large that gravity forces easily break the links between the particles in the settled floc. Hence coarse sands or suspensions of glass beads would settle to the same volume in both fresh and salt water.

The parallelism between sediment volume and the degree of stability of a suspension is demonstrated in Figure 28*a*. Curve 2 in this figure shows the sediment volume of a sodium montmorillonite suspension after ultracentrifugation as a function of the concentration of NaCl in the suspension. As pointed out before, the changes in the yield stresses of somewhat more concentrated suspensions with increasing NaCl concentration may be considered a measure of the stability changes in the system. These changes are represented by curve 1 of Figure 28*a*, which is taken from the previously discussed Figure 24*b*. The parallelism between the essential features of curves 1 and 2 is clear. The sediment volume of the pure sol decreases sharply upon the addition of a few milliequivalents per dm^3 of NaCl, as the

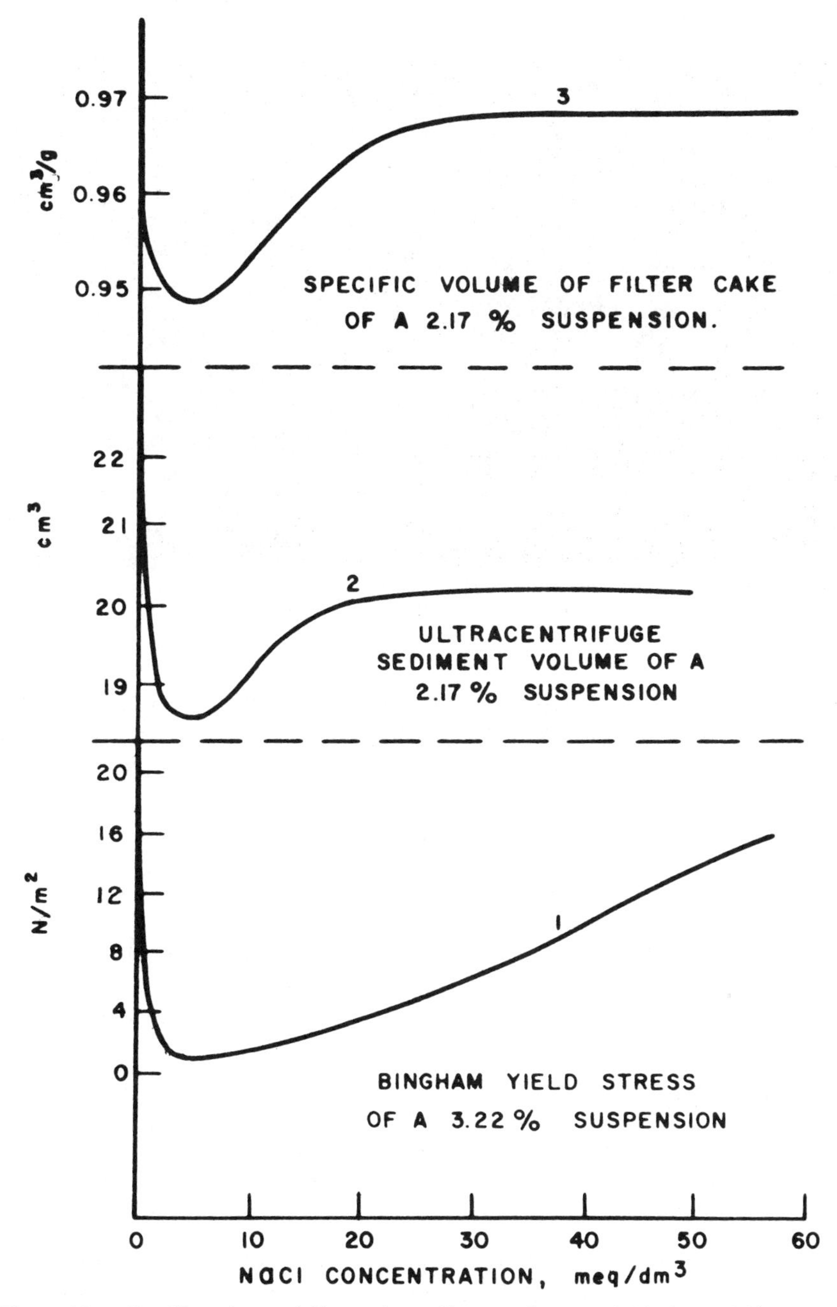

Figure 28*a*. Specific volume of filter cake, sediment volume, and Bingham yield stress of sodium montmorillonite sols as a function of the NaCl concentration.

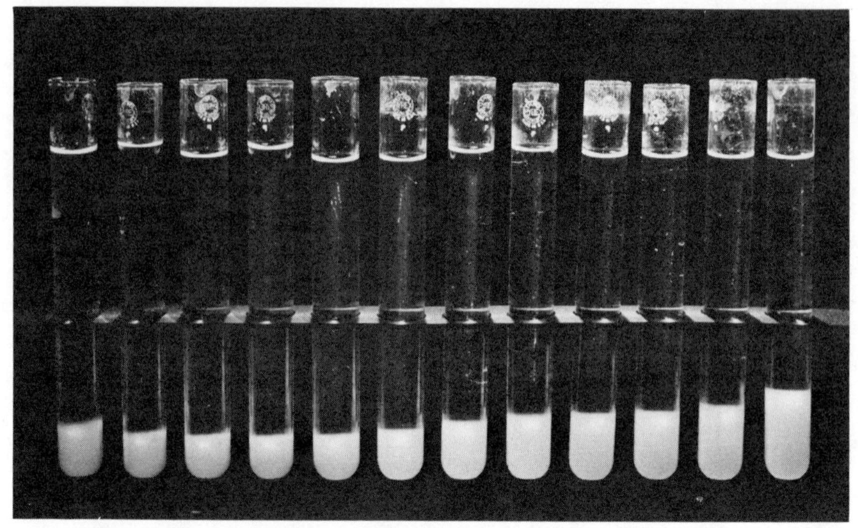

Figure 28*b*. Sediment volume of a sodium montmorillonite sol as a function of NaCl concentration (increasing from left to right). Sedimentation time: two months.

yield stress of the more concentrated suspension does, indicating a certain degree of deflocculation. With a further increase of the salt concentration, both the sediment volume and the yield stress increase because of progressing flocculation of the sols.

In the photograph of Figure 28*b*, the changes in sediment volume with salt concentration of sodium montmorillonite sols are clearly demonstrated. The photograph was made of a series of sols containing various amounts of salt, after several months' standing.

The change in sediment volume upon flocculation is sometimes used as the flocculation criterion in flocculation series experiments. Such experiments can be run quickly by centrifuging, with the test tubes replaced by calibrated centrifuge tubes.

B. Applications

Controlling the sediment properties by controlling the degree of stability or of flocculation of suspensions is very important in technology, as is illustrated by the following examples.

1. Separation of Dispersed Solids from a Suspension. When dispersed solids are separated from their suspension liquid by settling by gravity or by centrifuging, it is common practice to accelerate this process by the addition of flocculants.

2. Sedimentation Geology. In sedimentary geology it is of interest to note that the rate of sedimentation of clays which are entrained by rivers will be considerably increased when the river water comes into contact with sea water by which the clays are flocculated.

3. Paints. In a paint, the suspended pigment settles slowly on standing. Before use, the pigment must be resuspended by stirring. If the original suspension is stable, the bottom sediment will be rather compact. Such a dense sediment is difficult to stir, since the hydrodynamic dispersion forces will not get the proper grip on the pigment particles. If, on the other hand, the pigment suspension is in a flocculated condition, the sediment will be rather loosely packed and will be relatively easy to redisperse by stirring. Therefore, the ease of homogenization of a paint is controlled by the proper adjustment of the stability conditions of the pigment suspension.

4. Preparation of Thin Surface Coatings. Another technological application of stability control is encountered in the preparation of thin surface coatings by sedimentation or by electrodeposition, such as insulating films or fluorescent coatings. The compactness of these coatings can be varied by varying the stability of the suspension from which the coating is prepared (3).

5. Soils. A very important application of the control of sediment compaction is the chemical treatment of soils. In this application, filtration (drainage) also plays an important part. Therefore, the relation between stability and filtration will be discussed first.

II. FILTRATION OF SUSPENSIONS AND STABILITY

A. Principles

In the filtration of suspensions, the properties of the filter cake depend on the degree of peptization or flocculation of the suspension in the same way as the sediment properties do. When a stable suspension is filtered on a filter which retains the small individual particles, a thin and compact filter cake is obtained. Such a filter cake is rather impermeable, and soon after the first layers of the cake are deposited, the filtration process becomes very slow.

The filtration of a flocculated suspension yields a rather thick and porous filter cake. Usually, although not necessarily always, the permeability of such a porous filter cake is high, and a reasonable filtration rate is maintained. Of course, it is possible for the pore openings of the filter to be clogged by the particles, and this clogging effect may occur in stable as well as in flocculated suspensions. In that case, the filtration process is determined by the filter clogging, and the permeability may be independent of the degree of stability of the suspension (4).

Curve 3 in Figure 28*a* demonstrates the parallelism between filter cake and sediment properties for a sodium montmorillonite suspension at different salt concentrations. Curve 3 represents the specific volume of the filter cake, which appears to change with salt concentration in the same manner as the sediment volume of the suspensions does.

B. Applications

1. Analytical Chemistry. In chemical analytical work, the effect of peptization on the properties of the filter cake and on the rate of filtration is often encountered when the salt is washed out of a precipitate which is collected on a filter. As long as the suspension contains a flocculating amount of salt, filtration is rapid. However, when the salt is gradually removed, the precipitate peptizes, and the particles begin to run through the filter. If very fine filter paper is used, the filtration rate becomes very slow at this stage. In such cases, it is often helpful to keep the system flocculated by adding alcohol to the wash water. For example, in the determination of the cation exchange capacity of a clay, the clay is converted into the ammonium form with an excess of ammonium salt. The clay is then filtered and washed free from the ammonium salt to prepare it for a nitrogen determination. In the washing process, the addition of alcohol keeps the clay flocculated and thus aids in the filtration process by keeping the filtration rate high. (In addition, the use of alcohol in the washing procedure curbs the exchange of ammonium ions on the clay by hydrogen ions, which would occur to some extent if the clay were washed with water only.)

2. Management of Clay-Containing Soils (5). In the management of soils, one strives for good porosity and permeability characteristics in order to promote root growth, easy workability, and proper drainage. Obviously, these requirements will be met by a flocculated soil. Hence in a properly treated soil, the concentration of flocculating electrolytes should be kept at a flocculating level. Commonly, gypsum—or lime in acid soils—is used, thus taking advantage of the high flocculating power of the divalent calcium ions on clay and silt. Although gypsum is only moderately soluble in water, it is gradually washed out by rain water, and regular adjustments of the calcium ion level are required.

If the flocculating salt is allowed to be washed out to such an extent that the soil becomes peptized, undesirable changes in the soil will take place. The peptized clay and silt particles will be transported by the rain water and will form small, local filter cakes on narrow pore openings, thus reducing the permeability of the soil and hampering drainage. In this condition, the soil will become flooded by heavy rains, resulting in compaction of the submersed peptized soil by sedimentation, and drainage will be hampered still further. In this stage, the soil is more difficult to work to restore its porosity mechanically, which is a prerequisite for a renewed treatment with the flocculating calcium salts. Therefore, it is better policy to keep the soil in a flocculated condition by regular dosage of the flocculant.

Clayey soils which are flooded by sea water are in a flocculated condition because of the high salt content. The large excess of salt is readily washed out by rains, and the soil is suitable for raising a crop as long as a flocculating amount of salt is left. However, further leaching will soon remove the salt, and the clay will become peptized, resulting in a compact, impermeable, and hard-to-work soil, particularly since the clays will be primarily in the sodium form. It is common practice, therefore, to apply gypsum treatment before the condition of the soil becomes very poor.

It should be stressed that the flocculating action of the calcium ions on the clay in the soil is only part of the process of the improvement of soil structure. The complicated problem of the natural formation of a crumbly soil is not yet fully understood, particularly with respect to the effects of bacterial action or to the mechanism of the action of humus in the soil. In the light of the concept of a positive edge charge, it seems possible that the bridging of the clay particles by negatively charged colloidal sand particles or colloidal humus particles contributes to crumb formation. A difficulty inherent in experimenting with humus colloids is to extract them from the soil without physical or chemical alteration. [See Appendix V, reference B-25 (Chapter 10) for a review of colloid chemical studies on mixtures of clays and humus.]

An alternative method of soil flocculation is treatment with organic polyelectrolytes. The action of these so-called *soil conditioners* is discussed in Chapter Eleven (Section III–A–3b).

3. Permeability of Porous Formations. The permeability of a porous rock which contains a certain amount of clay is much higher for salt water than for fresh water. When salt water flows through the rock, the clay is in a flocculated condition and is probably fixed as a loosely built and permeable framework between the larger mineral grains. Hence the permeability of the rock in this condition will not be greatly affected by the presence of clays. On the other hand, when fresh water flows through the rock, the clays will be in a peptized condition. The individual peptized particles will be entrained by the fluid and will be deposited as microscopic filter cakes on narrow pore openings. Such filter cakes will reduce the

permeability of the rock considerably. The formation is called *water sensitive* (6–9).

It must be mentioned, however, that some authors interpret the difference between salt-water and fresh-water permeability on the basis of differences in swelling of the clay and do not assume that transportation of the clay takes place. Because of the greater degree of swelling of the clay in fresh water, they believe that the pores become more effectively clogged in fresh water than in salt water in which the swelling is repressed. This point of view was probably prompted by the observation that the largest differences between salt-water and fresh-water permeability occur when the rocks contain montmorillonite clays, which swell more than most other clays.* However, montmorillonites are also known as efficient filtration-rate reducers, and this fact would also support the first explanation.

The pronounced effectiveness of montmorillonite clays in reducing filtration rates is probably a result of the particular size and shape of the montmorillonite particles. The very thin, and possibly somewhat flexible, flakes may be expected to close small pore openings effectively by building a compact impervious sheath of flakes across the openings. Thicker clay particles, such as those of kaolinite, are far less effective in building impermeable filter cakes.

4. CONDITIONING OF DRILLING FLUIDS. One of the best known technical applications of stability control to achieve the desired filtration properties is the chemical conditioning of drilling muds. During the drilling operation, it is desirable to build a thin, impermeable filter cake on the faces of the porous formations which are encountered in the hole. The pressure operating in the filtration process is the difference between the hydrostatic pressure of the mud column and the hydrostatic pressure of the formation fluid (the connate water). Excessive losses of fluid which penetrates into the formation will be prevented if the deposited filter cake is rather impermeable. The suspended clay in the drilling fluid will be most effective in preventing such filtrate losses if the suspension is in the peptized condition, so that impermeable filter cakes will be rapidly deposited on the porous rocks. Therefore, the maintenance of a deflocculated condition of the mud by the addition of peptizing chemicals is important to meet the optimum filtration or so-called plastering requirements for the mud.

The plastering properties of drilling fluids also depend on the types of clay in the mud. It has been mentioned previously that montmorillonite

* One should think here in terms of osmotic swelling, since the interlayer swelling is not affected by the presence of salt unless the salt concentration is very high, that is, of the order of 1*M* and above.

clays are usually more effective in this respect than most other clays. Often, if the locally available clays for mud making are rather poor in their filtration behavior, these clays are blended with bentonite.

In order to maintain good plastering properties during drilling, frequent adjustments of the peptizer level must be made, and additional quantities of bentonite must sometimes be added. Often, clay formations are passed by the drill; hence mud is being made while drilling. It is this newly made mud which requires the further conditioning. In addition, peptizing chemicals may become less effective owing to decomposition in the borehole, for example, by hydrolysis. The adjustments are made partly on the basis of routine filtration tests at regular intervals.

The excellent plastering properties of peptized muds are partly due to the penetration of small individual peptized particles into the larger pores of the formation and subsequent trapping of the particles deeper in the formation. Also, the mud filtrate may peptize clays which are already present in the formation. The peptized particles may be entrained by the mud filtrate and be deposited elsewhere in the formation as micro filter cakes on certain pores. In these ways, the permeability of the rock itself is also reduced. This situation is undesirable, however, when the producing zone is being drilled. During the subsequent production of the oil, the trapped clay particles may not be removed, and thus the permeability reduction in the neighborhood of the borehole will hamper the flow of oil. Therefore, during completion of the hole in the producing zone, it is desirable to prevent particle penetration. This may be achieved by using a flocculated mud or an "aggregated" mud in which the particles are thicker. Examples of such drilling fluids which may be used as completion fluid will be given in Chapter Eleven.

The formation of filter cakes in boreholes is complicated by the dynamic conditions under which the cakes are deposited. Such conditions are not fulfilled in the standard routine tests for measuring plastering properties of muds. Laboratory and pilot tests under simulated drilling conditions have shown that the information obtained from standard static tests can be very misleading. Dynamic filtration processes are still under investigation, particularly in connection with the possible harm that penetrating solid particles can do to the formation in the producing zone.

5. CERAMICS. In the ceramic industry an exact control of the stability of clay suspensions is often very important. The following process is related with the filtration properties of the suspensions. Some ceramic objects are made by pouring a clay suspension in a porous mold (*slip casting*). Water is removed from the clay by capillary action, and the clay is deposited as a filter cake on the inside surface of the mold. The final properties of the deposited clay layers depend on the state of peptization or flocculation of

the suspension used and a delicate balance has to be maintained to achieve optimum properties of the final object.

III. RHEOLOGY AND STABILITY OF SUSPENSIONS (10–15)

The consequences of peptization and flocculation on the flow behavior or the rheology of suspensions, some of which have already been mentioned, are of utmost importance in technology and also in nature. The degree of stability of a dispersed system governs the pumpability of drilling fluids, the proper leveling of brush marks in painting, the smoothness of a paper coating, as well as the occurrence of landslides.

In view of the many technical implications of the relations between colloid chemical stability and rheology, some of the elementary principles of rheology will be discussed first.

A. Terminology

The flow and deformation of matter under the influence of mechanical forces is the subject of study in hydrodynamics, elasticity theory, and rheology. The first two deal with material behavior that can be described by simple linear models, whereas rheology is primarily concerned with the behavior of systems showing departures from such simple models. Many of these systems are colloidal systems.

1. NEWTONIAN FLOW. Consider a liquid contained between two parallel plates as shown schematically in Figure 29. Keeping one plate fixed, a tangential force is applied to the other plate, which imparts a *shear stress* σ (or τ) to the liquid. Through friction subsequent layers of liquid move parallel to each other in the direction of the applied stress (*laminar flow*), resulting in a velocity gradient between the plates when a steady state is reached. If the liquid behaves as a so-called *Newtonian liquid,* a constant velocity gradient or *rate of shear* $\dot{\gamma}$ is established, which is proportional to the applied shear stress. The proportionality constant is called the *shear viscosity,* or simply *viscosity* η, hence $\sigma = \eta\dot{\gamma}$. Curve 1 in Figure 30 shows the linear relationship between shear stress and rate of shear for a Newtonian liquid. The viscosity is determined by the slope of the line, or cot α.

(In hydrodynamics, stress is defined as a force acting across imaginary planes, and Newtonian behavior is expressed by a constitutive equation relating shear stress and rate of shear; it is an abstraction, and when one refers to a real liquid as a Newtonian liquid, it means that its flow behavior is consistent with that of a hypothetical liquid.)

Terminology describing Newtonian behavior, together with symbols and units, as proposed for international adoption by the International Union of Pure and Applied Chemistry, are summarized below:

Term	Symbol	SI = (cgs) (factor)
Shear stress	σ, τ	$N\ m^{-2}$ = (dyne cm^{-2}) (10)
Shear rate, rate of shear	$\dot{\gamma}$	$s^{-1} = (s^{-1})$
(Shear) viscosity	$\eta = \sigma/\dot{\gamma}$	$N\ s\ m^{-2}$ = (dyne s cm^{-2}, Poise) (10)
Fluidity	$\varphi = 1/\eta$	$N^{-1}\ s^{-1}\ m^2$ = ($dyne^{-1}\ s^{-1}\ cm^2$) (10^{-1})
Kinematic viscosity	$\nu = \eta/\rho$ (ρ = density)	$m^2\ s^{-1}$ = ($cm^2\ s^{-1}$) (10^4)
Viscosity of the solvent or the continuous medium	η_s, η_m	
Relative viscosity or viscosity ratio	$\eta_r = \eta/\eta_s$	dimensionless
Relative viscosity increment (formerly specific viscosity)	$\eta_i = (\eta - \eta_s)/\eta_s$	dimensionless
Reduced viscosity or viscosity number	η_i/ρ_D (ρ_D is the mass conc. of the dispersed phase)	$kg^{-1}\ m^3$ = ($g^{-1}\ cm^3$) (10^3)
Intrinsic viscosity or limiting viscosity number	$[\eta] = \lim_{\rho_D \to 0} (\eta_i/\rho_D)$	same
Inherent viscosity or logarithmic viscosity number	$\eta_{\ln}$ or $\eta_{\rm inh} = [\ln(\eta/\eta_s)]/\rho_D$	same

For dispersions of charged particles, that part of the viscosity connected with the charge of the particles is referred to as the *electroviscous effect*.

2. NON-NEWTONIAN FLOW. Several types of *non-Newtonian flow*, obeying different relations between shear stress and rate of shear, are illustrated in Figure 30, curves 2, 3, and 4. If flow occurs only above a certain finite stress, it is called *plastic flow* (curves 3 and 4). *Consistency* is a general term to describe the property of a material by which it resists permanent change of shape.

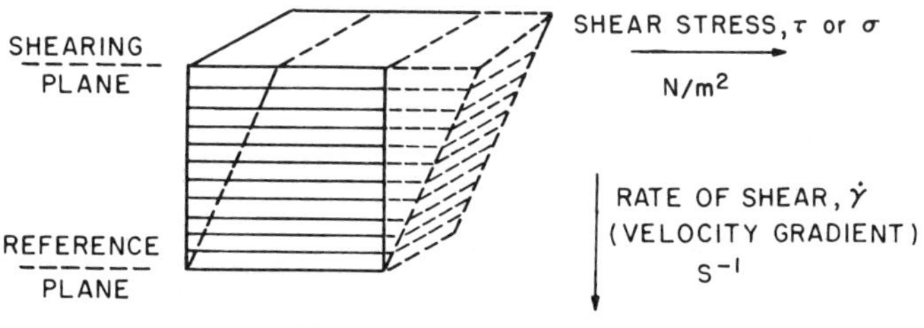

Figure 29. Laminar flow.

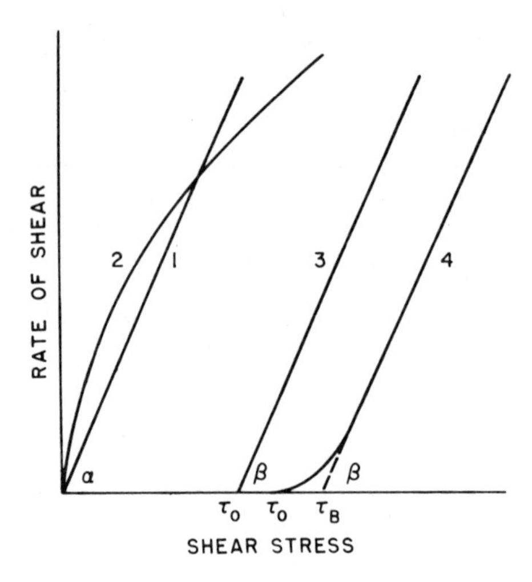

Figure 30. Relations between shear stress and rate of shear.

The *shear dependent viscosity* η is a coefficient equal to $\sigma/\dot{\gamma}$ at a given value of $\dot{\gamma}$; η_0 is the *limiting viscosity at zero shear rate;* η_∞ is the *limiting viscosity at infinite shear rate*; $[\eta_0]$ is the *limit of intrinsic viscosity at zero shear.*

The *differential viscosity* η_Δ is the derivative of stress with respect to the rate of shear at a given shear rate.

The *apparent viscosity* η_{app} is the ratio of shear stress and rate of shear calculated from measurements of forces and velocities in a given experiment, assuming that the liquid is Newtonian. The apparent viscosity of non-Newtonian liquids depends on the type and dimensions of the apparatus used.

If viscosity is a univalued function of the rate of shear, a reduction of the viscosity with increasing rate of shear is called *shear thinning,* and an increase of the viscosity with increasing rate of shear is called *shear thickening* (curve 2, Figure 30). Systems showing shear thickening often also show an increase of volume with increasing shear rate, or "dilatancy," and in older literature the term dilatancy was used at the same time for the rheological phenomenon instead of the term shear thickening.

The *yield stress* τ_0 or σ_0 is the shear stress at which yielding starts abruptly. Its value is arbitrary and depends on the subjective criterion of what is considered yielding to produce flow.

Many colloidal dispersions show so-called *Bingham flow,* which is characterized by curve 4 in Figure 30. For the linear part of the curve, the following relation applies: $\tau - \tau_B = \eta_\Delta\dot{\gamma}$ and τ_B (or σ_B) is called the *Bingham yield stress.*

The behavior characterized by curve 3 in Figure 30 is sometimes called *ideal plastic flow,* a type of flow which is only approximated by real systems.

At very low shearing stresses elastic and creep flow processes occur in many colloidal systems (See Chapter Seven, reference 32). These and some other variations of flow behavior will not be discussed here.

3. TIME-DEPENDENT PHENOMENA. A complication often encountered in dispersed systems is that their flow diagram is dependent on the previous shear history of the system. The application of a finite shear to a system may result in a decrease of the viscosity or consistency. If after discontinuing shear the decrease is permanent, this behavior is called *work softening* (or *shear breakdown*). If the original viscosity or consistency is recovered at an observable rate, this behavior is called *thixotropy* (from the Greek *thixis,* meaning touch, and *tropein,* meaning change). The time in which a certain yield stress is reached after discontinuing shear is called the *time of solidification* or the *time of thixotropic recovery. Rheopexy* describes the phenomenon in which the time of solidification after discontinuing a relatively high shear rate is shortened by applying a small shear rate, for example, by gently tapping the test tube containing the thixotropic system.

Terms to describe the opposite behavior in which shear results in either a permanent or a reversible increase of viscosity or consistency with time are, respectively, *work hardening* (opposite of work softening), and *antithixotropy* (opposite of thixotropy).

4. TURBULENT FLOW. The above terminology applies to conditions of low shear at which the streamlines are parallel, that is, at which laminar flow occurs. At high shear rates the laminar pattern of flow is disturbed, and so-called *turbulent flow* occurs. In this region of shear rates, when a volume element of the liquid changes direction because of an occasional disturbance, its inertia tends to make it proceed in the changed direction if the frictional forces are small compared with the inertia forces. The shear rate at which turbulence begins depends on the shape of the vessel, for example, on the shape of the entrance of a capillary tube, and on the ratio of the inertia and the frictional forces at that shear rate. This ratio is expressed in terms of a dimensionless quantity which is called the *Reynolds number.* For a certain vessel, turbulent flow is observed to set in when the Reynolds number reaches a certain critical value, the *critical Reynolds number,* which has about the same value for different liquids flowing through the same vessel.

In the turbulent region, the flow of liquids is no longer solely determined by rheological constants of the liquids but to a large degree by the inertia forces, which is reflected in the presence of the density of the liquid in the flow formulas.

B. Measurements of Flow Properties

Capillary viscometers of varied design are commonly used to measure the viscosity of Newtonian liquids. Although the capillary viscometric techniques are usually rather simple, in precise work several corrections must be applied, for which references may be made to monographs on viscosity.

In capillary-tube flow, the rate of shear and the shearing stress are not constant throughout the liquid; they vary between zero in the center of the tube and a finite value at the wall of the capillary. Since the viscosity is independent of the rate of shear or the shearing stress in the case of Newtonian liquids, the variation of these factors in the capillary tube is of no concern. However, in the case of non-Newtonian systems, the capillary viscometer is not always suitable. Although the interpretation of capillary-tube flow data in terms of a τ versus $\dot{\gamma}$ flow diagram is possible for relatively simple τ versus $\dot{\gamma}$ relations, viscometers which operate at uniform shear rates are generally preferred. Parallel-plate viscometers, in which the plates are moved with respect to each other in one direction, meet this requirement, but it is usually more practical to use a viscometer in which the plates are rolled up to form a set of concentric cylinders. In such "rotational-cylinder viscometers," the system is sheared between the cylinders by the rotation of one cylinder with respect to the other. At a given rate of rotation, the rate of shear in the sample between the cylinders is not strictly constant, but it varies only between narrow limits, particularly when the diameters of the cylinders are large and their clearance small. The shearing stress is derived from the torque on one of the cylinders, and the average rate of shear is calculated from the measured rate of rotation and the diameters of the two cylinders. In one type of coaxial-cylinder viscometer, the rate of rotation of the moving cylinder caused by applying a certain torque to that cylinder is measured (Couette type). In other instruments, one of the cylinders is rotated with a given speed, and the torque on the other cylinder is measured (MacMichael type).

It should be pointed out that for a system displaying ideal plastic flow (curve 3 of Figure 30), the curve relating rate of rotation and applied torque is straight only beyond a certain rate of rotation. Owing to the variation in shearing stress between the two cylinders as a function of the distance to the axis of rotation, the yield stress is not reached in all sections of the annular space simultaneously at a given torque. Hence a slight initial curvature is observed before the linear dependence is obtained. Such a deviation should not be confused with the initial curvature of the τ versus $\dot{\gamma}$ curve for a nonideal or Bingham plastic flow system. The torque at which the first rotation is observed (T_0) is lower than the value of the torque T_e obtained by extrapolation of the straight-line portion to the torque ordinate. The ratio of T_0 and T_e is fixed by the ratio of the radii of the two cylinders and is thus an apparatus constant. Only when this ratio of T_e and T_0 is larger than predicted from the apparatus dimensions is a deviation from ideal plastic flow indicated.

An analogous situation exists in tube flow. In this case, the shearing stress at a certain applied pressure difference varies even more, that is, from zero in the center of the tube to a finite value at the wall. The plot of pressure versus volume rate of flow for an ideal plastic system is nowhere a straight line, but it approaches asymptotically to a straight line at higher flow rates. When the asymptote is drawn, its intersection with the pressure ordinate occurs at a pressure P_e which is 4/3 times the pressure P_0 at which flow is first observed. Only when the ratio of P_e and P_0 is greater than 4/3 is a deviation from ideal plastic flow in the Bingham sense indicated.

In plastic systems, slip of the bulk of the mass at the walls of tubes or cylinders may occur. Slip may be prevented in a concentric-cylinder viscometer by thin, closely spaced vanes perpendicular to the cylinder surface. Then, shear occurs only in the plane enveloping the ends of the vanes, which is situated in the interior of the plastic material (5).

When a viscometer is used for analyzing certain technical flow problems or for control purposes in technical processes, the following considerations should be applied in choosing a viscometer and the conditions of testing: In the first place, it is essential that the viscometer chosen operates in the range of rates of shear which are prevailing in actual practice, at least when one is dealing with non-Newtonian liquids. For example, the prediction of the pumping pressures required to sustain a certain circulation rate of a drilling fluid in the mud circuit should be based on viscometric data which are obtained at rates of shear corresponding to those prevailing in the circuit at the given circulation rate. On the other hand, such data will be of no value for predicting the settling velocity of drill cuttings or sand in the mud, since this velocity is governed by the rheological behavior of the mud at the much lower rates of shear which prevail in the settling process. If one viscometer does not cover both ranges of shear rate with sufficient accuracy, two separate viscometers should be chosen to obtain data which can be used in the interpretation of either the pumping or the settling characteristics.

Another requirement which should be met in applied viscometry is that the flow pattern in the viscometer be laminar in the range of shear rates at which the flow pattern in practice is also laminar.

With regard to measurements in the viscometer, care should be taken to simulate the shear history in practice, if the rheological properties of the system are known to depend on this history, for example, in the case of thixotropy. Therefore, a prior analysis of the shear conditions in practice is necessary before the choice is made of viscometer and testing conditions to be used in connection with the particular practical problem.

C. Rheological Properties of Suspensions

In general, the rheological properties of dispersed systems depend on the following factors: (*1*) the viscosity of the fluid medium, (*2*) the concentration of the suspended matter, (*3*) the size and shape of the suspended particles, and (*4*) the forces of interaction between the particles, that is, the state of stability or flocculation of the suspension.

1. Dilute Sols and Suspensions. Dilute sols and suspensions behave as Newtonian liquids. The viscosity of the suspensions η is higher than that of the medium η_s; hence the relative viscosity η_r is greater than unity. Einstein has derived the following theoretical relation between the relative viscosity of a dispersion and the amount of dispersed solids: $\eta_r = 1 + 2.5\,\varphi$, in which φ is the relative volume of the dispersed solids (the total volume of the dispersed particles divided by the total volume of the dispersion). Hence the contribution of the dispersed solids to the relative viscosity can be written as $2.5\,\varphi = \eta_r - 1 = \eta_i$, the relative viscosity increment.

According to the theory, the contribution of the suspended matter to the viscosity is independent of the number of particles—only their total relative volume is important. The formula is valid for solid spheres which are far enough apart not to influence each other. They must be large with respect to the molecules of the medium but small compared with the dimensions of the viscometer. They must be wetted by the fluid in which they are suspended.

When the particles are anisometric, for example, are plates as in many clay suspensions, much larger proportionality constants than 2.5 apply. Theoretical expressions for the proportionality constant can be derived when the plates are idealized as ellipsoids of revolution or as discs. For ellipsoids of revolution with a small minor to major axis ratio p, one finds $\eta_i = (4/3\pi)(1/p)\varphi$ (Peterlin-Burgers), and for large thin discs with a small height to diameter ratio p, one finds $\eta_i = (32/15\pi)(1/p)\varphi$ (Simha). For expanding clays, the relative volume of the swollen particles may be calculated from $\varphi = m_C/2.85\,(100 - a)$, in which m_C is the concentration of the clay expressed as grams of dry clay present in 100 cm^3 suspension, 2.85 is the density of the dry clay, and a is the volume percent of water in the expanded clay as derived from the increase of the 001 spacing upon hydration. Therefore, from viscosity data of suspensions of such particles, their axis ratio can be derived. If, in addition, their volume is known, for example, from ultramicroscopic counting data and density, the height and the diameter of the plates can be obtained. This is an example of the combination of two indirect methods which enable the computation of the dimensions of anisometric dispersed particles.

Since the formulas are based on the assumption that particle interaction is absent, it is usually advisable to perform viscosity measurements at different particle concentrations and to extrapolate the data to zero concentration to find the initial slope of the plot of η_i versus ϕ.

2. Effect of Particle Interaction on the Flow Properties of Suspensions. When particle interaction cannot be neglected, the above formulas are no longer valid. For example, when attractive forces prevail between the particles, the viscosity of the suspension will be higher than that predicted from the formulas. If the suspensions are concentrated, and there is a net attraction between the particles, they will link to build a rigid matrix, and a certain yield stress is developed. The magnitude of the yield stress is related to the force required to break individual links and to the number of links. This also applies to the Bingham yield stress τ_B (curve 4, Figure 30). This yield stress is somewhat higher than the real or true yield stress τ_0. The linear part of the curve obeys the equation $\tau - \tau_B = n\dot{\gamma}$. With regard to the physical meaning of τ_B, Goodeve (Chapter Seven, references 23, 28, and 29) has presented the following analysis: Written in the form $\tau = \tau_B + n\dot{\gamma}$, the equation indicates that there are two contributions to the stress, one which is a shear dependent viscosity contribution ($n\dot{\gamma}$), and one which is independent of the rate of shear, the Bingham yield stress, τ_B. Goodeve argues that the shear independent contribution is due to the transmittal of impulses from one shearing layer to the next when the linked particles in adjoining layers are stretched until the link breaks. The impulse transmitted can be shown to be inversely proportional to the rate of shear. The links are constantly reestablished in the shearing process, and the instantaneous number of links per second per square centimeter can be shown to be proportional to the rate of shear. Hence the force due to shear making and breaking of links, or the product of the impulse and the number of impulses per second per square centimeter is independent of the rate of shear. In the nonlinear region of the flow curve at low shear rates, thermal making and breaking of links becomes important with respect to shear making and breaking.

The matrix of a clay gel in pure water (refer to Figure 24*b*) at a reasonably uniform particle size may be visualized as a cubic network of plates associated edge-to-face. The strength of the individual links in the matrix can then be derived from the Bingham yield stress and the number of links per unit volume, applying Goodeve's analysis and knowing particle dimensions and clay concentration.

Upon peptization of the originally stiff, rigid suspension, the particle links are weakened and are finally broken as soon as repulsion between the particles predominates. Consequently, the result of peptization is a reduction of the yield stress, which finally disappears, and the system obtains the character of a rather freely flowing liquid. These changes in the system are illustrated by the changing flow curves in Figure 31*a*. The flow behavior of the flocculated suspension is of the Bingham type, as shown by curve 3. Upon the addition of a peptizing chemical, the Bingham yield stress is reduced, and the flow curve shifts to position 2; finally, upon further addition of peptizer, the flow curve shifts to position 1. At this point, Bingham flow has practically become Newtonian flow. During the peptization process, the differential viscosity, as given by the slope of the curves, often decreases somewhat, but this decrease is only of secondary importance in the thinning effect, which is mainly due to the reduction of the yield stress. Therefore, the thinning action cannot correctly be referred to as a "viscosity reduction"—it is mainly a yield-stress reduction.

3. Thixotropy and Rheopexy. Clay suspensions which are more or less flocculated show the phenomenon of *thixotropy*. The stiff suspension becomes thinner on stirring and thickens again on standing. These changes in consistency appear to be due mainly to changes in the yield stress, as indicated by curves 1 and 2 in Figure 31*b*. Curve 2 represents the original flow behavior of the flocculated suspension. When stirred, the thinned suspension displays a flow behavior represented by curve 1, but after a period of rest, the flow curve 2 is again obtained. When the suspensions are very sensitive to shear, they will change materially during the determination of the flow diagram in the viscometer. In that case, a hysteresis loop is obtained when readings are taken subsequently at increasing and decreasing rates of shear (represented by curve 3 in Figure 31*b*).

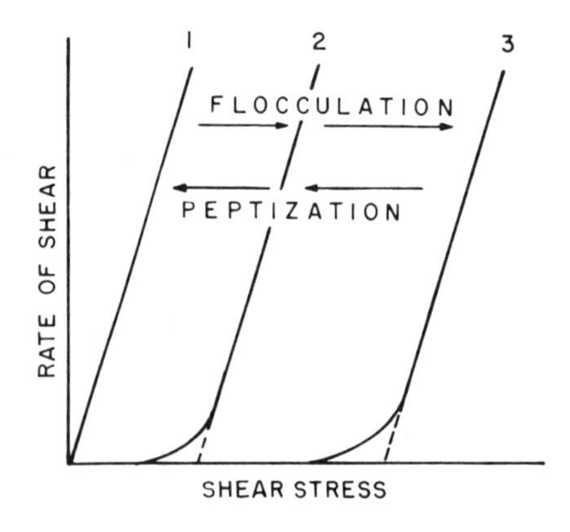

Figure 31*a*. Effect of peptization and flocculation on the flow behavior of a clay suspension (schematic).

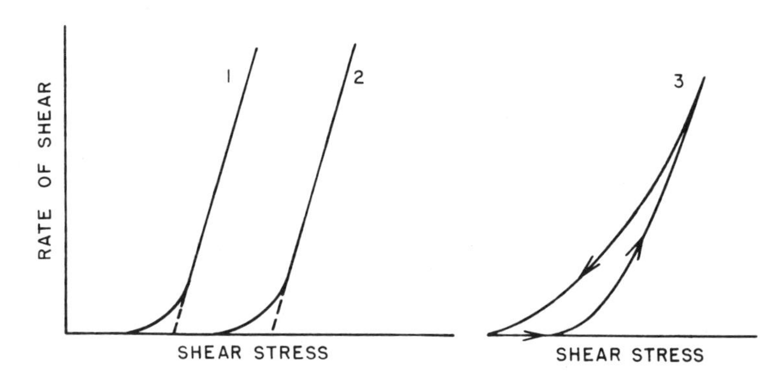

Figure 31*b*. Flow curves of thixotropic clay suspensions (schematic).

The reduction of the yield stress in a stirred system indicates that shear causes a breakdown of the particle links in the skeleton structure of the flocculated system. The restoration of the original yield stress indicates that during a period of rest, the links are re-established when Brownian motion brings the particles back together.

The process of thixotropic stiffening is essentially the same as slow coagulation in a dilute sol; therefore, in a given suspension the rate of stiffening will be dependent on the size of the residual energy barrier in the particle-interaction potential curve. When the salt concentration increases, the size of the barrier is reduced, and the rate of particle linking and stiffening of the suspension increases.

It is difficult to determine rates of recovery from measurements in viscometers since the system is disturbed by the measurement procedure. This would not occur if the measurement involved deformations below the elastic limit only. An elegant method is to measure the rate of propagation of very low amplitude shear waves through the gel as a function of time of recovery. The coefficient of shear elasticity which is proportional to the square of the propagation rate can then be calculated as a function of time of recovery, and this coefficient is a direct measure of the number of re-established links in the matrix. Applying the theory of coagulation kinetics, the size of the energy barrier in the system can be evaluated (see Chapter Seven, reference 28).

The increase of the rate of stiffening due to gentle shearing of the suspension, which is called *rheopexy,* is a consequence of the enhanced collision frequency in a stirred system. In dilute sols or suspensions, the increase of the rate of coagulation due to stirring of the sol corresponds to the phenomenon of rheopexy in more concentrated systems.

D. Applications

1. Drilling Fluids (37, 38). As an example of the practical consequences of yield-stress reduction of concentrated clay suspensions by the addition of chemical peptizers, we shall treat the problem of the pumpability of drilling fluids and drilling-fluid viscometry (10, 11).

In rotary drilling, a bit is attached to a hollow drill pipe which has a diameter smaller than the diameter of the bit. The drill pipe consists of sections which are screwed together by means of short "tool joints" which sometimes have a smaller internal diameter than that of the drill pipe. Drilling fluid is pumped from a reservoir (mud pit or mud tank) down the drill pipe; it passes through the openings in the bit (narrow "bit nozzles") and returns through the annular space between the drill pipe and the hole drilled, carrying up the drill cuttings. At the surface, the mud is freed from the cuttings by sieving and by settling in the mud pit.

In the mud circuit the drilling fluid must be circulated with a sufficiently high rate to bring the drill cuttings up and to cool the bit efficiently. For a given mud circuit, one likes to be able to anticipate the pumping pressures required to maintain the desired rate of circulation on the basis of the rheological behavior of the mud. In other words, one wants to relate viscometric data on the mud with its pumping characteristics.

The total pressure loss in the mud circuit when the mud is circulated is the sum of the pressure losses in the various parts of the circuit, that is, the surface connections, the drill pipe and tool joints, the bit nozzles, and the annulus between drill pipe and casing or the wall of the hole.

We shall consider the pressure losses in the drill pipe. Figure 32*c*, shows the pressure losses in the drill pipe as a function of the volume rate of circulation Q in the range of flow rates which are of practical interest. Curve 1 refers to a flocculated mud, and curve 2 refers to the same mud after the addition of a certain quantity of peptizer. At the smaller flow rates, laminar flow prevails in the pipe, and in the diagram the curves are practically straight lines in this region. (Actually, they are asymptotes to a straight line.)

If the fundamental τ versus $\dot{\gamma}$ diagrams of the two muds are known, the Q versus P curves can be calculated for the given dimensions of the pipe. In Figure 32*b*, the Bingham flow curves for the two muds are drawn; they are marked 1 and 2, corresponding to curves 1 and 2 in the flow diagram.

The fundamental τ versus $\dot{\gamma}$ curves can be obtained from viscometric data when a viscometer of suitable design is used. These data are plotted in Figure 32*a*. They were obtained with a rotational-cylinder viscometer and are presented as a plot of rate of rotation versus applied torque ($\omega - w$). The curves for the two muds considered are again marked correspondingly

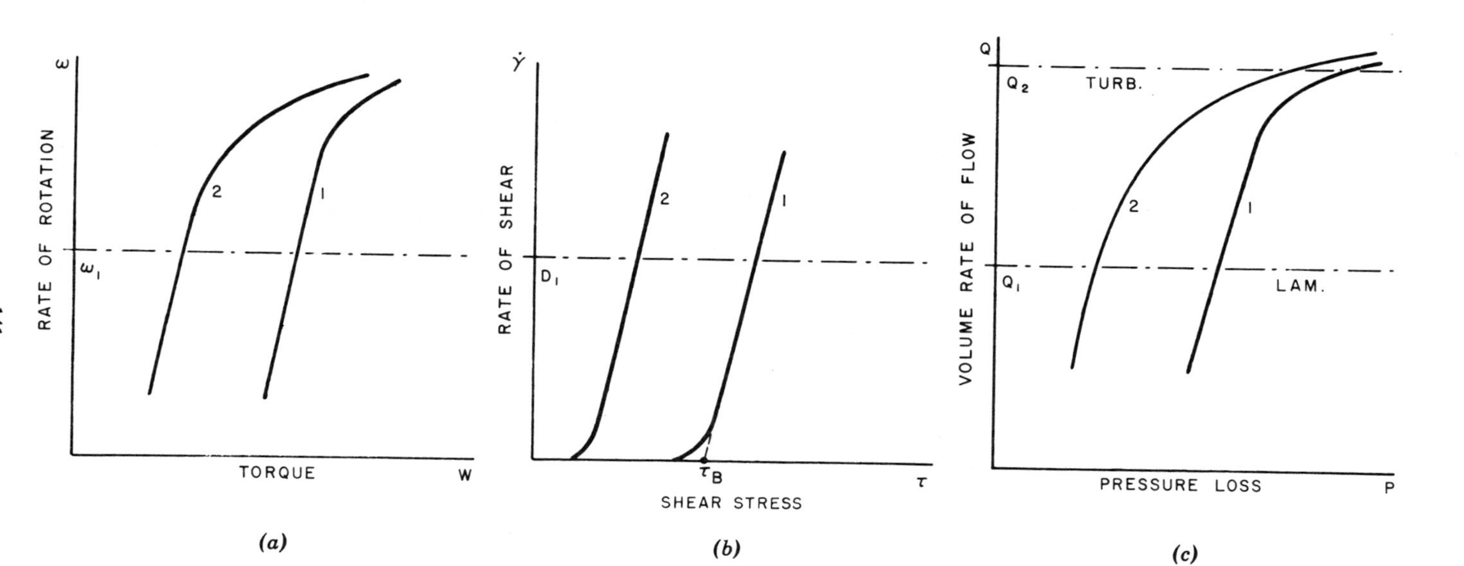

Figure 32. Relations between rheological, viscometer, and pipe-flow data for two drilling fluids showing Bingham behavior. (*a*) Rotational viscometer diagram. (*b*) Shear stress rate of shear diagram. (*c*) Pipe-flow diagram.

1 and 2. The complete procedure of calculating pressure losses as a function of rate of circulation is first to translate the viscometer data into the fundamental τ versus $\dot{\gamma}$ relation, which involves a knowledge of the apparatus constants for the conversion of rate of rotation into average shear rate and of torque into average shearing stress operating in the suspension between the two cylinders. The apparatus constant can be obtained either by calibration or by computation from the apparatus dimensions. The next step is to compute the Q versus P relation from the fundamental τ versus $\dot{\gamma}$ relation, as indicated above.

The rate of circulation Q_1 in the pipe-flow diagram *c* corresponds to an average rate of shear $\dot{\gamma}_1$ in the fundamental diagram *b* and to the rate of rotation ω_1 in the viscometer diagram *a*. Thus, the condition is fulfilled that the viscometer operates in the range of rates of shear which prevail in the mud circuit. The linearity of the curves in this region indicates that both in the viscometer and in the drill pipe, flow is laminar.

From the pipe-flow data, it is seen that the pumping pressure required to maintain a certain rate of flow Q_1 is primarily determined by the position of the steep curves in the diagram, and this position is determined by the magnitude of the Bingham yield stress. Therefore, the pressure requirements for the flow of the mud through the drill pipe in the considered region of flow rates decreases with decreasing yield stress. Slight changes in the slope of the curves, that is, in differential viscosity, do occur upon peptization, but such changes are only of secondary importance for the pressure requirements.

At the higher rates of circulation which are often practiced in drilling, the flow of the mud in the drill pipe becomes turbulent. In this region, the pressure losses increase considerably faster with increasing flow rates, as is demonstrated by the bending upper part of the curves in the pipe-flow diagram. Moreover, the pressure losses become less dependent on the flow characteristics of the muds, as is evident from the close approach of curves 1 and 2 in diagram *c* at high flow rates. In the turbulent-flow region, inertia forces become dominant, and the density of the mud becomes of primary importance.

In the turbulent region, pressure losses can no longer be predicted from rheological data for the mud on a theoretical basis, and one must rely on empirical relations. A considerable amount of work has been done to establish such empirical relations for turbulent flow of dispersed systems. In the hydraulics of these systems, a commonly used parameter is the *equivalent turbulent viscosity*. This is the viscosity of an imaginary Newtonian liquid which has the same density as that of the mud, and which would give the same pressure losses in the turbulent range as the mud. Attempts have been made to establish empirical relations between the two rheological constants of the mud, that is, differential viscosity and Bingham yield stress, with the equivalent turbulent viscosity. Such relations would enable the computation of turbulent-flow pressure losses from viscometric data obtained in the laminar-flow range.

No unique empirical relation of general validity exists, but an average relation can be found which is useful for the prediction of turbulent-flow pressure losses of drilling fluids with only moderate accuracy, although it is often accurate enough for engineering purposes. (Since in the viscometer, at high rates of rotation, flow also becomes turbulent, one might think that viscometer data obtained in the turbulent region would be useful to predict pressure losses in turbulent flow in pipes. However, the pattern of turbulence in the viscometer and in the pipes is entirely different; therefore, no useful correlations can be established.)

A different empirical approach to the problem of the prediction of turbulent-flow pressure losses is the establishment of a correlation between these pressure losses, that is, the equivalent turbulent viscosity, and the relative volume of the dispersed solids. Although in this approach the state of peptization or flocculation of the mud is completely neglected, even better accuracy is often obtained in pressure-loss predictions on the basis of this correlation than with the correlation with the rheological constants of the muds (11).

For a discussion of mud flow in the other parts of the mud circuit, a review of the hydraulic formulas, and a discussion of mud viscometry, reference 5 may be consulted.

With regard to the problem of the settling of drill cuttings and sand in drilling fluids, the following remarks can be made. When the drilling operation must be interrupted before the drill cuttings and sand have been circulated out of the hole, it would be desirable if the mud had some degree of thixotropic stiffening. Then, owing to the creation of a yield stress, the settling of the cuttings to the bottom of the hole would be prevented. Still, the necessary separation of the cuttings and the sand in the mud pit during the normal operations could be maintained by preventing thixotropic stiffening by stirring the mud in the pit. Since the development of thixotropy occurs when the mud is somewhat unstable, this requirement is incompatible with reaching optimum plastering and pumping characteristics, for which the suspension should be stabilized. In practice, however, a satisfactory compromise can be obtained.*

* In drilling-fluids practice, the terminology "stable" and "unstable" mud usually refers to whether the mud has a good or a bad cuttings-carrying ability. This usage of the terms "stable" and "unstable" should not be confused with their usage in the colloid chemical literature. Since a colloidally stable mud would have little rigidity and therefore a small cuttings-carrying ability, such a mud would be called an "unstable" mud in the field.

2. Paints (14, 15). A certain degree of thixotropy is a desirable property of paints for the following reasons: When the thixotropic paint is subjected to shear during brushing or spraying, the paint will become thin, enabling an easy application. Immediately following brushing or spraying, the yield stress should remain low enough to allow the brush marks or drops to level to a flat surface. Soon after the paint is leveled out, however, some buildup of yield stress should occur; otherwise, the paint would sag or drip. Of course, when volatile solvents are used in the paint, the buildup of yield stress is due largely to the evaporation of the solvent, by which the pigment concentration is increased.

Obviously, the control of the optimum degree of thixotropy required for proper application and leveling and nondripping properties will involve a rather subtle adjustment of the degree of flocculation of the paint.

Another desirable consequence of thixotropy in a paint is that during application, the paint in the container stiffens somewhat and thus retards the settling of the pigment. As mentioned previously, during storage of the container, the pigment settles to a comparatively loose sediment when the paint is colloidally somewhat unstable, and such a sediment is more easily stirred up than that of a stable pigment suspension. Therefore, a certain colloidal instability of a paint is desirable with regard to both application and ease of homogenization after storage.

3. Paper Fillers and Coatings (33–36). The paper industry is the largest user of kaolinite clays. It consumes about 50 percent of the production, as compared with 10 percent for refractories and 10 percent for pottery. The clays are applied as fillers and as coatings—8–20 percent in uncoated papers, and more in coated papers. The filler gives the paper a more homogeneous texture and thus makes it suitable for printing after calendering. The coating improves the brightness and the gloss and gives the paper better printing characteristics, as required in picture reproduction. (For the relation between gloss and degree of crystallinity see references 33 and 34).

The technique of application of the paper coating offers a rheological problem. A coating mixture or the so-called "color" consists of a clay pigment suspension containing a binder such as casein or a modified starch. The total solids content of the color is rather high. Depending on the particular technique used, the solids content may vary from 35 percent to as much as 60 percent. In application of the coating, ease of spreading and leveling to a thin layer on the paper requires a high degree of fluidity of the color, despite the rather high solids concentration. Therefore, peptizing agents are commonly used in the color formulation. In the case of starch colors, the most popular peptizing agents are phosphates. The casein colors do not require peptizing agents, since the casein itself has an appreciable thinning effect on the clay suspension.

Obviously, rheological control of coating formulations is important in the paper industry, and many studies have been made of the effects of particle size, particle shape, and types of adhesives and peptizers on the final properties of the papers prepared with different coating techniques. Commonly, conventional viscometers have been used in this work, but in the study of high-speed coating operations, other instruments should be used which operate at the high rates of shear prevailing in such operations.

E. Rheological Properties of Sediments

As mentioned previously, the sediments of stable suspensions are usually more densely packed than those of flocculated suspensions. The rheological properties of the two types of sediments are entirely different.

The voluminous sediment of a flocculated suspension behaves more or less like a Bingham system, as does a concentrated flocculated suspension. The closely packed sediment of a stable suspension, on the other hand, shows a very peculiar rheological behavior. At low shear stresses, the stable, closely packed sediment behaves like a liquid, and it can be poured slowly from a vessel. At high shear stresses, however, the sediment offers a strong resistance to deformation. For example, when a rod is suddenly pushed into it, the sediment behaves like a solid, but if the rod is left on top of the sediment, the rod sinks slowly. With a spatula, one can quickly break a lump out of the sediment, but the next moment it drips from the spatula. Apparently, we are dealing with a system which obeys the flow behavior represented by curve 2 of Figure 30.

This behavior of the stable sediment can be explained as follows: Since the particles repel each other, they are easily shifted with respect to each other by a small shearing stress. Simultaneously, since the particles are closely packed, liquid must flow through the interstitial spaces during the displacement of the particles. If the displacement is slow, the liquid is able to follow the change in geometry; during a rapid displacement, however, when a high shearing stress is suddenly applied, the capillary flow of the liquid becomes the bottleneck in the shearing motion of the sediment, and a high resistance against shear develops.

At the same time, the displacement of the particles involves a disturbance of the condition of close packing of the particles. Thus, the interstitial volume of the sediment increases, and water is drawn into the interior of the sediment: the system "dilates." Consequently, the sediment assumes a dry appearance when it is sheared suddenly, for example, around the point of impact of a rod which is rapidly pushed into the sediment. The increase of viscosity with shear rate, as well as the dilatant behavior, is primarily shown by systems containing more or less spherical particles in the narrow concentration range in which the proper close-packing conditions are realized and in which repulsion predominates. Among clays, the behavior can be observed in concentrated suspensions of rather coarse kaolinite particles. Upon the addition of flocculants, these systems are converted into Bingham systems. However, the latter does not apply in systems of very large particles, such as glass beads or sand grains, since in these systems interparticle forces are negligible with respect to gravity forces. Thus the behavior of wet coarse sand does not change when a flocculating salt is added, for example, when fresh water is replaced by sea water. Dilatancy is readily observed on the wet part of the ocean beach from the dry appearance of the sand around the foot when stepping on wet sand. When the foot is lifted, the sand regains its wet appearance.*

Finally, a special situation should be mentioned which gives rise to a certain rigidity of a stable suspension in which the particles are not closely packed. In systems containing rod- or needle-shaped particles, rigidity may be developed when these particles are haphazardly entangled like a heap of twigs. Often, in such systems, when a sufficiently high shear to overcome the yield stress is applied, the brush-heap structure is destroyed by orientation of the particles in a more or less parallel arrangement. Then, *work softening* or *shear breakdown* occurs. Of course, breakdown of individual rods or needles in the deformation process may also contribute to the shear breakdown.

F. Mechanical Properties of Soils (16–32)

Whenever clays are dominant constituents of soils, the structure and mechanical behavior of the soil is governed by the same principles as that of clay sediments. One very spectacular effect of relatively minor changes in the composition of the water phase in soils is the occurrence of sudden changes in mechanical properties as shown in landslide areas. Landslides usually occur as a result of shear breakdown (18), which may be explained as follows: Soils in landslide-prone areas often originate from sediments which were originally deposited under marine conditions. Clay and silt being in the flocculated condition under these circumstances, a coherent

* An interesting laboratory demonstration of this effect is the following experiment: A flexible plastic bottle is completely filled with closely packed glass beads and water. A glass tube is attached to the bottle through a hole in the cork. The tube is partly filled with water, and no air is included in the system. When the bottle is squeezed, the water level in the tube goes down, since the disturbance of the close packing of the glass beads causes an increase of the interstitial space. At the same time, the bottle bulges elsewhere owing to the force transmitted by the glass beads.

particle matrix exists which gives the soil rigidity. When the salt is gradually leached out by rain water, the system becomes deflocculated, and the particle matrix is no longer held together by attractive forces between the particles. Nevertheless, as long as the system is not disturbed, the matrix will hold together purely mechanically. The slightest disturbance will then cause a sudden collapse of the matrix and the system becomes fluid. In most typical situations, the liquefaction is permanent, hence the cause of the landslide is shear breakdown or work softening, and not thixotropy. In these areas, one can take cores with special tools designed not to disturb the soil while coring. The recovered core is solid, but it liquifies immediately upon stirring.

Landslides also may occur in areas where clays were originally deposited under fresh water conditions. In those areas, the deflocculation of the clay is due to leaching of organic anions from surface vegetation. These anions peptize the clays by edge adsorption and reversal of edge charge. Probably, this is also a contributing factor in deflocculation of the marine sediments.

Chemical methods for mechanical stabilization of these unstable soils include lime treatment in order to keep the soil in the flocculated condition, but it will often be necessary to build supporting concrete walls.

Shear breakdown in landslide areas and the thixotropic behavior of quick sands are problems of soil mechanics dominated by matrix building through EE and EF association, as is confirmed by fabric studies using the scanning electron microscope or impregnating techniques followed by examination of thin slides (24–29).

Another group of soil properties is related with the swelling behavior of clays, involving interactions between face surfaces and between face surfaces and water. These are discussed in the next chapter.

References

1. Hartwell, J. M. (1965), The diverse uses of montmorillonite, *Clay Minerals*, **6**, 111–118.
2. *Kaolin Clays and Their Industrial Uses*, J. M. Huber Corp., New York.

SEDIMENTATION

3. Hamaker, H. C., and Verwey, E. J. W. (1940), The role of the forces between the particles in electrodeposition and other phenomena, *Trans. Faraday Soc.*, **36**, 180–185.

FILTRATION

4. LaMer, V. K., and Smellie, R. H., Jr. (1962), Theory of flocculation, subsidence, and refiltration rates of colloidal dispersions flocculated by polyelectrolytes, *Clays, Clay Minerals*, **9**, 295–314.
5. Quirk, J. P., and Schofield, R. K. (1955), The effect of electrolyte concentration on soil permeability, *Soil Sci.*, **6**, 163–178.

6. Dodd, C. G., Conley, F. R., and Barnes, P. M. (1955), Clay minerals in petroleum reservoir sands and water sensitivity effects, *Clays, Clay Minerals*, **3,** 221–238.
7. Rollins, M. B., and Dylla, A. S. (1964), Field experiments on sealing permeable fine sand with bentonite, *Soil Sci. Soc. Am. Proc.*, **28,** 268.
8. Gray, D. H., and Rex, R. W. (1966), Formation damage in sandstones caused by clay dispersion and migration, *Clays, Clay Minerals*, **14,** 355–366.
9. Curry, R. B. (1967), Deposition of montmorillonite from suspension during flow through porous media, *Clays, Clay Minerals*, **15,** 331–344.

RHEOLOGY

10. van Olphen, H. (1950), Pumpability, rheological properties, and viscometry of drilling fluids, *J. Inst. Petrol.*, **36,** 223–234.
11. Havenaar, I. (1954), The pumpability of clay-water drilling fluids, *J. Petrol. Technol.*, **6,** 49–55.
12. von Engelhardt, W., and Lubben, H. (1956), Absolute Viskositätsmessungen mit einem Rotationsviskosimeter nach Couette-Hatschek, *Kolloid-Z.*, **147,** 1–6.
13. Langston, R. B., and Pask, J. A. (1958), Analysis of consistencies of kaolin-water systems below the plastic range, *Clays, Clay Minerals*, **5,** 4–22.
14. Jordan, L. A. (1958), Changing patterns, *J. Oil and Colour Chemists' Assoc.*, **41,** 273–287.
15. Doherty, D. J., and Hurd, R. (1958), The preparation, properties, and rheological investigation of thixotropic paint systems, *J. Oil and Colour Chemists' Assoc.*, **41,** 42–84.

See also Chapter Seven, references 20–41.

SOIL STRUCTURE, AND SOIL ENGINEERING

16. Lambe, T. W. (1958), The structure of compacted clay, *J. Soil Mech. Found. Div., Am. Soc. Civil Engrs.*, **SM 2,** Paper 1654, 1–34.
17. Lambe, T. W. (1958), The engineering behavior of compacted clay, *J. Soil. Mech. Found. Div., Am. Soc. Civil Engrs.*, **SM 2,** Paper 1655, 1–35.
18. Rosenqvist, I. T. (1958), Remarks on the mechanical properties of soil-water systems, *Reports of the Swedish Society of Clay Research*, No. **10,** pp. 433–457.
19. Symposium on the Engineering Aspects of the Physico-Chemical Properties of Clays 1962, *Clays, Clay Minerals*, **9,** 12–218.
20. Bradfield, R. (1950), Soil Structure, *Trans. Intern. Congr. Soil Sci., 4th, Amsterdam*, Vol. II, 9–19.
21. Koenigs, F. F. R. (1961), The mechanical stability of clay soils as influenced by the moisture condition and some other factors, Thesis, Wageningen, The Netherlands.
22. Williamson, W. O. (1955), Effects of deposition and deformation on the microstructure of clays, *Research* (*London*), **8,** 276–281.
23. Geuze, E. C. W. A., and Rebull, P. M. (1966), Mechanical force fields in a clay mineral particle system, *Clays, Clay Minerals*, **14,** 103–116.
24. Martin, R. T. (1966), Quantitative fabric of wet kaolinite, *Clays, Clay Minerals*, **14,** 271–288.
25. Sloane, R. L., and Kell, T. R., (1966) The fabric of mechanically compacted kaolin, *Clays, Clay Minerals*, **14,** 289–296.
26. Diamond, S. (1970), Pore size distributions in clays, *Clays, Clay Minerals*, **18,** 7–24.
27. Diamond, S. (1971), Microstructure and pore structure of impact-compacted clays, *Clays, Clay Minerals*, **19,** 239–250.

28. Lincoln, J., and Tettenhorst, R. (1971), Freeze-dried and thawed clays, *Clays, Clay Minerals*, **19,** 103–108.
29. O'Brien, N. R. (1971), Fabric of kaolinite and illite floccules, *Clays, Clay Minerals*, **19,** 353–360.
30. McKyes, E., and Yong, R. N. (1971), Three techniques for fabric viewing as applied to shear distortion of a clay, *Clays, Clay Minerals*, **19,** 289–294.
31. Keller, W. D., and Hanson, R. F. (1975), Dissimilar fabrics by scan electron microscopy of sedimentary versus hydrothermal kaolins in Mexico, *Clays, Clay Minerals*, **23,** 201–204.
32. Martin, R. T., and Ladd, C. C. (1975), Fabric of consolidated kaolinite, *Clays, Clay Minerals*, **23,** 17–26.

PAPER

33. Murray, H. H., and Lyons, S. C. (1960), Further correlations of kaolinite crystallinity with chemical and physical properties, *Clays, Clay Minerals*, **8,** 11–17.
34. Hemstock, G. A. (1962), Effect of clays upon the optical properties of paper, *Tappi* **45,** 158A–159A.
35. Bundy, W. M., Johns, W. D., and Murray, H. H. (1965), Physico-chemical properties of kaolinite and relationship to paper coating quality, *Tappi*, **48,** 688–696.
36. Bundy, W. M. (1967), Kaolin properties and paper coating characteristics, *Chem. Engrg. Progress*, **63,** 57–64.

DRILLING FLUIDS

37. Jones, G. K. (1960, 1961), Drilling fluids, a current review, *Petroleum* (*London*), **23,** 43, 101 (1960); **24,** 53 (1961).
38. Larsen, D. H. (1955), Use of clay in drilling fluids, *Clays, Clay Minerals*, **1,** 269–281.

CHAPTER TEN

Interlamellar and Osmotic Swelling—Applications

I. CLAY-WATER RELATIONSHIPS: SWELLING AND COMPACTION IN SOIL ENGINEERING AND SEDIMENTARY GEOLOGY

Swelling and compaction of clays and soils are often very troublesome in the foundations of roads and buildings. Without going into engineering details, we shall summarize the principles of spontaneous swelling and shrinking of clays as a guide to the understanding of the often tremendous forces which are operating in clays and soils in contact with water. The same principles apply to certain aspects of the formation of sedimentary basins, in which the geologist is interested.

The often spectacular changes which take place when a dry clay is contacted with water suggest that clays interact directly with water in a very special manner. However, there are no significant differences in the order of magnitude of the energy of adsorption of water on clay surfaces and on surfaces of other inorganic materials, nor are there important differences in the charge density of these surfaces. Nevertheless, the behavior of the clay-water system is very different from that of a quartz-water system, for example. The causes of these differences become clear when both the unusual shape of the clay particle and the dual character of its surface are considered. These features of the clay particles strongly influence the effectiveness of interparticle forces in creating large swelling pressures and in building gel structures.

In many clay systems, the clay plates assume a parallel position by which large expanses of surface are able to interact at comparatively short distances. This applies to the layers in a stack of a montmorillonite particle, as well as to individual particles of both expanding and nonexpanding clays.

Two ranges of interaction may be distinguished. In the first stage of hydration of the dry clay particles, water is adsorbed in successive monolayers on the surfaces and pushes the particles or the layers of a montmorillonite clay apart. In this stage, the principal driving power is the adsorption energy of the water layers on the clay surface, as will be discussed later. The volume changes accompanying this stage of swelling may be as high as 100 percent of the original volume of the dry clay when four monolayers of water enter between the layers of a montmorillonite clay. When the hydration takes place on the exterior surfaces of nonexpanding clays, the volume changes become smaller with increasing thickness of the particles.

The second stage of swelling is due to double-layer repulsion by which the particles or the layers may be pushed further apart. In this stage one usually speaks about osmotic swelling. Large volume changes accompany this stage of swelling, but there is often a limit to the swelling caused by the formation of cross links by EF associated particles leading to the formation of gels, as discussed in Chapter Seven (Section II-C).

Although the surface hydration and the double-layer repulsion are the principal driving forces in the two stages respectively, these are not the only forces which operate between the particles or the layers in the two stages. We shall discuss these forces in some detail and consider their quantitative evaluation.

A. Short-Range Particle Interaction—Swelling Due to Hydration Energy

The short-range interaction between clay surfaces can be conveniently studied by analyzing the separation of layers in an expanding clay mineral. However, this analysis also applies to the separation of the flat surfaces of individual clay particles.

The expanding clays offer a unique possibility for a quantitative determination of the net potential curve of interaction of the layers of the clay as a function of the distance between them when the clay is submersed in water. This interaction curve can be derived by thermodynamic manipulation of the results of the determination of the water-vapor adsorption and desorption isotherms of the clay and simultaneous layer-distance measurements by X-ray diffraction.

It is interesting to note that the derived net work to remove the last monolayer of water when the plates are brought together is of the order of 0.1 J/m^2 (or 100 erg/cm^2). The pressure required to remove this monolayer of thickness, 0.25 nm, will then be as follows:

$$P = \frac{10^{-1}}{(2.5 \times 10^{-10})} = 4 \times 10^{8} \text{ Pa (or 4000 bar)}$$

Therefore, it may be concluded that in sedimentary clay beds overburden pressures will commonly be not high enough to squeeze out the last adsorbed water monolayers from between grain surfaces, or from the interlayer space in expanding clays, although at higher temperatures the required pressures will be less. Conversely, the lifting power of clays in the first stages of hydration is enormous.

1. Thermodynamic Analysis of Intracrystalline Swelling. The intracrystalline swelling phenomenon is also of theoretical interest for the analysis of adsorption of water on silicate surfaces. The arrangement of alternating silicate layers and adsorbed water layers in the layer package offers some unique possibilities for a direct experimental access to properties and structure of adsorbed water as a function of surface coverage. By measuring the *001* spacing of the clay during the adsorption process, the successive build-up of water layers can be followed. Information on the bonding mechanism between water and the clay surface can be obtained from infrared absorption spectra and other physical techniques. The possibility to vary the angle between infrared beam and the clay plate allows the bond direction to be determined in favorable cases. Since the relative amount of adsorbed water with respect to that of the solid is large, the magnitude of effects of adsorption is large and makes their determination relatively easy.

In the thermodynamic treatment of the adsorption isotherm, the system is usually treated as a one-component system—water—assuming that the solid is inert. The latter is certainly not true for swelling clays since the layers part during the adsorption process and cations originally located in the tetrahedral holes in the dry clay may leave these holes during water adsorption. Still another effect has been observed on the substrate: during adsorption of water the b dimension of the unit cell changes somewhat. This change is attributed to forces exerted by hydrogen-bonded water molecules in epitaxial arrangement with tetrahedral oxygen atoms. The effect is strongest at monolayer coverage, and it becomes rapidly smaller with the adsorption of subsequent layers. This is a rare example of a directly observable structural change of a solid surface due to adsorption of water. Therefore, when treating the system as a one-component system, the changes in the solid phase must be taken into consideration when interpreting the thermodynamic data derived from the adsorption isotherm.

The Gibbs energy of adsorption, hence the work done in separating the layers during the adsorption process, can be evaluated from the adsorption isotherm:

$$V_R = \frac{RT}{18{,}000\,\Sigma}\left[\int_{p_1}^{p_2} \frac{n_p}{p}\,dp + n_p \ln\left(\frac{p_2}{p_1}\right)\right]$$

in which n_p is the amount of water in milligrams which is adsorbed per gram of dry clay at a relative vapor pressure p, and Σ is the total area of opposite layer surfaces (cm^2) per gram of dry clay, or half the total layer surface area per gram neglecting the external surfaces. The integral can be evaluated graphically by plotting n/p versus p.

The net energy of layer interaction derived from the isotherm data is the sum of the following three contributions: van der Waals attraction between the layers, electrostatic interaction between the layers, and the hydration energy. The electrostatic interaction contribution depends on the layer charge and the position of the cations with respect to the layers. At any given stage of hydration, the cations may either remain embedded in the hexagonal holes of the silica sheet, as they are in the dry condition of the clay, or they may leave the holes and assume positions midway between the layers. In the first case, the electrostatic interaction is a repulsion, in the second case an attraction. The computation of these electrostatic terms is given in Appendix IV, Sections G-1 and G-2.

When the ions remain associated with the layer surfaces, the electrostatic repulsion appears to be rather small, and to some extent will compensate the van der Waals attraction, which is also small (see Appendix V). Hence the calculated net repulsion must be primarily due to hydration energy which in this case would have to be the hydration energy of the surface. Surface hydration energy may be partly due to hydrogen bond formation between adsorbed water molecules and surface oxygen atoms, and partly due to interaction of the water dipoles with the cations buried in the tetrahedral holes. If, on the other hand, the cations assume midway positions, the attractive energy will be substantial, hence the hydration energy should be rather high, that is, equal to the electrostatic attraction plus the (relatively small) van der Waals attraction. But in this arrangement, the relatively large cation-dipole interaction energy will be available through cation hydration, although not even the full-bulk cation hydration energy is involved in the confinement of the interlayer space.

Since both alternatives yield consistent pictures of the energy balance, it is obviously impossible to decide from this analysis alone whether the cations move out of the holes during the hydration process.

There are, however, a number of methods available to determine the position of the cation in the hydrates. When the charge density of the clay is high, for example, in some vermiculites, the position of the cations can be determined from X-ray diffraction patterns (applying one-dimensional Fourier analysis). For clays of lower charge densities, the position of the ions can be derived from electron spin resonance information (69-71), from surface acidity determination (Chapter Eleven, reference 133), and from statistical considerations (19). The results show that the cation position varies with such factors as ion size, surface charge, and specific cation-to-surface forces.

From adsorption isotherms taken at different temperatures, or by combining isotherm data with calorimetric data on heats of wetting at different initial degrees of surface hydration, the integral entropy can be evaluated as a function of surface coverage. From the entropy data, certain conclusions may be drawn regarding the mobility of the water molecules on the clay surface. For example, the entropy of water adsorbed on the exterior surfaces of kaolinite particles was found to be higher than that of liquid water, but less than that of water vapor (38). On the other hand, the entropy of interlayer water in a high-charge vermiculite was found to be less than that of liquid water for a one-layer and a two-layer hydrate (53, 54), indicating a higher degree of order in the system, that is, in the adsorbed phase and the solid phase together. Since parting of the layers and the movement of the cations away from the tetrahedral holes are likely to be accompanied by an increase of entropy, the entropy of the adsorbed phase is probably even lower when taking the changes in the solid phase into account. This result is not surprising considering that, for example, at monolayer coverage the interlayer consists of cations and water molecules in a 1:2 ratio, so that this assembly will have the character of a two-dimensional solid, In the two-layer hydrate, the cation-water ratio is 1:6 with the water molecules in octahedral coordination with the cations. The entropy for this configuration appears to be not as low as for the monolayer hydrate; apparently, the water molecules have some more freedom to move.

These observations are supported by infrared absorption behavior and by nuclear magnetic resonance (61). The infrared absorption data are discussed in the following section.

2. Infrared Absorption Spectra of Hydrated Clay Minerals. An important tool for analyzing the bonding state of adsorbed molecules is infrared spectroscopy. The absorption of infrared radiation is due to characteristic vibrational or rotational modes of atomic groups, and their frequencies determine the position of absorption bands in the infrared spectra. Since these frequencies are affected by the bonding state of the atomic groups, the position of the absorption bands can give a clue regarding the bonding state of the atomic groups in the adsorbed condition. For example, the vibrational stretching frequency of OH groups is lowered by hydrogen bonding of the OH group to oxygen. The magnitude of the frequency shift gives a measure of the bonding energy of the group.

Frequencies of the infrared spectra are usually expressed in cm^{-1}, and the frequencies attributed to structural OH groups in clay minerals are somewhat lower, but close to that of the unperturbed OH group. The structural OH bands are rather narrow in many clay minerals, and for these minerals, they can be easily distinguished from the broad bands observed for the hydrated minerals in the range from about 3200–3650 cm^{-1}, and which are attributed to the presence of hydrogen-bonded water molecules on the surfaces (62).

Absorption in the higher frequency range of this band is attributed to water molecules which are hydrogen bonded to oxygen in the silica surface whereas the frequencies near the maximum of the band are attributed to water molecules which are hydrogen bonded to other water molecules. Such assignments of the observed frequency ranges in the spectra are in part based on determination of the bond direction from the angular dependence of the absorption intensity. However, changes of absorption with angle can only be observed in systems in which the adsorbed water is sufficiently immobilized. Actually, infrared spectroscopy, using the variable angle techniques, confirms entropy calculations (61).

The small changes of frequency attributed to water molecules hydrogen-bonded to the oxygen of the surface indicate that such bonding is rather weak, perhaps of the order of a few kcal/mol at most, since frequency shifts of some 100–500 cm^{-1} roughly correspond to hydrogen bonding energies in the range of 1–5 kcal mol (36).

The infrared spectrum of adsorbed water on clays also shows a band around 1600 cm^{-1} which is due to the OH bending frequency. This rather isolated band is useful as an indication of the presence of adsorbed water.

B. Long-Range Particle Interaction—"Osmotic Swelling" or Electrical Double-Layer Repulsion

At plate distances beyond about 10 A (equivalent with four monomolecular layers of water), the surface hydration energy is no longer important, and the electrical double-layer repulsion becomes the major repulsive force between the plates. The swelling pressure of a clay paste can be determined by measuring the confining pressure which is required to prevent further uptake of water by the paste which is in contact with water via a porous frit or semipermeable membrane (18–21). Swelling pressures of clays are found to decrease with increasing plate separations (increasing water content of the pastes) in a range from several times 10 atm to less than 0.1 atm. This region is of interest in foundation engineering as well as in the study of the first stages of compaction of sediments in nature.

This stage of swelling is also called the stage of "osmotic swelling." With regard to this term it will be desirable to define precisely the connection between osmotic pressure and swelling pressure. Referring to the following sketch, we consider a fixed clay plate, and another plate at some distance from the fixed plate and parallel with it.

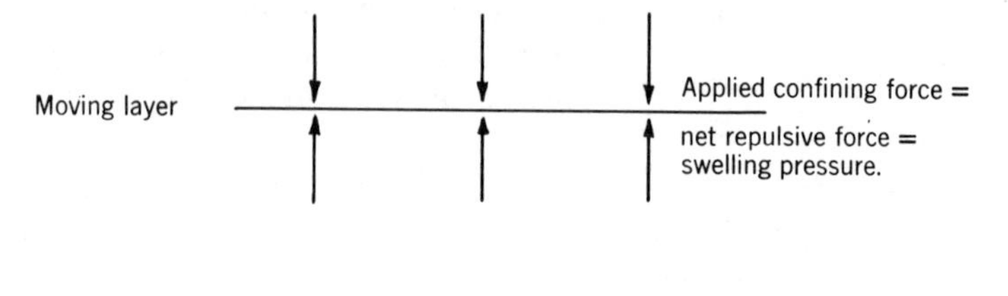

If, at a given layer or plate distance, the plates tend to part still further, the clay is said to have a positive swelling pressure. The swelling pressure is defined as the *net* repulsive force per unit area at the given distance. The swelling pressure is determined by measuring the equal confining force which must be applied externally to the moving layer to keep the layers at the given distance.

As mentioned in Chapter Seven (Section II-F-2), the swelling pressure is also called the osmotic pressure of the clay-water system. This statement may lead to misunderstanding and needs further elaboration. The term osmotic pressure is identified with pressures arising from concentration differences of dissolved ions or molecules in different parts of a system. Therefore, this term can legitimately be applied to a pressure arising from the differences in ion concentrations between the unit layers or between the clay plates and in the external submersion liquid. From the theoretically derived distribution of ions in the double layers between the clay plates, and in the equilibrium liquid, the osmotic pressure can be computed. This pressure is identical with the double layer repulsion of the two plates as computed by other methods (see Appendix III). Consequently, the swelling pressure as given by the applied confining pressure will be identical with the osmotic pressure only when the double-layer repulsion is the single operating force, or by far the dominant force between the plates.

In general, measured swelling pressures as a function of water content (or plate distance) are of the magnitude of the calculated double-layer repulsion, which therefore apparently dominates the swelling process. Under commonly prevailing conditions, the repulsion easily amounts to several atmospheres, or pressures which are indeed high enough to lift moderate-size buildings. Of course, the thickness of the clay layer should be such that the total volume change generated in the swelling process is large enough to cause an observable uplift. The proper procedure to avoid these effects is to remove the clay layers and to replace them with coarse sand and provide proper drainage.

For the foundation engineer, the swelling pressure as measured by the confining force is the important parameter, whether it is equal to the osmotic pressure, or whether other forces are involved. However, for a better understanding of the behavior of foundations and sediments in the early stages of compaction, it is important to consider whether forces other than the osmotic (or double-layer) force contribute to the swelling force as measured by the confining force.

Some more insight in this problem was awarded by a different kind of experiment in which flakes of montmorillonite clays were allowed to swell freely without applying any confining force. In the flakes, particles are oriented in a parallel position. From X-ray observations it appeared that in some montmorillonite clays the layers become parted by more than about four monolayers of water and reach large equilibrium distances. These distances could be measured up to about 120 A. The equilibrium distances obtained decrease with increasing salt content of the system (30). This observation supports the concept that the repulsive force is the double-layer repulsion since this repulsion will be of shorter range due to compression of the double layers when the electrolyte concentration of the system is increased.

At each equilibrium distance, an energy minimum must occur in the net potential curve of interaction between the layers. Therefore, an attractive force must operate between the layers which just compensates the double-layer repulsion at the equilibrium distance.

If it is assumed that the van der Waals attraction between the layers is the force which compensates the double-layer repulsion, the latter can be evaluated from an estimate of the van der Waals attraction. It appears that in this way a double-layer repulsion is computed which is much smaller than would be expected on the basis of the Gouy model of the double layer. Hence these observations would indicate that the Gouy repulsion is considerably reduced due to specific adsorption of the cations at the clay surface.

However, it also appears that the attractive forces required to compensate the double-layer repulsion (corrected for specific cation adsorption) at the different equilibrium spacings observed at different electrolyte concentrations would have to change much less with the layer distance than would be expected from the behavior of van der Waals forces in general. Therefore, a different attractive force has been considered. It has been proposed that this attractive force is a cross-linking force between the layers, established by positive edge-negative face association of occasional nonparallel particles. From an evaluation of the magnitude of individual cross-linking forces based on rheological observations of gels, it can be shown that a relatively small number of such cross links could compensate the full Gouy double-layer repulsion. Thus, the case for specific cation adsorption is weakened. However, since the number of cross links is not known, the attractive force at each equilibrium distance cannot be evaluated exactly. Hence the repulsive force cannot be evaluated, and no estimate can be made of the magnitude of the specific cation adsorption potential, if any exists (29).

An argument in favor of the picture of cross-linking by positive edge-negative face association is that an increase of the swelling distances is observed when the degree of cross-linking is reduced by reversal of the edge charge by phosphate addition (29, 31).

Returning to the experiments on swelling and deswelling by varying the confining pressures on the systems, the consequences of the picture of cross linking are the following:

At the equilibrium distances observed in the freely swelling flakes, no confining pressures would be needed since the particles are held together by cross links. Nevertheless, since the matrix of the clay system will have a certain rigidity due to cross linking, the matrix may support some external pressure without losing water. In general, therefore, part of the confining pressure is grain pressure, and only part is fluid pressure conteracting particle repulsion. Thus, the confining pressure is not an appropriate measure of the particle repulsion.

The picture of cross linking also explains why the swelling and deswelling of soils and clays by applying decreasing and increasing confining pressures is often not reversible. Such behavior is indicative of the occurrence of irreversible changes in the structural geometry of the system as a function of compression history. On the other hand, the reversibility of swelling and deswelling in certain systems does not necessarily guarantee that cross linking is absent, since the matrix may show elastic deformation and recovery. When the degree of cross-linking is reduced by the addition of phosphates, the swelling and deswelling becomes more reversible (31).

References

RANGE OF HYDRATION FORCES

1. Anderson, D. M., and Low, P. F. (1958), The density of water adsorbed by lithium-, sodium-, and potassium bentonite, *Soil Sci. Soc. Am. Proc.*, **22,** 99–103.
2. Bradley, W. F. (1959), Density of water sorbed on montmorillonite, *Nature,* **183,** 1614–1615.
3. Low, P. F. (1961), Physical chemistry of clay-water interaction, *Advan. Agron.*, **13,** 269–327.

4. Deeds, C. T., and van Olphen, H. (1961), Density studies in clay-liquid systems. Part I—The density of water adsorbed by expanding clays, *Advan. Chem. Ser.*, **33** (Solid surfaces and the gas-solid interface), 332–339.
5. Deeds, C. T., and van Olphen, H. (1963), Density studies in clay-liquid systems. Part II—Application to core analysis, *Clays, Clay Minerals*, **10**, 318–328.
6. Graham, J. (1962), Density of clay-water systems, *Nature*, **196**, 1124.
7. Graham, J. (1964), Adsorbed water on clays, *Rev. Pure. Appl. Chem.*, **14**, 81–88.
8. Oster, J. D., and Low, Ph. F. (1964), Heat capacities of clay and clay-water mixtures, *Soil Sci. Soc. Am. Proc.*, **28**, 605–609.
9. Leonard, R. A., and Low, P. F. (1964), Effect of gelation on the properties of water in clay systems, *Clays, Clay Minerals*, **12**, 311–325.
10. Davidtz, J. C., and Low, P. F. (1970), Relation between crystal-lattice configuration and swelling of montmorillonites, *Clays, Clay Minerals*, **18**, 325–332.
11. Davey, B. G., and Low, P. F. (1971), Physico-chemical properties of sols and gels of Na-montmorillonite with and without adsorbed hydrous aluminum oxide, *Soil Sci. Soc. Am. Proc.*, **35**, 230–236.
12. Low, P. F., and White, J. L. (1970), Hydrogen bonding and polywater in clay-water systems, *Clays, Clay Minerals*, **18**, 63–66.
13. Ravina, I., and Low, P. F. (1972), Relation between swelling, water properties and b-dimension in montmorillonite-water systems, *Clays, Clay Minerals*, **20**, 109–123.
14. Kay, B. D., and Low, P. F. (1972), Pressure induced changes in the thermal and electrical properties of clay-water systems, *J. Colloid Interface Sci.*, **40**, 337–343.
15. Kay, B. D., and Low, P. F. (1975), Heats of compression of clay-water mixtures, *Clays, Clay Minerals*, **23**, 266–271.

HYDRATION, SWELLING

16. Hendricks, S. B., and Jefferson, M. E. (1938), Structure of kaoline and talc-pyrophyllite hydrates and their bearing on water sorption of clays, *Am. Mineralogist*, **23**, 863–875.
17. Hendricks, S. B., Nelson, R. A., and Alexander, L. T. (1940), Hydration mechanism of the clay mineral montmorillonite saturated with various cations, *J. Am. Chem. Soc.*, **62**, 1457–1464.
18. Macey, H. H. (1942), Clay-water relationships and the internal mechanism of drying, *Trans. Brit. Ceram. Soc.*, **41**, 73–121.
19. Mering, J. (1946), The hydration of montmorillonite, *Trans. Faraday Soc.*, **42B**, 205–219.
20. Forslind, E. (1948), The crystal structure and water adsorption of the clay minerals, *Trans. Intern. Ceram. Congr. 1st, Maastricht, Neth.*, 98–110.
21. Barshad, I. (1949), The nature of lattice expansion and its relation to hydration in montmorillonite and vermiculite, *Am. Mineralogist*, **34**, 675–684.
22. Williamson, W. O. (1951), The physical relationship between clay and water. *Trans. Brit. Ceram. Soc.*, **50**, 10–34.
23. Mering, J., and Glaeser, R. (1954), Sur le rôle de la valence des cations échangeables dans la montmorillonite. *Bull. Soc. Franç. Mineral. Crist.*, **77**, 519–530.
24. Walker, G. F. (1956), The mechanism of dehydration of Mg-vermiculite, *Clays, Clay Minerals*, **4**, 101–115.
25. Walker, G. F. (1956), Diffusion of interlayer water in vermiculite, *Nature*, **177**, 239–240.
26. Bradley, W. F., and Serratosa, J. M. (1960), A discussion of the water content of vermiculite, *Clays, Clay Minerals*, **7**, 260–270.
27. van Olphen, H. (1956), Forces between suspended bentonite particles, *Clays, Clay Minerals*, **4**, 204–224.

28. van Olphen, H. (1954), Interlayer forces in bentonite, *Clays, Clay Minerals*, **2,** 418–438.
29. van Olphen, H. (1962), Unit layer interaction in hydrous montmorillonite systems, *J. Colloid Sci.*, **17,** 660–667.
30. Norrish, K. (1954), The swelling of montmorillonite, *Discussions Faraday Soc.*, **18,** (Coagulation and Flocculation), 120–134.
31. Norrish, K., and Rausell-Colom, J. A. (1963), Low angle diffraction studies of the swelling of montmorillonite and vermiculite, *Clays, Clay Minerals*, **10,** 123–149.
32. Nahin, P. G. (1955), Swelling of clay under pressure. *Clays, Clay Minerals*, **3,** 174–185.
33. Bolt, G. H., and Miller, R. D. (1955), Compression studies of illite suspensions, *Soil Sci. Soc. Am. Proc.*, **19,** 285–288.
34. Warkentin, B. P., Bolt, G. H., and Miller, R. D. (1957), The swelling pressures of montmorillonite, *Soil Sci. Soc. Am. Proc.*, **21,** 495–497.
35. Warkentin, B. P., and Schofield, R. K. (1960), Swelling pressures of dilute Na-montmorillonite pastes, *Clays, Clay Minerals*, **7,** 343–349.
36. Lippincott, E. R., and Schroeder, R. (1955), One-dimensional model of the hydrogen bond, *J. Chem. Phys.*, **23,** 1099–1106.
37. Martin, R. T. (1959), Water vapor adsorption on kaolinites: hysteresis, *Clays, Clay Minerals*, **7,** 259–278.
38. Martin, R. T. (1960), Water vapor adsorption on kaolinite: entropy of adsorption, *Clays, Clay Minerals*, **8,** 102–114.
39. Zettlemoyer, A. C., Young, G. J., and Chessick, J. J. (1955), The surface chemistry of silicate minerals, III—Heats of immersion of bentonite in water, *J. Phys. Chem.*, **59,** 962–966.
40. Wade, W. H., and Hackerman, N. (1959), Heats of immersion, II—Calcite and kaolinite—The effect of pretreatment, *J. Phys. Chem.*, **63,** 1639–1641.
41. Slabaugh, W. H. (1959), Heats of immersion of preheated homoionic clays, *J. Phys. Chem.*, **63,** 1333–1335.
42. Hutchinson, E., and Longwell, D. I. (1959), Measurements of heats of wetting of montmorillonites, Division of Colloid Chemistry, 136th Meeting, ACS, Atlantic City, N.J. (abstract).
43. Brooks, C. S. (1960), Free energies of immersion of clay minerals in water, ethanol, and *n*-heptane, *J. Phys. Chem.*, **64,** 532–537.
44. Barshad, I. (1960), Thermodynamics of water adsorption and desorption on montmorillonite, *Clays, Clay Minerals*, **8,** 84–101.
45. Kolaian, J. H., and Low, P. F. (1962), Thermodynamic properties of water in suspensions of montmorillonite, *Clays, Clay Minerals*, **9,** 71–84.
46. Scholz, A. (1960), Neue Ergebnisse über die innerkristalline Quellung, *Kolloid-Z.*, **173,** 61–63.
47. Greene-Kelly (1962), Charge densities and heats of immersion of some clay minerals, *Clay Min. Bull.*, **5,** 1–8.
48. Norrish, K., and Rausell-Colom, J. A. (1962), Effect of freezing on the swelling of clay minerals, *Clay Min. Bull.*, **5,** 9–16.
49. Rowell, D. L. (1963), Effect of electrolyte concentration on the swelling of orientated aggregates of montmorillonite, *Soil Sci.*, **96,** 368–374.
50. Yong, R., Taylor, L. O., and Warkentin, B. P. (1963), Swelling pressures of sodium montmorillonite at depressed temperatures, *Clays, Clay Minerals*, **11,** 268–281.
51. van Olphen, H. (1963), Compaction of clay sediments in the range of molecular particle distances, *Clays, Clay Minerals* **11,** 178–187.
52. Osterman, J. (1964), Studies on the properties and formation of quick clays, *Clays, Clay Minerals*, **12,** 87–108.

53. van Olphen, H. (1965), Thermodynamics of interlayer adsorption of water in clays. I—Sodium vermiculite, *J. Colloid Sci.*, **20,** 822–837.
54. van Olphen, H. (1969), Thermodynamics of interlayer adsorption of water in clays. II—Magnesium vermiculite, *Proc. Intern. Clay Congress,* Tokyo, **1,** 649–657.
55. Andrews, D. E., Schmidt, P. W., and van Olphen, H. (1967), X-ray study of interactions between montmorillonite platelets, *Clays, Clay Minerals,* **15,** 321–330.
56. Mering, J., and Brindley, G. W. (1967), X-ray diffraction band profiles of montmorillonite—Influence of hydration and of the exchangeable cations, *Clays, Clay Minerals,* **15,** 51–62.
57. van Olphen, H. (1968), Particle interaction and particle-water interaction in clay-water systems, *Tappi,* **51,** 145A–148A.
58. Touillaux, R., Salvador, P., Vandermeersche, C., and Fripiat, J. J. (1968), Study of water layers adsorbed on Na- and Ca-montmorillonite by the pulsed nuclear magnetic resonance technique, *Israel J. Chem.*, **6,** 337–348.
59. Taylor, T. R., and Schmidt, P. W. (1969), Interparticle potential energies in Na-montmorillonite clay suspensions, *Clays, Clay Minerals,* **17,** 77–82.
60. Woessner, D. E., and Snowden, B. S., Jr. (1969), NMR doublet splitting in aqueous montmorillonite gels, *J. Chem. Phys.*, **50,** 1516–1523.
61. Hougardy, J., Serratosa, J. M., Stone, W., and van Olphen, H. (1970), Interlayer water in vermiculite: Thermodynamic properties, packing density, nuclear pulse resonance, and infra-red absorption, *Far. Soc., Special Disc.*, **1,** 187–193.
62. Farmer, V. C. (1971), The characterization of adsorption bonds in clays by infrared spectroscopy, *Soil Sci.*, **112,** 62–68.
63. Nayak, N. V., and Christensen, R. W. (1971), Swelling characteristics of compacted, expansive soils, *Clays, Clay Minerals,* **19,** 251–262.
64. Murat, M., and Gielly, J. (1970), Determination des isothermes d'adsorption de vapeur d'eau par mesures électriques et gravimétriques simultanées, *Bull. Soc. Chim. de France,* **1970,** 1262–1266.
65. Quinson, J. -F., Escoubes, M., and Blanc, R. (1971), Profil énergétique de l'adsorption d'eau a 25°C sur montmorillonites homoioniques (K^+, Na^+, H^+, Ca^{++}), *Bull. Groupe Franç. Argiles,* **24,** 49–67.
66. Escoubes, M., Quinson, J. -F., Gielly, J., and Murat, M. (1972), Interaction de la vapeur d'eau avec quelques minéraux argileux: étude par couplage thermogravimétrie-calorimétrie, *Bull. Soc. Chim. de France,* **1972,** 1689–1698.
67. Keren, R., and Shainberg, I. (1975), Water vapor isotherms and heat of immersion of Na/Ca-montmorillonite systems—I: Homoionic clay, *Clays, Clay Minerals,* **23,** 193–200.
68. van Olphen, H. (1975), Water in soils, *Soil Components, Volume 2: Inorganic Components,* J. E. Gieseking, Editor, Springer, New York, pp. 497–527.
69. McBride, M. B., Mortland, M. M., and Pinnavaia, T. J. (1975), Exchange ion positions in smectite: Effects on electron spin resonance of structural iron, *Clays, Clay Minerals,* **23,** 162–163.
70. McBride, M. B., Pinnavaia, T. J., and Mortland, M. M. (1975), Perturbation of structural Fe(3+) in smectites by exchange ions, *Clays, Clay Minerals,* **23,** 103–107.
71. Berkheiser, V., and Mortland, M. M. (1975), Variability in exchange ion position in smectite: Dependence on interlayer solvent, *Clays, Clay Minerals,* **23,** 404–410.

CHAPTER ELEVEN

Interaction of Clays and Organic Compounds

I. INTRODUCTION

We have thus far dealt mainly with entirely inorganic systems of clay, water, and inorganic salts acting as flocculants and peptizers. In the present chapter, we shall discuss clay-organic systems, some of which are of considerable technological interest. Certain organic compounds act as peptizers for clay-water systems, and their application as such appreciably widens the scope of chemical-treatment procedures for clay suspensions. Other organic compounds, by their adsorption on the clay surface, modify the wetting characteristics of the clay. They may change the originally water-wet clay surface into one which is preferentially wet by oil. Such a modification is important in the secondary recovery of petroleum, the making and breaking of emulsions, and the preparation of clay-in-oil dispersions.

A. Terminology—Surface Activity

A few remarks on terminology should be made at this point. In principle, any compound adsorbed in an interface and changing the properties of that interface might be called a *surface-active agent.* However, this term is normally restricted to *amphipolar* or *amphipathic* molecules, consisting of an oleophilic and a hydrophilic part. The adsorption of the surface-active agent or *surfactant* changes the interfacial tension, and it may change the double-layer structure at the interface or the wetting characteristics of the solid for different liquids. Depending on their effect on a particular system, the compounds may be labeled *peptizers* or *flocculants, emulsifiers* or *demulsifiers,* or *wetting agents* for oil or for water. The fact that certain compounds may act in different ways on different systems must be stressed. For example, a surfactant may change the wettability of the surface of one solid but not that of another solid; or an emulsifier for one liquid-liquid system may act as a demulsifier for another system. Moreover, a compound may act both as a surfactant in a solid-liquid system and as an emulsifier in a liquid-liquid system. Therefore, the classification on the name tag of the compound should not be taken at face value, but it should be used with reference to a definite system.

The typical amphipolar surfactant molecule has a hydrophobic hydrocarbon part and a hydrophilic polar or ionic group. Well-known examples are quaternary bases ("cationics"), soaps ("anionics"), and long-chain alcohols ("nonionics"), as well as cationic, anionic, and nonionic polymers. Because of the dual character of the molecules, they feel most happy in the water-oil interface where both parts of the molecules can satisfy their preference for either medium. Or, to put it more scientifically, though not more clearly, their accumulation in the interface is energetically and entropically preferred. The tendency to reduce the contact with water and to seek relatively nonpolar environments is summarized in the catch-all term *"hydrophobicity."*

The amphipolar surfactants also have a tendency to self-association; they are examples of "association colloids." The nonpolar parts of the molecules seek the company of each other in preference to remaining in contact with water, another manifestation of hydrophobicity. This "micellization" is considered the result of *"hydrophobic bonding."* This bonding mechanism is still not completely understood quantitatively since it is the result of a complicated set of both attractive and repulsive interactions between the molecules.

B. Wetting

The wetting ability of a liquid for a solid surface is measured by the value of the contact angle between the surface and the drop of liquid resting on that surface (Figure 33). The smaller the contact angle, the better the liquid wets the surface. When the contact angle is zero, the surface is said to be perfectly wetted by the liquid. (Still, for different liquids which form zero contact angles with a certain solid surface, the energy of adhesion between liquid and solid may vary.) The contact angle is determined by the three vectors of interfacial tension as indicated in Figure 33.

The relative wetting ability of two immiscible liquids on the same surface is measured by the contact angle between the solid surface and the contact plane between the two liquids (Figure 33*b*).

C. Classification

In the following discussion of the action of organic compounds on clays, a distinction will be made between compounds of low to moderate molecular weight and macromolecular compounds. The latter, when dissolved in water, represent the class of "hydrophilic colloids"; therefore,

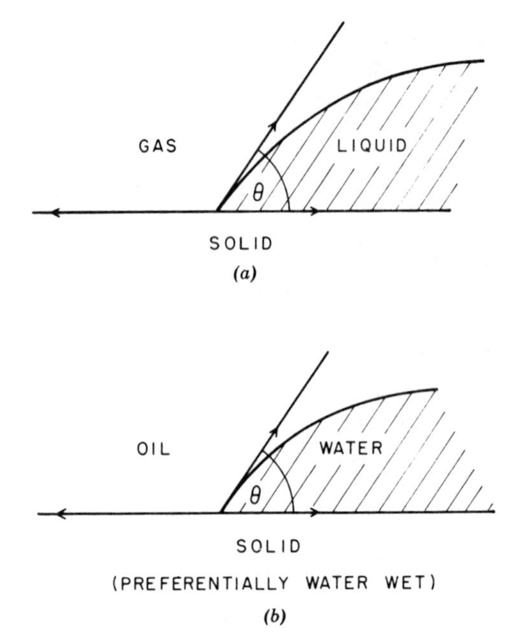

Figure 33. Wetting of solids by liquids—Contact angles.

their action on clay suspensions falls in a special category which may be called "interaction between hydrophilic and hydrophobic colloids."

In each category, we shall discuss subsequently the action of anionic, cationic, and nonionic compounds. These compounds contain organic anions (phenolates, carboxylates, sulfonates, etc.), organic cations (amine salts, quaternary ammonium salts), and polar groups (alcohols, ketones, etc.), respectively.

Because of the difference in the polarity of charge on edge and face surfaces of clay particles, one can expect that organic cations will be adsorbed at the negative faces, and organic anions at the positive edges. In analogy with water, polar organic compounds will be adsorbed on both hydrophilic surfaces.

In addition, chemical reactions between clays and organic compounds will be reviewed briefly. Finally, clay stabilized emulsions and clay dispersions in oil are discussed with particular reference to the stability factors prevailing in organic dispersions.

D. X-Ray Observations and Infrared Absorption Spectra

The face adsorption of organic cations and polar compounds is not limited to the exterior surfaces of the layer packets of expanding clays (montmorillonites), for interlamellar adsorption also takes place. This adsorption process is accompanied by an increase of the basal spacing of the clay with respect to that of the dry untreated clay. These interlamellar adsorption complexes are usually called *clay-organic complexes*.

The orientation and the packing density of the organic molecules between the layers can be derived from X-ray data combined with quantitative adsorption measurements. However, it is often not possible to arrive at definite conclusions regarding the precise arrangement of the interlayer molecules owing to certain limitations in the interpretation of these data.

To a certain extent, the basal spacing of an organic complex of montmorillonite is a measure of the van der Waals dimensions of the organic molecules in a direction determined by their orientation between the layers. The clay may be considered as a "micromicrometer" for measuring the dimensions of organic molecules, with the X-ray pattern as the scale. In this micromicrometer, the reference planes for measuring the gap occupied by the organic molecules are the planes touching the oxygen atoms of the layers. A "defect" of the micromicrometer is that these planes do not represent surfaces which are perfectly flat on a molecular scale. There are small indentations between triplets of oxygen atoms in these surfaces and deeper holes in the center of hexagonal rings of oxygen atoms (Figure 17). Organic molecules between the layers may be partly keyed to these indentations and holes, and may thus protrude beyond the planes touching the oxygen atoms. Actually, when these planes are used as reference planes, the van der Waals dimensions of the organic molecules are usually somewhat smaller than those observed in crystals of these compounds. Keying of the organic molecules may indeed be partly responsible for the apparent decrease of the van der Waals dimensions. Perhaps, as several authors have pointed out, the van der Waals dimensions of the organic molecules may be reduced owing to an effect of the specific adsorption forces, acting between the layers and the organic molecules, on the bond lengths in the layer—organic molecule—layer assembly. We may add to such an effect a possible influence of attractive forces between the layers on the van der Waals dimensions of the organic molecules between the layers. Particularly, when the exchangeable cations are located midway between the layers, the electrostatic attraction between the negative layers and the layers of midway cations may be appreciable. Finally, the apparent decrease of the van der Waals dimensions of the interlayer organic molecules may also be due to some tilting of these molecules.

Although a one-dimensional Fourier analysis is often helpful in establishing the general orientation of the interlayer molecules, the resolution of the electron density sketches is usually insufficient to yield a precise information on the angle of tilt, the bond lengths, and keying in the clay-organic complex.

In the sixties, the application of infrared absorption spectroscopy to clay-organic complexes (and as discussed before, to clay-water systems as well) was a major breakthrough for the understanding of the bonding mechanisms between clays and organic molecules. In the adsorbed state, the vibrational modes of functional groups in the organic molecules are modified. From the corresponding spectral band shifts and absorption intensity variations conclusions can be drawn regarding the character of the bonds with the clay surface. Also, the bond directions can be evaluated from the angular dependence of the absorption intensity when the oriented clay plates are rotated with respect to the infrared beam.

X-ray diffraction and infrared spectroscopy are powerful complementary tools for the study of clay-organic complexes. Additional information can be obtained from ultraviolet absorption, electron spin resonance, and nuclear magnetic resonance spectroscopy for appropriate systems.

A discussion of these techniques and the analysis of the spectra is outside the scope of this book. The reader is referred to references 4 through 7 and the detailed presentations in the monograph by Theng, Appendix V, reference B-32.

E. Adsorption Measurements

Quantitative information on the amount of organic compound adsorbed in the interlamellar space may be obtained in different ways, but there are several limitations in the interpretation of the data.

In adsorption from the vapor phase, there is the difficulty that multilayer adsorption may already occur to some degree on the exterior surface, when between the layers only one monomolecular layer of the organic compound is adsorbed. When the plates are rather thin, the required correction of the amount adsorbed for exterior multilayer adsorption may be appreciable.

The same applies to measurements of adsorption from solution. An additional uncertainty in the interpretation of adsorption data from solution is that solvent adsorption may occur. If a polar compound is adsorbed from its solution in a hydrocarbon liquid, the clay may obtain a certain adsorption capacity for the hydrocarbon as soon as part of its surface becomes covered with the polar compound. It is difficult to account for such solvent adsorption.

Quantitative adsorption measurements of polar organic liquids from the liquid phase are impossible since the adsorption complex cannot be separated from the liquid phase for analysis. However, apparent density measurements of the clay in such liquids offer a possibility to derive the packing density of the adsorbed interlayer molecules by comparing the apparent density of the clay in the liquid with its crystallographic density. (See Chapter Ten, reference 4.)

II. COMPOUNDS WITH LOW TO MODERATE MOLECULAR WEIGHTS

A. Organic Anions—Specifically Tannates (11–14)

Tannates are widely used as peptizers in drilling fluids. The most popular tannate in this application is derived from quebracho tannin, a red-colored tannin extracted from the heartwood of the South American quebracho tree. This particular tannin belongs to the group of the catechins. According to Freudenberg (11), it is a polycondensation product of the monomer:

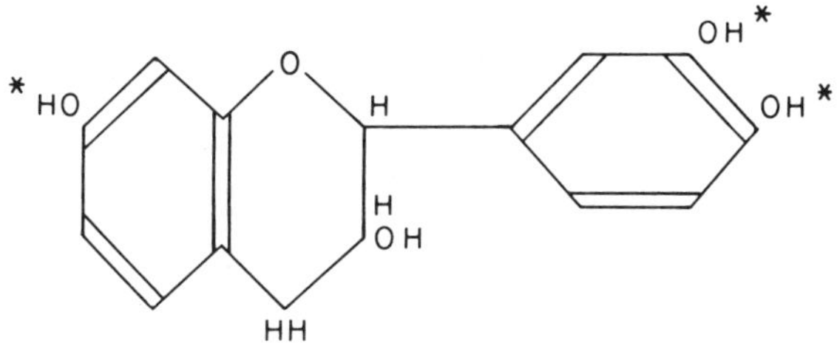

The three phenolic groups, marked with an asterisk, are not involved in the condensation reaction; they are preserved in the final condensation product. Thus, the equivalent weight of the tannin derived from the above monomer formula amounts to 91. The lower-molecular-weight fraction of the tannin is soluble in water, the higher-molecular-weight fraction is soluble in an alkaline solution in which the alkali-phenolate is formed. Possibly, this salt is partially colloidally dissolved. Since the tannin is a weak acid, the sodium salt solution shows an alkaline reaction.

1. Effect of Tannates on Clay Suspensions. Clay suspensions are usually peptized by the addition of a few tenths of one percent of a tannate. In this respect, the action of the tannate is comparable with that of the inorganic peptizing agents discussed earlier. Systematic experiments have shown that the mechanism of peptization is the same for both groups of chemicals. The tannate annions are adsorbed at the edge surfaces of the clay particles by complexing with the exposed octahedral aluminum ions. Consequently, the edge charge is reversed and a negative double layer is created by which EF and EE association is prevented. This analysis of the peptizing mechanism is supported by the following observations:

A small but measurable anion adsorption which occurs on the clay does not take place on the faces of the clay plates, since in montmorillonites no increase in basal spacing is observed after the adsorption of the tannate. Furthermore, positive alumina sols become negative and peptized when tannate is added.

The large size of the condensed natural tannin molecule is not essential for its peptizing activity, since the sodium salts of some simple low-molecular-weight phenols act in the same way on clays. In order to be effective, a phenol must contain at least three phenol groups in the molecule, two of which should be adjacent. Thus, mono- and divalent phenols are inactive; and of the three-valent phenols, pyrogallol (1,2,3-hydroxybenzene) in the form of the sodium salt peptizes a clay, but Na-phloroglucinolate (1,3,5-hydroxybenzene) is inactive in this respect. Possibly the two adjacent phenolic groups are in the right position to complex with an aluminum ion at the clay particle edge, while the third phenolic group gives the surface its negative charge.

Another group of phenolates which act as peptizing agents for clay suspensions are the synthetic alizarin dyes. These are alumina-mordant dyes, and since such a mordant is, in effect, present at the edges of the clay plates, the dye will be adsorbed on the edge surfaces. The resulting peptized clay suspensions are intensively colored. Although these dyes are too expensive to be applied in drilling fluids they might, for example, find application in such systems as tracers.

Because of the rather high valence of the polycondensed natural tannate anions, their charging effect on the edges may be greater than that of the simple phenolates. Also, the activity reduction of the sodium ions by the polyvalent anions will result in a greater contribution to the increased salt tolerance of the clays. The tannates are indeed much more effective than the simple phenolates, as is demonstrated by the two curves in the flocculation diagram presented in Figure 34. Actually, even at rather high concentrations of the tannate, no flocculation of the clay by the tannate alone is observed, wheras the pyrogallol-salt flocculates the clay suspension at a moderate concentration.

2. Application of Tannates in "Red Muds" and "Lime Red Muds." Although drilling fluids are peptized by tannates of the quebracho type at a concentration level of a few tenths of one percent, usually as much as 1–2 percent of the quebracho with a slight excess of alkali is compounded in the mud. Such a treatment is possible, since the quebrachate does not flocculate the clay at high concentrations. Because of the red color of the quebracho-treated drilling fluids, they are called *red muds*. The special feature of these muds is that they have good *formation-conserving* properties; that is, they prevent the dispersal of the clays in the wall of the borehole or the disintegration of the drill cuttings from these formations.

The reduced dispersion tendency of formation clays and drill cuttings in contact with red muds seems rather paradoxical in view of the peptizing

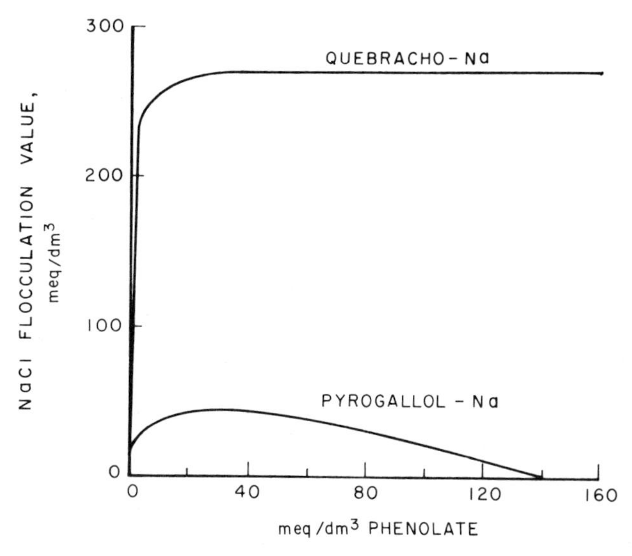

Figure 34. Peptization of sodium bentonite suspensions by quebracho-Na and by pyrogallol-Na.

action of the tannins on clays. This effect is easily explained when the mechanism of disintegration of lumps of clay is considered. The first stage in the disintegration of a lump of clay in the presence of water is osmotic swelling of the lump owing to the double-layer repulsion between the faces of the individual clay plates. The edge surfaces of the plates thereby become accessible for the peptizing agents, and further disintegration takes place. In the presence of the high electrolyte concentration prevailing in the red muds, the face double layers will be compressed, hence the osmotic swelling will be suppressed. (In other words, the osmotic swelling is reduced because of the smaller difference in ion concentration in the interior of the clay lump and in the mud.) Hence the lump of clay swells very little and the clay particles in the interior do not become accessible for the peptizing tannate anions, particularly since they cannot diffuse easily through the narrow pores with negatively charged walls.

The applicability of red muds is limited to the drilling of relatively shallow wells—at the higher temperatures in deeper holes, considerable stiffening of the mud occurs. This effect of heat on the consistency of the mud is not completely understood. Certain soluble aluminates or other products of a chemical reaction between alkali and the oxides exposed at the particle edges of the clay are probably responsible for the stiffening because of the flocculating power of these reaction products.

This adverse effect of heating can largely be prevented by the addition of solid calcium hydroxide to the red mud. Possibly the added lime precipitates the reaction products of alkali and clay, and thus makes them harmless. For example, sodium aluminate, which is a possible reaction product of alkali and clay, causes stiffening of a red mud when added to a freshly prepared red mud at room temperature. The subsequent addition of lime appeared to have a thinning effect on the mixture, probably because of the precipitation of the aluminate in the form of the calcium salt.

Soon after the introduction of the lime-treated red muds, it appeared that there is also a limit to the temperatures at which these muds can be used. At high temperatures in very deep holes, the lime red muds stiffen to such an extent that the term solidification is more appropriate (12). The solidification is assumed to be a result of the formation of cement-type reaction products of clay, alkali, and lime. The substitution of lime by barium hydroxide extends the range of applicability of the muds to somewhat higher temperatures (13).

The fact that a lime-treated red mud as such is colloidally stable seems rather contradictory in view of the powerful flocculating action of divalent calcium ions on clays. In order to resolve this paradox, some systematic work was done on the complicated system of clay, water, polycondensed tannin, alkali, and lime. In these studies, the complex and ill-defined components of a real mud were substituted by simpler model components. Instead of raw clays, a well-defined sodium montmorillonite clay was used, as well as the model sols alumina and quartz. The quebracho tannin was substituted by simple phenols such as pyrogallol. (With the latter component, air must be excluded to prevent oxidation which makes the phenol ineffective. Possibly the quebracho-treated muds also lose some of their effectiveness by oxidation during the drilling operation.) The simple model systems are considered representative, since their behavior appears to be completely analogous to that of practical muds with respect to the effect of changes in the alkali and lime content and their response to the addition of flocculating salts. It was concluded from this work that the flocculating effect of the divalent calcium ions in these systems is eliminated by the complexing of these ions by the tannate or phenolate anions. Probably these complex anions are adsorbed at the clay particle edges and thus contribute to the creation of a stabilizing negative charge on these surfaces.

Simultaneously, part of the calcium ions will take up exchange positions on the faces of the clay particles. Because of this exchange and because of the high electrolyte concentration of the mud, FF association of the particles is promoted. Consequently, the average thickness of the clay particles in the lime red mud is greater than that in the red muds. This effect of "aggregation" of the clay particles in a deflocculated system is clearly demonstrated by the greater settling tendency of the particles and the reduced tendency to gel. Higher clay concentrations can therefore be maintained in lime red muds. Also, the thicker particles of the lime red mud do not easily penetrate into porous formations. Therefore, in spite of the better filtration properties of the red muds, the lime red muds are more suitable for drilling into the oil-bearing formations, at least if the latter do not contain clays which will be peptized by the mud filtrate (see Chapter Nine, Section II-B-3).

The behavior of red muds and lime red muds was discussed in some detail as a demonstration of the complexity of a colloid chemical analysis of a practical system. The interpretations are bound to remain somewhat speculative, although we believe that the substitution of model systems for the complex and rather ill-defined practical systems has thrown some light on the mechanism of the effects of the various chemicals in these systems. On the basis of such an understanding of the mechanism of chemical treatment rests the hope for efficient development of new and better muds.

Another group of anionic organics frequently used in drilling muds are various lignosulfonates. The mechanism of their thinning action on clay suspensions is probably analogous to that of the tannates (14).

B. Organic Cations—Specifically Amine Salts (15–36)

Organic anions are adsorbed at the edges of the clay particles. Organic cations, on the contrary, are adsorbed on the negative face surfaces of the clay. This fact is evident from the much larger adsorption capacity of the clay for these cations and also from the increase of the basal spacing of montmorillonite clays after treatment with organic cations.

When an amine salt, for example, $R—NH_3^+Cl^-$, or a quaternary ammonium salt or base ($R_4N^+Cl^-$ or $R_4N^+OH^-$) is added to a clay-water suspension, the organic cation replaces the cations which were originally present on the clay surfaces; in other words, "exchange adsorption" takes place. There appears to be a strong preference of the clay for the organic cation, which is often practically quantitatively adsorbed until all the exchange positions are occupied by the organic cation. The amino groups become strongly attached to the clay surface. Simultaneously, the hydrocarbon chains attach themselves to the clay surface and displace the previously adsorbed water molecules. The resulting "clay-organic complex" is schematically represented in Figure 35*a*. In this figure, it is assumed that sufficient space is available on the clay surface to accommodate the hydrocarbon chains. When the chains are too long to lie flat in the available space they may tilt, or, in swelling clays, two layers of the organic cations may become superimposed.

The surface of the exchange complex is now covered by hydrocarbon

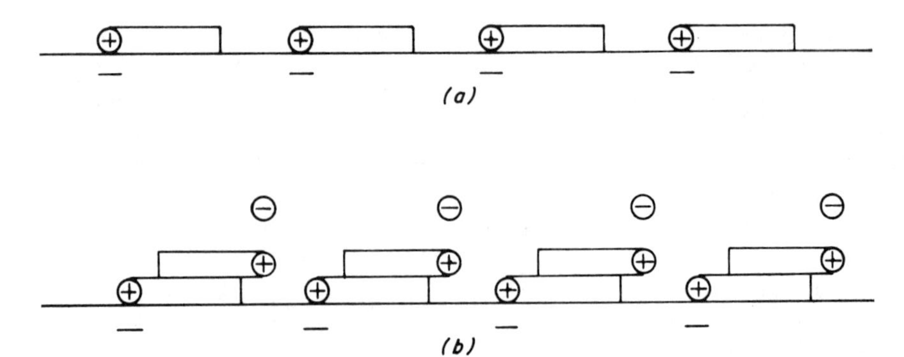

Figure 35. Schematic representation of adsorption of an amine salt on a clay layer surface. (*a*) Exchange adsorption of an equivalent amount of amine salt. (*b*) Physical adsorption of amine salt on adsorbed amine cations when amine salt in excess of the cation exchange capacity of the clay is added.

chains; consequently, it is no longer wetted by water, and the clay precipitates. At the same time, the modified surface has become preferentially wet by oil, and the clays can be homogeneously dispersed in hydrocarbon fluids. The exchange complex of the clay also promotes emulsification. We shall return to applications of the treated clays in oil dispersions and in emulsions in the later sections of this chapter.

The adsorption of organic cations is usually not limited to an amount equivalent to the cation exchange capacity of the clay. For example, for a quaternary ammonium compound with a long chain and three methyl groups, and a montmorillonite clay, adsorption of about $2\frac{1}{2}$ times the cation exchange capacity of the clay has been observed. The excess of the organic salt is physically adsorbed by van der Waals linking of the hydrocarbon chains of the exchange-adsorbed cations and those of the excess organic molecules. The ionized groups of the physically adsorbed molecules will point towards the water phase. The organic cations and the anions of the ionic groups together will create a more or less diffuse electric double layer, and the adsorption complex now assumes a positive charge, as sketched in Figure 35*b*. This arrangement is evident from the fact that the clay again becomes peptized and that the particles behave as positive particles in an electrophoresis experiment. In this condition, the clay is no longer oleophilic; it is unsuitable for the preparation of clay in hydrocarbon dispersions or for the preparation of emulsions.

Since the optimum emulsifying power of the clay-organic complexes is displayed at the equivalence point of exchange, an interesting possibility for determination of the cation exchange capacity (CEC) by means of a simple test-tube experiment is offered:

In a series of test tubes, a given amount of a clay suspension of known concentration is treated with increasing amounts of a standard solution of pure cetyltrimethylammonium bromide. The contents of each tube are vigorously shaken with a given amount of an oil. The most stable and finest emulsion is obtained in the tube in which an equivalent amount of quaternary ammonium salt was added to the clay (72). Visual inspection of the contents of the tubes appears to be a sharp enough criterion for establishing the optimum quality of the emulsion. The CEC of the clay is then computed from the amount of clay present and the number of milliequivalents of the quaternary ammonium salt which was added in the best emulsion.

The orientation of the interlayer cations, and therefore the basal spacing of the complexes, is usually determined by the requirement that the projected surface area of the cation in a given orientation should be equal to, or smaller than, the available surface area per exchange site. A proposed method of estimating the surface charge density is based on the measurement of the basal spacing of complexes with mono- and di-alkylammonium cations of different chain lengths (25).

Aromatic cations assume either a parallel or an upright position between the layers, depending on the available space.

C. Polar Organic Compounds (37–51)

Many organic compounds with a dipole character are adsorbed on the layer surfaces and probably also on the edge surfaces of a clay, in analogy with the behavior of water. The adsorption energy of many of these compounds is probably of a magnitude comparable with that of water: they can displace adsorbed water from clays, but frequently the organic compounds can be removed from the clay by washing with water.

As in the case of the adsorption of water on clays, we do not know exactly to what extent the polar groups of the organic molecules associate with the counter-ions of the clay, and to what extent they are hydrogen bonded with the oxygen surfaces. However, infrared absorption indicates that the latter is less important.

Organic clay complexes with polar molecules can be prepared either by contacting the dry clay with polar organic liquids or vapors, or by mixing the polar liquid and a clay-water suspension.

When montmorillonite clays are treated with polar organic compounds, such as alcohols, glycols, or amines, the organic molecules penetrate between the layers and displace the interlayer water. The basal spacing of the complex depends on the size of the organic molecules and on their orientation and packing geometry.

Alternatively, when certain organic compounds are added to an unknown mixture of clays, a basal spacing typical for the complex of that compound will indicate the presence of montmorillonite in the mixture. Ethylene glycol, which gives a basal spacing of about 17 A with most montmorillonites, is commonly used for this identification purpose.

When the clay surface becomes covered with polar molecules containing a substantial proportion of hydrocarbon groups, the surface becomes oleophilic, and in this condition, clay-oil dispersions can be prepared with the treated clay. However, since the exchange cations are still present in the complexes, they are usually rather sensitive to water. For example, a pyridine complex of bentonite in the presence of water initially hydrates stepwise, as shown by X-ray diffraction at low water concentrations, and disperses completely when a large amount of water is added (47). Pyridinium ion (or other organic cation) exchange complexes are not affected by water. Exceptions are low molecular weight alkane ammonium complexes (e.g., of *n*-butylammonium) which show a spectacular interlayer swelling with water (46).

The addition of a polar organic compound to a clay-water suspension often causes flocculation of the clay. It has been pointed out before (Chapter Four, Section III) that this effect is predicted by the Gouy theory. Sometimes, a peptizing effect is observed, which might be explained by changes occurring in the Stern layer. The effect of polar organic compounds on the stability of clay suspensions is therefore rather unpredictable. (See Chapter Four, Section VII.)

III. MACROMOLECULAR COMPOUNDS (52–67)

A. Polyelectrolytes

A technically very important and powerful method of controlling the stability of hydrophobic colloids is the addition of water-soluble polyelectrolytes, that is, hydrophilic or macromolecular colloids, to the hydrophobic sols.

Polyelectrolytes consist of long-chain molecules with ionizing groups which are usually located along the entire length of the chain. The polyion may be a polycation with amino groups, or a polyanion with carboxyl, sulphate, sulphonate, or other negative groups. A single-polymer molecule may contain both positive and negative groups, for example amino and carboxyl groups, as is the case in proteins.

Many polyelectrolytes occur in nature—gum arabic, gelatin, alginates, pectin, and so on. A growing number of synthetic polyelectrolytes which are becoming available are either wholly synthetic, that is, made by the polymerization of monomers, or they are chemically altered natural macromolecular compounds, for example, oxidized starch and carboxymethylcellulose.

When the polyelectrolytes are dissolved in water, their functional groups become more or less dissociated. As a result of the mutual repulsion of the charged groups along the chain, the flexible chain molecule will assume a stretched configuration. The rather high viscosity of the polyelectrolyte solutions is a consequence of the presence of the stretched molecules. This effect of charge on viscosity is referred to as an *electroviscous effect.* An appreciable viscosity reduction is observed when small amounts of salt are added to the polyelectrolyte solutions. This effect is typical for hydrophilic colloidal systems. The concentrations of various salts at which a certain viscosity reduction is achieved are smaller when the valence of the ion of opposite sign increases. Apparently, the reduction of the viscosity of hydrophilic colloids upon the addition of salts is analogous to the flocculating effect of salts on hydrophobic colloids, which obeys the Schulze-Hardy rule.

The electroviscous effect in hydrophilic colloids can be explained as follows: Opposite a charged functional group of the chain, a counter-ion atmosphere is created which has a net charge equal to the charge on the functional group. The mutual repulsion between the charged groups, which causes the chain to stretch, is analogous to the mutual repulsion of the diffuse double layers of the particles in hydrophobic sols. In the presence of salt, the diffuse atmospheres around the charged groups are compressed, and the range of mutual repulsion is reduced. As in the hydrophobic sols, the double-layer compression is dictated by the valence of the ions of opposite sign of the added salts. For the dissolved chain molecules, a randomly coiled condition is statistically the most probable configuration. Therefore, from the point of view of entropy, the chains will have a tendency to assume the coiled condition. In the salt-free solution, the mutual repulsion of the charged groups prevents coiling, but a reduction of the mutual repulsion by the addition of salt allows the chains to assume the more probable coiled condition. Because of the coiling of the chains, the contribution of the large molecules to the viscosity decreases. From this explanation it is clear why the effect of salts on the electroviscous effect follows a rule which is analogous to the Schulze-Hardy rule.

(Previously, the high viscosity of hydrophilic colloids was thought to be due to the formation of large hydration shells around the colloidal units. The electroviscous effect was attributed to dehydration by the added salts. However, many experimental facts appeared to be incompatible with this concept and the theory was abandoned.)

At very high electrolyte concentrations in the region of molar solutions, the hydrophilic colloid is sometimes precipitated from the solution. The precipitation is considered a consequence of a reduction of the solubility of the polyelectrolyte; it is "salted-out." However, in many polyelectrolyte solutions salting-out does not occur, and the macromolecules remain in solution at high salt concentrations.

1. Effect of Polyelectrolytes on Clay Suspensions. The effect of polyelectrolytes or hydrophilic colloids on the stability of hydrophobic sols, including clay suspensions, depends on their relative concentrations.

When the polyelectrolyte content of a hydrophobic sol amounts to a few tenths to several percent, the salt tolerance of the hydrophobic sol is considerably increased, even to the extent that the sol remains stable in concentrated salt solutions. Because of this effect, the polyelectrolyte is called a *protective colloid* for the hydrophobic sol.

In the presence of very small amounts of polyelectrolyte, however, the hydrophobic sol becomes more sensitive toward flocculation by salt, although the polyelectrolyte alone does not flocculate the hydrophobic sol, if they carry the same charge. This effect is called *sensitization.* When the polyelectrolyte and the hydrophobic sol particles carry opposite charges, the hydrophobic sol is flocculated by extremely small amounts of the polyelectrolytes, as may be expected from the high valence of the latter. The protective action of the polyelectrolyte, on the other hand, usually is independent of the sign of the charges of the hydrophobic and hydrophilic colloid.

The sensitizing and the protective actions of polyelectrolytes on hydrophobic sols are illustrated by the flocculation diagram (Figure 36). The salt-flocculation value of a sodium montmorillonite sol is plotted as a function of the concentration of the polyanion of sodium carboxymethylcellulose. Because of the widely different concentration ranges in which either effect prevails, the region of sensitization is enlarged in the lower part of the figure.

The protective action of a hydrophilic colloid on a hydrophobic sol has been known for a long time. A well-known example is the protection of a gold sol against flocculation by salts. The protection of gold sols is used as the basis for a routine test of the protecting power of a polyelectrolyte. The so-called "gold number" of a hydrophilic colloid is the number of milligrams of polyelectrolyte which just prevents the flocculation of 10 ml of a gold sol by 1 ml of a 10 percent NaCl solution. A smaller gold number indicates a greater protective power.

2. The Mechanism of the Protective and Sensitizing Action. Despite the fact that both the protective and the sensitizing

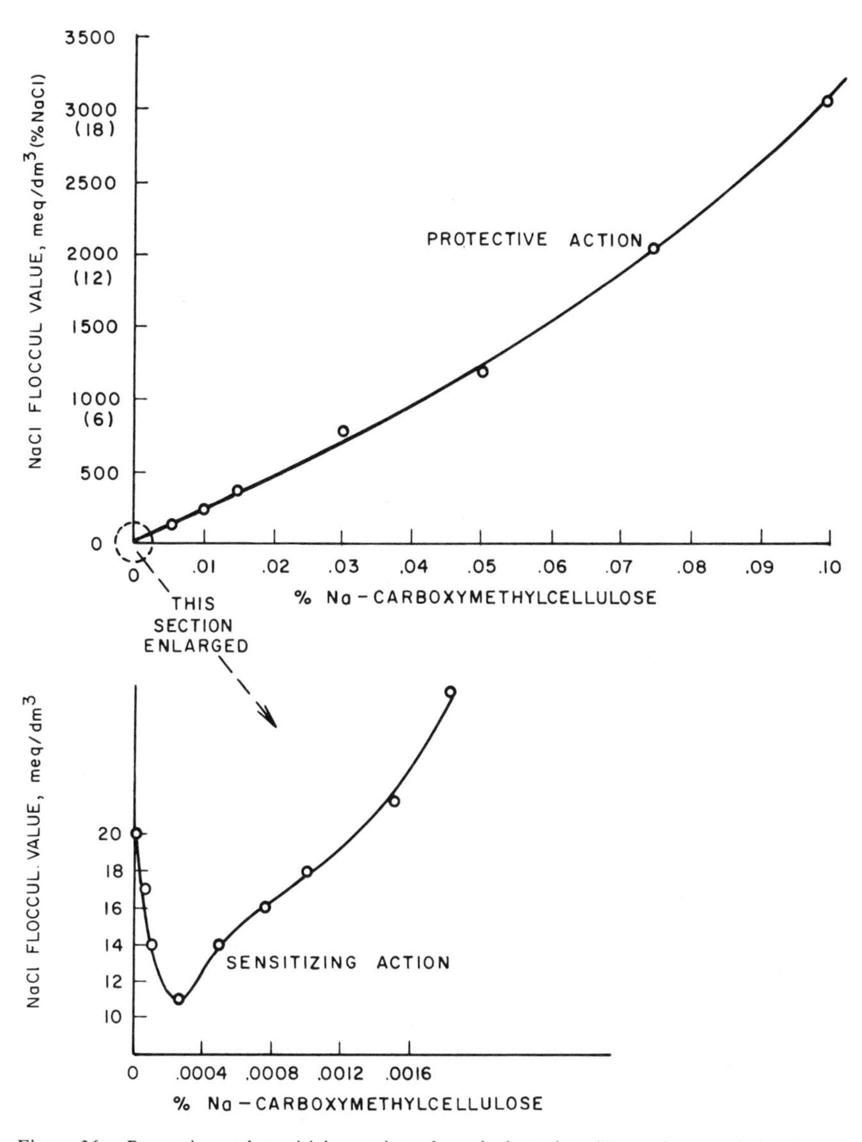

Figure 36. Protective and sensitizing action of a polyelectrolyte (Na-carboxymethylcellulose) on a sodium bentonite suspension.

effect of hydrophilic colloids on hydrophobic sols have been known for a long time, no systematic study leading to a full understanding of the mechanism of these actions has yet been carried out. Fragmentary investigations have indicated that the problem is a rather complex one. It has been observed that the polyelectrolytes are adsorbed by the hydrophobic particles and that the adsorption equilibrium depends on the salt content of the system. The adsorption is probably governed by entropy effects, although chemisorption may also contribute, for example, in the case of adsorption of polyanions on the edge surfaces of clays. In the adsorption complex, both double-layer repulsion and entropy repulsion may contribute to an improved stability of the hydrophobic sol. Moreover, in the coagulation processes, the relative magnitudes of the rate of coagulation and the rate of polymer adsorption appear to be important.

Nevertheless, as a practical guide, the following rough picture of the mechanism of protection or sensitization is usually accepted:

In the range of polyelectrolyte concentrations in which protection is observed, the polyelectrolytes envelop the hydrophobic sol particles and lend them the special characteristics of the hydrophilic sol, that is, their stability and high salt resistance. This point of view is supported by the observation that the hydrophobic sol particles obtain an electrophoretic mobility which is characteristic for the particular hydrophilic sol used.

In the region of sensitization, in which a relatively small amount of polyions is available, the polyion is thought to become adsorbed on several hydrophobic sol particles at the same time, thus forming a bridge between them. In other words, the sol particles are adsorbed on different spots of the same chain. In this way, particle linking or flocculation of the hydrophobic sol is promoted. It has been shown that in the absence of salt, polyion adsorption does not occur, and this is one of the reasons why the polyion alone does not flocculate the sol. A certain amount of salt must therefore be present to establish particle bridging by the simultaneous adsorption of polyelectrolyte on different sol particles. Another function of the salt is to allow the sol particles to come closer together by reduction of the range of double-layer repulsion and thus to promote bridging of the particles by the polyelectrolyte chains. The reality of such an effect is indicated by the following observations on the effects of changing the sequence of addition of polyelectrolyte and salt which demonstrates the importance of the relative rates of coagulation and polymer adsorption. As follows from the above considerations, in order to promote bridging, salt should be added to the sol before the polyelectrolyte is added. If the polyelectrolyte is added before the salt, there is a greater chance that the polyelectrolyte ions will become adsorbed on individual particles when salt is added. Then, in the subsequent flocculation process, the particles will probably become attached by those

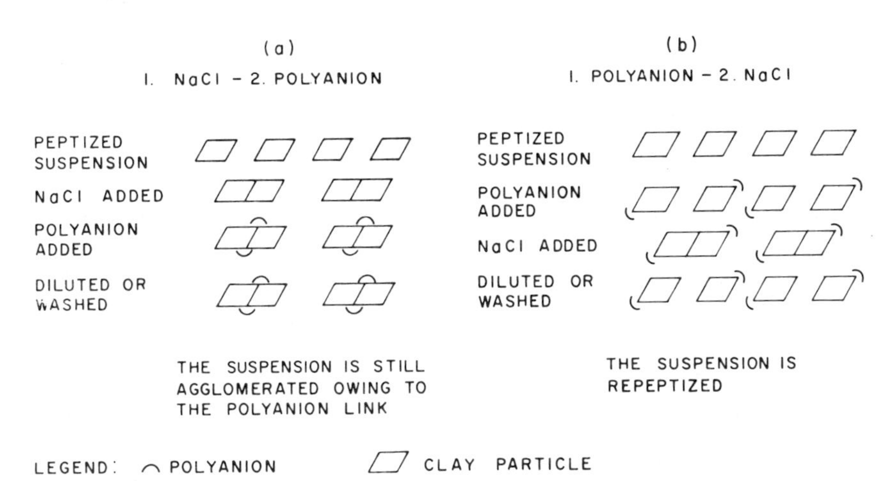

Figure 37. Schematic representation of the interaction of clay particles with polyanions and salt as a function of the sequence of addition. (*a*) Salt added before adding polyanions. (*b*) Salt added after the addition of polyanions. [After R. A. Ruehrwein and D. W. Ward (56).]

parts of their surfaces which are not covered by the polyions, as is indicated in Figure 37. When the salt is removed from this system by washing, the suspension disperses easily owing to the stabilizing effect of the adsorbed polyions. The suspensions to which the salt was added first do not repeptize when washed owing to the linking of the particles by the polyelectrolyte chains.

3. Applications

a. *Flocculation aids.* The addition of very small amounts of polyelectrolytes to tenacious suspensions containing some flocculating salt often yields flocs which are much easier to separate than those obtained by salt flocculation alone. In such applications the polyelectrolyte is called a *flocculation aid.* They are often used to treat "slimes" containing clays, for example, slimes from mining operations or slimes in waste-water treatment plants. The polyelectrolyte flocculated system has improved sedimentation and filtration properties.

b. *Soil conditioners.* Polyelectrolytes may be used to keep soils in the proper flocculated condition required for root growth and drainage. In this application, the polyelectrolytes are referred to as *soil conditioners.* Although the same effect can be obtained by gypsum treatment of the soil, the soil conditioners are not easily washed out by rain water, since they are strongly adsorbed on the clay surface. However, dispite their greater permanence as well as the small amounts required, their use still seems to be limited because of their comparatively high price. As with other flocculants, the soil conditioners are most effective if they are thoroughly worked into an aerated soil.

c. *Drilling-mud extenders.* Drilling muds are made up primarily from locally available clays as well as from the underground clays dispersed in the mud when the hole is drilled. If these local clays happen to be rather "lean," it is difficult to obtain the proper plastering and cutting-carrying ability. Lean clays, such as kaolinite, consist of rather thick particles; their gel-forming ability is therefore small, and their filter cakes are rather permeable. In such circumstances, a good grade (i.e., fine particle size) bentonite is added to the mud to improve the "yield" and plastering properties of the clay. (A clay is said to have a higher "yield" if the amount of clay required to give a suspension of a given yield stress is smaller.) Since good grade bentonites are relatively expensive, means to increase their yield as well as the yield of poorer (i.e., coarser) bentonites have been sought. Obviously, an increased yield could be obtained by adding flocculating salts. However, flocculation with the aid of small amounts of polyelectrolytes is preferred. The latter give higher yields, probably because less "aggregation" takes place than in flocculation with inorganic salts. Moreover, the plastering properties of the muds, although usually not as good as those of the original mud, are affected less by polyelectrolyte treatment than by flocculation with inorganic salt. The polyelectrolytes, for example polyacrylates, are therefore preferred for increasing the "yield" of a clay and thus "extending" the beneficial effect of a certain amount of bentonite when added to a lean mud. Hence the term *mud extender* was coined for the polyelectrolyte in this application.

d. *Protection of salt-water drilling fluids.* The applications of polyelectrolytes in colloidal systems so far mentioned were based on their sensitizing action when applied at low concentrations. The protective action of polyelectrolytes at higher concentrations is the basis for their use in salt-water drilling fluids. These muds are used in drilling through salt domes or in offshore drilling. The stabilizing effect of polyelectrolyte solutions on clay suspensions is very large and surpasses the effect of other types of peptizers (Figure 36). As long as the polyelectrolyte is not salted out, it may be used in saturated salt solutions. Modified starches and carboxymethylcellulose salts are most widely used in drilling muds. The higher price of the latter is offset by their greater permanence compared with starches, which are more sensitive to bacterial action.

e. *Miscellaneous applications.* Freeze-drying of clay gels yields aerogels in which some of the coherence of the matrix existing in the hydrogel is preserved, although during the freezing step the matrix is affected by the growth of the ice crystals. When polyelectrolytes are present in the hydrogel, the mechanical strength of the resulting aerogel is significantly increased. Such aerogels might find application as insulating material or in gas chromatography, the latter in particular since the clay surfaces may be pretreated with organic molecules through which considerable variety in chromatographic separations can be achieved (106–108).

The adsorption capacity of clays for proteins is put to use in the clarification of wine (105).

B. Nonionic Polymers

1. Interaction with Clay Suspensions. Nonionic polymers are macromolecules with a number of polar functional groups which are distributed along the chains. Purely hydrocarbon polymers are not included in this group. Examples are polysaccharides and synthetic polyoxyethylenes. At relatively high concentrations they usually have a protective action on hydrophobic sols; at small concentrations they sometimes flocculate a sol and sometimes do not, as is the case with low-molecular-weight polar compounds.

Rather little effort has been put into systematic research on the interaction of hydrophobic sols and nonionic polymers. At present, little understanding has been obtained about the mechanism of these interactions. Polyoxyethylenes have been found to become adsorbed on the flat surfaces of clay particles, as is evident from the increased basal spacings observed for the polyoxyethylene complex with montmorillonite. Phenomena not understood are how the adsorbed macromolecules affect the ion distribution in the Stern and Gouy layer, whether edge adsorption takes place simultaneously, or to what extent entropy stabilization plays a part in peptization reactions between nonionic polymers and clay suspensions.

2. Applications

a. *Drilling fluids.* So-called "surfactant muds" have been introduced in drilling-fluid technology. These muds consist of a clay suspension, a nonionic polymer of the polyoxyethylene type, the salt of an anionic polyelectrolyte such as sodium carboxymethylcellulose, and a strongly flocculating electrolyte such as calcium sulphate. Owing to the high electrolyte content of the mud, osmotic swelling of formation clays encountered in drilling is repressed. These muds therefore have good formation-conserving properties. It is claimed that the nonionic polymer aids in the repression of the swelling. Although the clay is in a flocculated condition, the plastering properties of the mud are kept on a reasonable level by the carboxymethylcellulose. Since the mud components do not readily react with clays, like those in the formation conserving lime red muds, high-temperature solidification is avoided.

IV. EMULSIONS CONTAINING DISPERSED CLAYS (68–72)

An emulsion is a dispersion of fine droplets of a liquid in another liquid which is the *continuous* or *external phase,* the droplets being the *dispersed* or *internal phase.* Usually the droplets are larger than about 0.1 μm. Emulsions consisting of water and a variety of organic liquids may be prepared. The organic liquid phase is usually referred to as the "oil phase." The customary notation for water-in-oil emulsions is W/O, and O/W indicates oil-in-water emulsions.

In order to obtain a stable emulsion, a third component, the *emulsifier,* must be present. Very small amounts of such emulsifiers are required for a stable emulsion. Some emulsifiers promote the formation of O/W emulsions, others are specific emulsifiers for W/O emulsions. For example, an alkali soap makes an O/W emulsion, whereas a soap of a cation of higher valence makes a W/O emulsion.

When an emulsion becomes unstable, the droplets coalesce, and two bulk phases are created. The emulsion is said to *break* (*demulsification*). Two separate processes should be distinguished in the breaking of emulsions: (*1*) an agglomeration of the droplets, which is analogous to flocculation in a hydrophobic sol, and (*2*) the coalescence of the agglomerated droplets, which is governed by a reduction of the total interfacial free energy of the system.

Under certain conditions, an O/W emulsion can be *inverted* and become a W/O emulsion. A W/O emulsion is often broken by the addition of a compound which acts as an emulsifier for the O/W emulsion, and vice versa.

Many different theories have been advanced to explain emulsification and the stability of emulsions. In view of the complexity and variety of emulsion compositions, it is not surprising that no single theory is applicable to all systems. A detailed discussion of the proposed emulsion theories is given in a monograph by Becher (69).

An important group of emulsifiers are surface-active agents which are accumulated in the oil-water interface. Emulsification of oil in water by surfactants should not be confused with the process of *"solubilization"* of oily meterials in the presence of micelles of surfactants. For example, some surfactants form spherical micelles in water with the hydrocarbon tails turned inside. Oil will then be accommodated in the hydrocarbon interior of these micelles, and "solubilized."

Another group of emulsifiers are finely divided solids, such as carbon, certain basic sulfates, or hydroxides. These particles, which should preferably be in the colloidal size range, are accumulated in the oil-water interface, since they are wetted to some extent both by water and by oil. It has already been mentioned that clays may become effective emulsifiers when an organic compound is adsorbed on their surface. For example, an equivalent-exchange complex of a clay and an organic cation appears to have an optimum emulsifying ability. In this condition, the face surfaces of the clay have become oleophilic, whereas the edge surfaces will still be hydrophilic, since no adsorption of the organic cation on these surfaces will occur.

A remarkable rigidity is sometimes displayed by an oil-water interface in which clay-organic exchange complexes are accumulated. When a pendant drop of one liquid in the other is made to recede, the interface obtains the appearance of a shrunken wrinkled bag.

A. Applications

In drilling for oil, emulsion-type drilling fluids are often used. They may be emulsions of oil in a clay water-base mud, or emulsions of water in an oil-base mud. In the preparation of these emulsions, surface-active emulsifiers are sometimes added, but it is possible that adsorption complexes of these emulsifers and the solids in the muds are the actual stabilizers of the emulsions.

Clay-organic complexes may act as the emulsifying agents in natural crude-oil emulsions.

V. CLAY DISPERSIONS IN OIL

Clays can be dispersed in oil when the particle surface has been made oleophilic by the adsorption of suitable organic compounds. For example, exchange adsorption of an organic amine cation or a quaternary ammonium cation usually yields an eleophilic clay which is dispersible in oils. Alternatively, certain polar organic compounds adsorbed on the clay may make the clay oil-dispersable.

Several different organo-clays are commercially available. They can be dispersed directly in oil while applying shear.

Often, the best way to prepare an oil dispersion is to make a water dispersion first and then transfer the dispersed clay from the water phase into an oil phase. Three different methods of transfer can be distinguished:

1. The organic compound is added directly to the clay suspension in water. The clay precipitates. If the water has been displaced effectively from the clay surface, it will be possible to separate the precipitate from the water phase, for example, by filtration, and to disperse the precipitate in oil. This process is sometimes called the *flushing process*. A homogeneous

dispersion is obtained when high shear rates are applied in the dispersion equipment.

2. After the addition of the organic compound to the hydrous dispersion, the precipitate is washed with a water-miscible alcohol; this transfers the hydrous precipitate or hydrogel into an "alcogel." Subsequently, the alcogel is washed with the hydrocarbon oil, which must be miscible with the alcohol used in the previous step. In this way the alcogel is transferred into an oleogel.

3. The organic-treated hydrogel or alcogel is transferred into an aerogel by freeze-drying or by evaporation of the liquid above the critical temperature and pressure (Kistler process). The aerogel, which retains a good deal of the structure of the original liquid gel, converts to an oleogel by imbibition of oil.

The stability of the resulting dispersions in oil depends on the balance between interparticle repulsion and attraction. It was mentioned earlier that in oil dispersions, the repulsion which has to counteract the van der Waals attraction may be of an entropic nature, but if ions are present in the dispersion double-layer repulsion is the most likely stabilizing factor.

A detailed study of the stability conditions in clay-oil dispersions has not been made. Normally these dispersions are in the flocculated or gelated condition when they are prepared with cationic organic compounds. These compounds are probably exclusively adsorbed on the negative layer surfaces and not on the (positive) edge surfaces. Therefore, the edge surfaces will not carry a protective organic coating that would prevent van der Waals association. Hence EE association and possibly also EF association will occur and the system flocculates or gelates.

A. Applications (100–103)

Oil-dispersible clays are used in the manufacture of lubricating greases. In these "inorganic gel" greases, in which the clay is in the associated (flocculated) condition, the clay skeleton replaces the soap skeleton of the conventional lubricating greases. The advantage of the clay-base greases is their smaller susceptibility to heating.

Organophilic clays can also be used in oil-base drilling fluids. Oil-base "muds" consist of oil, a colloidally suspended compound, and weighting materials such as ground limestone, or other material. Blown asphalt is generally used as the colloidal material. The function of the colloidal material is twofold: It imparts a certain yield stress to the suspension by which the weighting material is prevented from settling, and it gives the fluid good plastering properties.

As with the lubricating greases, the substitution of the organic colloidal constituent by the inorganic oleophilic clay may be expected to improve the high-temperature performance of the oil-base muds. This is indeed the case, but it appears that the colloidal organophilic clay alone does not impart sufficiently good plastering properties to the drilling fluid. When used in conjunction with blown asphalt, however, the organophilic clay seems to improve the mud properties at elevated temperatures, and thus it extends the range of applicability of oil-base muds to deeper wells in which higher temperatures prevail.

Some other applications in the field of petroleum exploration and production which are related to a modification of the wetting properties of mineral surfaces are the release of hydrocarbons from source rocks, and the use of surfactants in secondary recovery processes. In source rocks, hydrocarbons are generated from occluded organic material. They are released under suitable circumstances by overburden pressure, transported through pores and eventually trapped, and accumulated in porous rocks (reservoirs) from which they can be produced. It is known that considerable amounts of hydrocarbons are permanently retained in the source rocks. One possible reason for this retention is that part of the hydrocarbons is adsorbed on the pore surfaces, since these surfaces will have obtained an oleophilic character due to the adsorption of polar or ionic organic molecules, which are the precursors of the hydrocarbons. From adsorption isotherms of model systems, it can be shown that commonly existing overburden pressure differentials are too low to provide the needed work of desorption of the hydrocarbons (104).

As a general note, we may stress that all uses of additives which are adsorbed on the faces of the clay particles require a rather high dosage of the additive, particularly when montmorillonite clays are involved. For example, when the exchange capacity of a montmorillonite clay with a long-chain organic cation is to be satisfied, almost equal weights of clay and organic additive are required. Therefore, the applications of these additives from an economic point of view are much more limited than those of additives which are adsorbed at the edge surfaces of the clay particles, such as the "thinners" for clay water-base drilling muds.

In summary, the surface properties and the colloid chemical behavior of clay systems can be varied widely by the incorporation of organic compounds, providing considerable latitude in tailoring a system to suit a specific application. Clay-organic interactions are important in many cosmetic and household products in which clays are frequently used as thickeners or fillers. In the formulation of these products, application of the principles discussed in this chapter will help in guiding the development of a suitable formula and should minimize the amount of empirical product development work.

VI. INTERCALATION AND INTERSALATION IN KAOLINITES (79–99)

Halloysite used to be the only member of the kaolinite group known to have a capacity to expand: in nature it occurs as an interlayer hydrate. However, the hydrate is not very stable; it easily loses its water and collapses irreversibly.

In the late fifties, Wada found that the hydrated halloysite intercalates certain salts such as potassium acetate (81, 82). Subsequently, he observed that such salts also intercalate with other members of the kaolinite group under suitable conditions. This phenomenon, called *intersalation,* is generally observed for salts of low molecular weight carboxylic acids and large cations (K, Rb, Cs, NH_4).

Intersalation was originally achieved by grinding the mineral with moist crystals of the salts, or by contacting the mineral with saturated solutions. Apparently, a condition for intersalation is a low degree of ionization of the salt. It was found later that intersalation is also accomplished with dilute solutions of the salts in water-alcohol mixtures with a high alcohol content (87). Once the layers of the kaolinites have been parted by the salts, they can be replaced with other types of salts which are not able to expand the mineral directly.

The interlayer complex contains the complete salt as well as water, which can be reversibly desorbed by degassing. For example, a typical kaolinite-potassium acetate intersalation complex has a *001* spacing of 14 A. By degassing, the spacing is reduced to 11.7 A, but the 14-A spacing is restored by exposure to water vapor.

The intersalated material can be removed by washing with water. Then, the mineral either collapses and the original 7.2-A spacing is obtained, or an interlayer hydrate is formed. Usually, such hydrates show a mixed-layer X-ray pattern. However, in the case of a nacrite, a very regular hydrate with a spacing of 8.4 A is observed. In this hydrate, the water molecules are keyed with the tetrahedral cavities at less than the monolayer level (87).

In addition to these salts, another class of compounds is able to intercalate. These comprise urea, hydrazine, amides, dimethylsulfoxide, pyridine-*N*-oxide, and others. The resulting *001* spacings vary with the type of compound. Once such intercalation complexes have been formed, the interlayer material can be exchanged for other materials which would not intercalate directly. In this way, a whole new field of clay-organic complexes has been opened.

Weiss (92) has proposed that intercalation was probably behind the secret of ancient Chinese porcelain manufacture. The common practice was to bury kaolins for a while with urine; therefore, the kaolin must have slowly expanded by intercalation of urea. Subsequently, the expanded mineral could be ground in a special way to produce thinner individual particles which would be a much improved base for porcelain manufacture.

All kaolinite intercalating materials have in common that they have a certain hydrogen bonding capacity. Hence it has been assumed that the driving force in intercalation was for these compounds to break hydrogen bonds between the kaolinite layers and to establish hydrogen bonds between the compounds themselves and the kaolinite surface. However, quantitative estimates of the cohesive energy between kaolinite layers show that the hydrogen bonding accounts for only a small fraction—perhaps 10 percent—of the electrostatic bonding energy, that is, about equal to the van der Waals contribution to the cohesive energy. Although the hydrogen bonding energy between the intercalating compounds and the clay surface—as estimated from the infrared spectral shift—appears to be considerably larger than the hydrogen bonding between the layers, it is still not large enough to compete with the electrostatic contribution to cohesion. Therefore, it has been proposed that another contribution may be the decrease of the electrostatic attraction caused by an increase in the dielectric constant when the compounds penetrate between the layers from the periphery of the particles. This idea is supported by the observation that the addition of small amounts of water enhances intercalation (see Chapter Five, reference 26).

VII. SUMMARY OF PARTICLE INTERACTION (73–78)

In the first few chapters, in the discussion of interparticle forces and sol stability, the competition between double-layer repulsion and van der Waals attraction was emphasized. As an alternative source of repulsion, the so-called "entropic repulsion" was mentioned. In the present chapter, which deals with organic systems, some additional mechanisms of particle interaction were introduced. A summary of the many factors determining flocculation and peptization therefore seems useful at this point.

A. Factors Promoting Deflocculation

1. Electric Double-Layer Repulsion. Electric double-layer repulsion is not limited to hydrous systems. It may exist in dispersions in organic media if additives are present which form ions in such systems.

2. "Entropic" Repulsion. "Entropic" repulsion may also occur in both hydrous and organic dispersions. This repulsion is due to steric hindrance in a layer of adsorbed surface-active molecules or polymers, which

occurs when the particles approach each other to within a distance at which the adsorbed molecules begin to interfere. Since this type of repulsion occurs at only moderately large distances, it is likely to be effective only for small particles, for which the total van der Waals attraction at these distances is not too large. Large particles, for which the van der Waals attraction is stronger, will in general need the longer-range double-layer repulsion to achieve stability.

3. SHORT-RANGE HYDRATION OR "LYOSPHERE" REPULSION. A short-range repulsion between particles exists if the particles adsorb solvent molecules to the extent of one or two monolayers. The repulsion is determined by the work required to desorb these molecules when the particles come very close to each other. This short-range repulsion tends to make less deep the attraction minimum that is due to strong van der Waals attraction. (In organic dispersions, which are stabilized by adsorbed surface-active molecules, the solvent molecules are often the same size as the molecules of the additive. In that case, one treats the changes in configuration and adsorption of the solvent and solute molecules simultaneously.)

4. BORN REPULSION. When the crystals come into contact, Born repulsion prevents the interpenetration of the crystal structures. It is important to note that extruding points on the surfaces, which are seldom perfectly smooth, will prevent close approach of large parts of the particle surfaces.

B. Factors Promoting Flocculation

1. VAN DER WAALS ATTRACTION. The magnitude of the attraction depends on the type of dispersion medium, but in hydrous systems there is little dependence of the total force on the electrolyte concentration. The magnitude of the attraction further depends on the size of the particles and on their composition, namely, the packing density of the atoms. If the particle is surrounded by a shell of adsorbed molecules of either solvent or solute, or both, and the density of atom packing in this shell is less than in the particle itself, the effective van der Waals attraction between two particles will be reduced. Hence a smaller repulsion would be required to achieve stability in this case.

The attraction is also affected by the shape of the particles: for anisodimensional particles, the attraction as a function of the distance will vary with the positioning of the particles with respect to the path of mutual approach.

2. ELECTROSTATIC ATTRACTION. In dispersions of two kinds of particles carrying double layers of opposite charge, these particles attract each

other electrostatically. The ensuing agglomeration is called "mutual flocculation." This type of flocculation can occur in both hydrous and organic systems. In hydrous clay systems, "internal mutual flocculation" occurs because of the presence of double layers of opposite sign on the edge and flat surfaces of the clay particles.

3. BRIDGING OF PARTICLES BY POLYFUNCTIONAL LONG-CHAIN COMPOUNDS. Polyfunctional long-chain compounds may become adsorbed on more than one particle simultaneously and thus bridge several particles. This effect explains the sensitization of inorganic hydrosols by small amounts of polyelectrolytes.

4. BRIDGING OF PARTICLES BY A SECOND IMMISCIBLE LIQUID COMPONENT. A mechanism of particle agglomeration which has not been mentioned so far is one which explains the rather dramatic flocculating effects of traces of water on an organic dispersion. This effect is shown by water, and other polar liquids as well. If these liquids are immiscible with the organic dispersion medium, and if the dispersed particles are preferentially wetted by the added polar liquid, the added water is believed to envelop the particles with a thin film. When two particles come together, their water films will flow together at the juncture points since this arrangement results in a reduction of the interfacial area and thus to a reduction of the total free interfacial energy of the system. A quantitative treatment of this mechanism has been presented by Kruyt and van Selms (73, 75).

These are the principal factors which are of direct importance to dealing with particle interactions in clay systems. Of more peripheral interest are the following phenomena related to aggregation and dispersion of particles and molecules which have been briefly mentioned in the text: hydrophobic bonding, primarily important for micellization of surfactants; solubilization of oils in micelles; emulsification of liquids and demulsification; coacervation, that is, the separation of a liquid phase of higher particle concentration (e.g., tactoids); the precipitation of polymers by salts due to lowering of the solubility.

In view of the large variety of possible mechanisms of particle agglomeration and dispersion, it is often difficult to give even a qualitative analysis of the factors operating in a given practical system. However, a knowledge of the possible factors and the conditions under which they are operative will be a good guide for research on such systems to improve formulation. Further complications in dealing with practical systems are that the components are often ill-defined, and also that the establishment of the adsorption equilibrium at the particle surfaces is often slow which gives rise to metastable systems and aging phenomena.

Finally, it should be kept in mind that the particles in a dispersion are subject not only to interparticle forces, but also to gravitational, diffusion, and convection forces.

VIII. CHEMICAL REACTIONS BETWEEN CLAYS AND ORGANIC COMPOUNDS

A full discussion of the chemical reactions which take place on clay surfaces is outside the scope of this book. However, a few general notes may be helpful to introduce the reader to some of the literature on this rather complex and not yet fully developed subject. The industrial interest in catalytic activity of clays and of "activated" (acid-treated) clays is mostly historical since the clays have been replaced by synthetic oxide catalysts. However, a growing interest in clay-organic reactions relates to the roles that clays probably have played in the generation of prebiotic materials of some complexity. In addition, a knowledge of clay-organic interaction is important for the study of the fate of pollutants in soils.

The individual surface sites at which reactions take place or at which catalytic reactions are initiated are of several different kinds. Their nature may be derived from the kinds of reactions that occur on clay surfaces, and in favorable circumstances, the presence of these sites may be confirmed by physical (primarily spectral) and chemical methods of surface analysis.

In studying the reaction mechanisms, it is again important to distinguish between reactions taking place on face surfaces or on edge surfaces. There are a number of ways to assess the location where a particular reaction takes place. For example, edge surfaces may be blocked by prior adsorption of phosphates to observe whether the reaction is inhibited. If during the reaction interlayer expansion is observed, the possibility of the reaction taking place on the faces should certainly be considered, but this observation by itself is not a definitive indication that the face surfaces provide the reaction sites. The expansion may be due to interlayer adsorption of the unreacted material, whereas the reaction takes place on the edge surfaces. An example is the reaction that occurs between dry bentonite and dry pyridine, resulting in a deep blue colored product. The clay expands since pyridine is adsorbed between the layers, but upon removal of the interlayer pyridine by heating, the expansion is eliminated while the color remains unaffected. Hence the color reaction actually takes place on the edge surfaces. Another clue to the location of the reaction sites is the yield of the reaction product. If the yield is very low, the reaction probably will take place at edge surfaces. In the above example, the yield of the blue product was small and of the order of the adsorption capacity of edge surfaces (35).

Chemical reactions on clay surfaces include acid-base and oxidation-reduction reactions, as well as many reactions which are catalyzed by acid-base or oxidation-reduction sites (e.g., polymerization, hydrocarbon cracking, peptide bond formation, and others).

The following sites have been identified on clay surfaces:

1. *Lewis acid sites,* that is, electron-pair accepting sites. These sites occur in the absence of water. They are associated with such surface features as incompletely coordinated Al, for example, tetrahedrally coordinated Al in the silica sheet, or incompletely coordinated aluminum atoms resulting from dehydration of edge sites. In the presence of water, Lewis sites become Brönstedt sites.

2. *Brönstedt acid sites,* that is, proton donor sites. These sites not only occur in acid-treated clays, but also in clays with other cations in exchange position. The cations are associated with water molecules which they polarize so that protons become available. The acid strength depends on the type of cation, and on its position on the surface: inside or outside the tetrahedral holes. The acid strength increases with decreasing water content, and for acid catalysis nearly dry clays are required (129–133).

3. *Oxidizing sites,* that is, electron-accepting sites, and

4. *Reducing sites,* that is, electron-donor sites. These sites are respectively associated with ferric and ferrous ions in octahedral position, and can affect reactions on both edge and face surfaces. Oxidation may also be the result of the presence of adsorbed oxygen on surfaces.

The one-electron accepting oxidation sites are sometimes referred to as Lewis acid sites in the widest sense.

Many color reactions of clays and organic compounds have been studied, and they are often used for identification purposes. Some are of the acid-base, or the oxidation-reduction (electron-donor-acceptor), type (110–115).

5. *Surface hydroxyl groups.* These occur at crystal edges, bound to Si, Al, or other edge-exposed octahedral ions. In kaolinites they also occur on the aluminum-hydroxide faces. Their number may be determined by methylation with CH_3I, correcting for the reaction of CH_3I with (separately determined) adsorbed water. The fraction of OH groups with a strong acid character may be determined by reacting the clay with diazomethane (118, 119).

It has been debated whether surface hydroxyl groups bound to silicon (silanol groups) occur on the face surfaces of montmorillonites, since several reactions by which silanol groups would be esterified resulted in increased *c* spacings (116). Such face OH groups are present in alternative structures for montmorillonites, as proposed by Edelman and Favejee and by McConnell (See Chapter Five, References 5 and 6). However, these proposals cannot be accepted on the basis of X-ray and other evidence, and the alleged esterification reactions are probably due to reactions with residual adsorbed water or to polymerization reactions (117).

When considering reactions with OH groups and with acid sites on kaolinites, it is important to consider the possible presence of strongly adsorbed alumino silicates on kaolinite face surfaces (see Chapter Five, reference 76).

IX. CLAYS AND THE ORIGIN OF LIFE (134–157)

Clays have long been considered of having played an important role in the generation of prebiotic compounds of some complexity, and experimental support for this hypothesis is growing. Clays may have played several different roles in biological evolution. These are the following:

1. Through the large adsorption capacity of clays, reactions between components present in natural waters at very low concentrations could be significantly accelerated when the components are concentrated in clay sediments. Owing to the dual character of the edge and the face surfaces, both anionic and cationic compounds can be adsorbed, in addition to a variety of nonionic polar molecules. Apolar compounds can be adsorbed as soon as the surfaces obtain an oleophilic character due to prior adsorption of cationics, anionics, and nonionics with suitable hydrocarbon moities. A certain selectivity in these processes can be expected because of variations in adsorption energy, particularly during leaching of sediments by solutions of competing compounds.
2. Increased reactivity of adsorbed compounds may occur selectively, due to activation of adsorbed molecules by the clay surface. Such catalytic action will be different for edge and for face surfaces, and will depend strongly on environmental conditions, particularly the presence or absence of water.
3. The steric constraints on adsorbed species imposed by the clay surfaces, and particularly by confinement in the interlayer space, can result in the generation of stereospecificity of reaction products, for example, of polymerization products of adsorbed monomers.
4. A key element in the creation of prebiotic materials is replication, and it has been proposed that possibly the most important role of clays in biological evolution could be the directing effect of patterns of crystal imperfections in clay minerals on adsorption and reaction processes on their surfaces, analogous to the "information transfer" inherent in crystal growth.

Again, in studies of this kind, it will be important to establish the site of the adsorption and reaction processes, whether they occur on edge or on face surfaces, or possibly on both.

References

GENERAL

See also the monograph in Appendix V, reference B-32.

1. MacEwan, D. M. C. (1962), Interlamellar reactions of clays and other substances, *Clays, Clay Minerals,* **9,** 431–443.
2. Cowan, C. T., and White, D. (1962, 1963), Adsorption by organo-clay complexes, *Clays, Clay Minerals,* **9,** 459–467 (I); **10,** 226–234 (II).
3. Brindley, G. W., Suito, E., and Koizumi, M. (1972), Clay-organic complexes (Report of U.S.-Japan seminar), *Clays, Clay Minerals,* **20,** 189–191.

SPECTRA

4. McAtee, J. L., Jr. (1963), Organic cation exchange on montmorillonite as observed by ultraviolet analysis, *Clays, Clay Minerals,* **10,** 153–162.
5. Little, L. H. (1966), *Infrared Spectra of Adsorbed Species,* Academic, New York.
6. Mortland, M. M. (1970), Clay-organic complexes and interactions, *Advan. Agron.,* **22,** 75–117.
7. Farmer, V. C. (1971), The characterization of adsorption bands in clays by infrared spectroscopy, *Soil Sci.,* **112,** 62–68.

ANIONIC, CATIONIC, AND NONIONIC SURFACTANTS

8. Law, J. P., and Kunze, G. W. (1966), Reactions of surfactants with montmorillonite: adsorption mechanism, *Soil Sci. Soc. Am. Proc.,* **30,** 321.
9. Schott, H. (1968), Deflocculation of swelling clays by nonionic and anionic detergents, *J. Colloid Interface Sci.,* **26,** 133–139.
10. Hower, W. F. (1970), Adsorption of surfactants on montmorillonite, *Clays, Clay Minerals,* **18,** 97–106.

ANIONS

11. Freudenberg, K., and Maitland, P. (1934), Der Quebracho Gerbstoff, *Ann. Chem. Liebigs,* **510,** 193–205.
12. Gray, G. R., Neznayko, M., and Gilkeson, P. W. (1952), Solidification of lime treated muds at high temperature, *World Oil,* **134,** 101–106.
13. Darley, H. C. H. (1956), The use of barium hydroxide in drilling muds, *J. Petrol. Technol., Trans. A.I.M.E.,* **207,** 252–255.
14. Browning, W. C. (1955), Lignosulfonate stabilized emulsions in oil well drilling fluids, *J. Petrol. Technol.,* **7,** 9–15.

CATIONS

15. Smith, C. R. (1934), Base exchange reaction of bentonites and salts of organic bases, *J. Am. Chem. Soc.,* **56,** 1561–1563.

16. Gieseking, J. E., and Jenny, H. (1936), Behavior of polyvalent cations in base exchange, *Soil Sci.*, **42,** 273–280.
17. Gieseking, J. E. (1939), Mechanism of cation exchange in the montmorillonite-beidellite-nontronite type of clay minerals, *Soil Sci.*, **47,** 1–14.
18. Hendricks, S. B. (1941), Base exchange of the clay mineral montmorillonite for organic cations and its dependence upon adsorption due to van der Waals forces, *J. Phys Chem.*, **45,** 65–81.
19. Jordan, J. W. (1949), Organophilic bentonites, I—Swelling in organic liquids, *J. Phys. & Colloid Chem.*, **53,** 294–306.
20. Jordan, J. W., Hook, B. J., and Finlayson, C. M. (1950), Organophilic bentonites, II—Organic liquid gels, *J. Phys. & Colloid Chem.*, **54,** 1196–1208.
21. Franzen, P. (1954), X-ray analysis of an adsorption complex of montmorillonite with cetyltrimethylammonium bromide (Lissolamine), *Clay Minerals Bull.*, **2,** 223–225.
22. Barrer, R. M., and Reay, J. S. S. (1957), Sorption and intercalation by methylammonium montmorillonites, *Trans. Faraday Soc.*, **53,** 1253–1261.
23. Cowan, C. T., and White, D. (1958), The mechanism of exchange reactions occurring between sodium montmorillonite and various *n*-primary aliphatic amine salts, *Trans. Faraday Soc.*, **54,** 691–697.
24. Slabaugh, W. H., and Kupka, F. (1958), Organic cation exchange properties of calcium montmorillonite, *J. Phys. Chem.*, **62,** 599–601.
25. Weiss, A., and Kantner, I. (1960), Uber eine einfache Möglichkeit zur Abschätzung der Schichtladung glimmerartiger Schichtsilikate, *Z. Naturforsch.*, **15b,** 804–807.
26. Bergman, K., and O'Konski, C. T. (1963), A spectroscopic study of methylene blue monomer, dimer, and complexes with montmorillonite, *J. Phys. Chem.*, **67,** 2169–2177.
27. McAtee, J. L., Jr. (1962), Cation exchange of organic compounds on montmorillonite in organic media, *Clays, Clay Minerals*, **9,** 444–450.
28. Greenland, D. J., and Quirk, J. P. (1962), Adsorption of 1-*n*-alkylpyridinium bromides by montmorillonites, *Clays, Clay Minerals*, **9,** 484–499.
29. Brindley, G. W., and Hoffman, R. W. (1962), Orientation and packing of aliphatic chain molecules on montmorillonite, *Clays, Clay Minerals*, **9,** 546–556.
30. Davidson, D. T., Demirel, T., and Rosauer, E. A. (1962), Mechanism of stabilization of cohesive soils by treatment with organic cations, *Clays, Clay Minerals*, **9,** 585–591.
31. Rosauer, E. A., Handy, R. L., and Demirel, T. (1963), X-ray diffraction studies of organic cation-stabilized bentonite, *Clays, Clay Minerals*, **10,** 235–243.
32. Weiss, A. (1963), Mica-type layer silicates with alkylammonium ions, *Clays, Clay Minerals*, **10,** 191–224.
33. Diamond, S., and Kinter, E. B. (1963), Characterization of montmorillonite saturated with short chain amine cations. I: Interpretation of basal spacing measurements; II: Interlayer surface coverage by the amine cation, *Clays, Clay Minerals*, **10,** 163–173 (I); 174–190 (II).
34. Walker, G. F. (1967), Interactions of *n*-alkylammonium ions with mica-type layer lattices, *Clay Minerals*, **7,** 129–143.
35. van Olphen, H. (1968), Modification of the clay surface by pyridine-type compounds, *J. Colloid Interface Sci.*, **28,** 370–375.
36. Lagaly, G., Fitz, St., and Weiss, A. (1975), Kink block structures in clay-organic complexes, *Clays, Clay Minerals*, **23,** 45–54.

POLAR MOLECULES

37. MacEwan, D. M. C. (1944), Identification of the montmorillonite group of minerals by X-rays, *Nature*, **154,** 577–578.

38. MacEwan, D. M. C. (1948), Complexes of clays with organic compounds, I, *Trans. Faraday Soc.*, **44,** 349–367.
39. Bradley, W. F. (1945), Molecular associations between montmorillonite and some polyfunctional organic liquids, *J. Am. Chem. Soc.*, **67,** 975–981.
40. Glaeser, R. (1949), On the mechanism of formation of montmorillonite-acetone complexes, *Clay Minerals Bull.*, **3,** 88–90.
41. Greene-Kelly, R. (1955), Sorption of aromatic organic compounds by montmorillonite, Part I—Orientation studies; Part II—Packing studies with pyridine, *Trans Faraday Soc.*, **51,** 412–424 (I), 425–430 (II).
42. Haxaire, A., and Bloch, J. M. (1956), Sorption de composés organiques azotés par la montmorillonite. Etude de mécanisme, *Bull. Groupe Franc. Argiles,* **8,** 17–22.
43. Brindley, G. W., and Rustom, M. (1958), Adsorption and retention of an organic material by montmorillonite in the presence of water, *Am. Mineralogist,* **43,** 627–640.
44. Hoffman, R. W., and Brindley, G. W. (1960), Adsorption of nonionic aliphatic molecules from aqueous solutions on montmorillonite, Clay-organic studies, II, *Geochim. Cosmochim. Acta,* **20,** 15–29.
45. Tensmeyer, L. G., Hoffman, R. W., and Brindley, G. W. (1960), Infrared studies of some complexes between ketones and calcium montmorillonite, Clay-organic studies, III, *J. Phys. Chem.*, **64,** 1655–1662.
46. Garrett, W. G., and Walker, G. F. (1962), Swelling of some vermiculite-organic complexes in water, *Clays, Clay Minerals,* **9,** 557–567.
47. van Olphen, H., and Deeds, C. T. (1962), The stepwise hydration of clay-organic complexes, *Nature,* **194,** 176–177.
48. Tettenhorst, R., Beck, C. W., and Brunton, G. (1960), Montmorillonite-polyalcohol complexes—Part I, *Clays, Clay Minerals,* **9,** 500–519.
49. Brunton, G., Tettenhorst, R., and Beck, C. W. (1963), Montmorillonite-polyalcohol complexes—Part II, *Clays, Clay Minerals,* **11,** 105–116.
50. Szanto, F., and Varkonyi, B. (1963), Concerning the sediment volume of montmorillonite suspensions. Part I: Sediment volumes of montmorillonite suspensions in water-alcohol mixtures, *Koll. Zeit.*, **191,** 123–230.
51. van Assche, J. B., van Cauwelaert, F. H., and Uytterhoeven, J. B. (1972), Sorption of organic polar gases on montmorillonite, *Proc. Intern. Clay Conf.*, Madrid, **2,** 327–338.

POLYELECTROLYTES AND NONIONIC POLYMERS

52. Mattson, S. (1932), The laws of soil colloidal behavior, VII—Proteins and proteinated complexes, *Soil Sci.*, **23,** 41–72.
53. Ensminger, L. E., and Gieseking, J. E. (1941), Adsorption of proteins by montmorillonite clays and its effect on base exchange capacity, *Soil Sci.*, **51,** 125–132.
54. Talibudeen, O. (1955), Complex formation between montmorillonite clays and amino-acids and proteins, *Trans. Faraday Soc.*, **51,** 582–590.
55. Hauser, E. A. (1950), Modified gel forming clay and process of producing same, U.S. Patent, No. 2,531, 429.
56. Ruehrwein, R. A., and Ward, D. W. (1952), Mechanism of clay aggregation by polyelectrolytes, *Soil Sci.*, **73,** 485–492.
57. Michaels, A. S., and Morelos, O. (1955), Polyelectrolyte adsorption by kaolinite, *Ind. Eng. Chem.*, **47,** 1801–1809.
58. Warkentin, B. P., and Miller, R. D. (1948), Conditions affecting formation of the montmorillonite-polyacrylic acid bond, *Soil Sci.*, **85,** 14–18.
59. Montgomery, R. S., and Hibbard, B. B. (1955), Theoretical aspects of the soil conditioning activity of polymers, *Soil Sci.*, **79,** 283–292.

60. Foster, W. R., and Waite, J. M. (April, 1956), Adsorption of polyoxyethylated phenols on some clay minerals, Symposium on Chemistry in the Exploration and Production of Petroleum, Division of Petroleum, ACS, Dallas, 129th Meeting (preprinted).
61. Mortensen, J. L. (1962), Adsorption of hydrolyzed polyacrylonitrile on kaolinite, *Clays, Clay Minerals,* **9,** 530–545.
62. Greenland, D. J. (1963), Adsorption of polyvinylalcohols by montmorillonite, *J. Colloid Sci.,* **18,** 647–664.
63. Dekking, H. G. G. (1964), Preparation and properties of some polymer-clay compounds, *Clays, Clay Minerals,* **12,** 603–616.
64. Healy, T. W., and La Mer, V. K. (1964), The energetics of flocculation and redispersion by polymers, *J. Colloid Sci.,* **19,** 323–332.
65. La Mer, V. K., and Healy, T. W. (1963), Adsorption-flocculation reaction of macromolecules at the solid-liquid interface, *Reviews of Pure and Applied Chemistry,* **13,** 112–133.
66. Hesselink, F. Th. (1971), On the theory of the stabilization of dispersions by adsorbed macromolecules—I, *J. Phys. Chem.,* **75,** 65–71.
67. Hesselink, F. Th., Vrij, A., and Overbeek, J. Th. G. (1971), On the theory of the stabilization of dispersions by adsorbed macromolecules—II, *J. Phys. Chem.,* **75,** 2094–2103.

EMULSIONS

68. Clayton, W. (1954), *The Theory of Emulsions and Their Technical Treatment,* Fifth Edition, C. G. Sumner, Editor, Chemical Publishing Co., Inc., New York.
69. Becher, P. (1957), *Emulsions, Theory and Practice,* American Chemical Society Monograph Series, Reinhold, New York.
70. van den Tempel, M. (1953), *Stability of Oil-in-Water Emulsions,* Rubberstichting, Delft, The Netherlands, Communication No. 225. (See also *Rec. Trav. Chim.,* **72,** 419–432 (I), 433–441 (II), 442–461 (III).
71. Proc. *2nd Intern. Congr. Surface Activity,* Vol. I, 417–481.
72. van Olphen, H. (1951), A tentative method for the determination of the base exchange capacity of small samples of clay minerals, *Clay Minerals Bull.,* **1,** 169–170.

PARTICLE INTERACTION

73. Kruyt, H. R., and van Selms, F. G. (1943), The influence of a third phase on the rheology of suspensions, *Rec. Trav. Chim.,* **62,** 415–426.
74. Howe, P. G., Benton, D. P., and Puddington, I. E. (1955), The nature of the interaction forces between particles in suspensions of glass spheres in organic liquid media, *Can. J. Chem.,* **33,** 1189–1196.
75. Farnand, J. R., Smith, H. M., and Puddington, I. E. (1961), Spherical agglomeration of solids in liquid suspension, *Can. J. Chem. Eng.,* **39,** 94–97.
76. Varkonyi, B., and Szanto, F. (1964), Concerning the sediment volume of montmorillonite suspensions. Part II: The effect of water on the sediment volume in organic fluids, *Koll. Zeit.,* **199,** 52–56.
77. Poland, D. C., and Scheraga, H. A. (1966), Hydrophobic binding and micelle stability: The influence of ionic head groups, *J. Colloid Interface Sci.,* **21,** 273.
78. Tanford, C. (1975), *The Hydrophobic Effect,* Wiley, New York.

INTERCALATION AND INTERSALATION IN KAOLINITES

79. Garrett, W. G., and Walker, G. F. (1959), The cation exchange capacity of hydrated halloysite and the formation of halloysite-salt complexes, *Clay Minerals Bull.,* **4,** 75–80.

80. Andrew, R. W., Jackson, M. L., and Wada, K. (1960), Intersalation as a technique for differentiation of kaolinite from chloritic minerals by X-ray diffraction, *Soil Sci. Soc. Am. Proc.*, **24,** 422–424.
81. Wada, K. (1959), Oriented penetration of ionic compounds between the silicate layers of halloysite, *Am. Mineralogist,* **44,** 153–165.
82. Wada, K. (1959), An interlayer complex of halloysite with ammonium chloride, *Am. Mineralogist,* **44,** 1237–1247.
83. Wada, K. (1961), Lattice expansion of kaolin minerals by treatment with potassium acetate, *Am. Mineralogist,* **46,** 78–91.
84. Wada, K. (1963), Quantitative determination of kaolinite and halloysite by NH_4Cl retention, *Am. Mineralogist,* **48,** 1286–1299.
85. Wada, K. (1965), Intercalation of water in kaolin minerals, *Am. Mineralogist,* **50,** 924–941.
86. van Olphen, H., and Deeds, C. T. (1963), *Proc. Intern. Clay Conf.*, Stockholm, **2,** 380–381 (short communication).
87. Deeds, C. T., van Olphen, H., and Bradley, W. F. (1966), Intersalation and interlayer hydration of minerals of the kaolinite group, *Proc. Intern. Clay Conf.*, Jerusalem, **2,** 183–199; **1,** 295–296.
88. Ledoux, R. L., and White, J. L. (1964), Infrared study of the OH-groups in expanded kaolinite, *Science,* **143,** 244–246.
89. Ledoux, R. L., and White, J. L. (1966), Infrared studies of hydrogen bonding between kaolinite surfaces and intercalated potassium-acetate, hydrazine, formamide and urea, *J. Colloid Interface Sci.*, **21,** 127–152.
90. Cruz, M., Laycock, A., and White, J. L. (1969), Perturbation of OH groups in intercalated kaolinite donor-acceptor complexes. I. Formamide-, methylformamide-, and dimethyl formamide-kaolinite complexes, *Proc. Intern. Clay Conf.*, Tokyo, **1,** 775–789.
91. Weiss, A. (1961), Eine Schichteinschlussverbindung von Kaolinit mit Harnstoff, *Angew. Chem.*, **73,** 736.
92. Weiss, A. (1963), A secret of Chinese porcelain manufacture, *Angew. Chem., Int. Ed. Engl.*, **2,** 697–703.
93. Weiss, A., Thielepape, W., Göring, G., Ritter, W., and Schäfer, H. (1963), Kaolinit Einlagerungsverbindungen, *Proc. Intern. Clay Conf.*, Stockholm, **1,** 287–305.
94. Weiss, A., Thielepape, W., and Orth, H. (1966), Kaolinit-Einlagerungsverbindungen, *Proc. Intern. Clay Conf.*, Jerusalem, **1,** 277–294.
95. Range-J., K., Range, A., and Weiss, A. (1969), Fire-clay type kaolinite or fire-clay mineral? Experimental classification of kaolinite-halloysite minerals, *Proc. Intern. Clay Conf.*, Tokyo, **1,** 3–13.
96. Wiewiora, A., and Brindley, G. W. (1969), Potassium acetate intercalation in kaolinite and its removal: Effect of material characteristics, *Proc. Intern. Clay Conf.*, Tokyo, **1,** 723–733.
97. Olejnik, S., Aylmore, L. A. G., Posner, A. M., and Quirk, J. P. (1968), Infrared spectra of kaolin mineral-dimethyl sulfoxide complexes, *J. Phys. Chem.*, **72,** 241–249.
98. Olejnik, S., Posner, A. M., and Quirk, J. P. (1971), The infrared spectra of interlamellar kaolinite-amide complexes. I: The complexes of formamide, *N*-methyl formamide and dimethyl formamide, *Clays, Clay Minerals,* **19,** 83–94.
99. Olejnik, S., Posner, A. M., and Quirk, J. P. (1971), The infrared spectra of interlamellar kaolinite-amide complexes. II: Acetamide, *N*-methylacetamide and dimethylacetamide, *J. Colloid Interface Sci.*, **37,** 536–547.

APPLICATIONS

100. Jordan, J. W. (1963), Organophilic clay base thickeners, *Clays, Clay Minerals,* **10,** 299–308.
101. Nahin, P. G. (1963), Perspectives in applied organo-clay chemistry, *Clays, Clay Minerals,* **10,** 257–271.
102. Boner, C. J. (1954), *Manufacture and Application of Lubricating Greases,* Reinhold, New York.
103. Bollo and Woods, (1962), *Advances in Petroleum Chemistry and Refining,* Kobe, K. A., and McKetta, J. J., Editors, Interscience, New York.
104. van Olphen, H. (1963), Clay-organic complexes and the retention of hydrocarbons by source rocks, *Proc. Intern. Clay Conf.,* Stockholm, **1,** 307–317.
105. Rankine, B. C., and Emerson, W. W. (1963), Wine clarification and protein removal by bentonite, *J. Sci. Food, Agr.,* **14,** 685–689.
106. Mortimer, J. V., and Gent, P. L. (1964), Use of organo-clays as gas chromatographic stationary phase, *Anal. Chem.,* **36,** 754–755.
107. White, D. (1964), The structure of organic montmorillonites and their adsorption properties in the gas phase, *Clays, Clay Minerals,* **12,** 257–267.
108. van Olphen, H. (1967), Polyelectrolyte reinforced aerogels of clays—Application as chromatographic adsorbents, *Clays, Clay Minerals,* **15,** 423–436.
109. van Olphen, H. (1966), Maya Blue: A clay-organic pigment, *Science,* **154,** 645–646.

COLOR REACTIONS

110. Hasegawa, H. (1961, 1962), Spectroscopic studies on the color reaction of acid clay with amines, *J. Phys. Chem.,* **65,** 292–296, Part II—Reaction with aromatic tertiary amines, *J. Phys. Chem.,* **66,** 834–835.
111. Theng, B. K. G. (1971), Mechanisms of formation of colored clay-organic complexes. A review, *Clays, Clay Minerals,* **19,** 383–390.
112. Furukawa, T., and Brindley, G. W. (1973), Adsorption and oxidation of benzidine and aniline by montmorillonite and hectorite, *Clays, Clay Minerals,* **21,** 279–287.
113. Thompson, T. D., and Moll, W. F., Jr. (1973), Oxidative power of smectites measured by hydroquinone, *Clays, Clay Minerals,* **21,** 337–350.
114. Lahav, N., and Anderson, D. M. (1973), Montmorillonite-benzidine reactions in the frozen and dry states, *Clays, Clay Minerals,* **21,** 137.
115. Besson, G., Estrade, H., Gatineau, L., Tchoubar, C., and Mering, J. (1975), A kinetic survey of the cation exchange and the oxidation of a vermiculite, *Clays, Clay Minerals,* **23,** 318–322.

CLAY-ORGANIC REACTIONS

116. Deuel, H. (1952), Organic derivatives of clay minerals, *Clay Minerals Bull.,* **1,** 205–213.
117. Brown, G., Greene-Kelly, R., and Norrish, K. (1952), Organic derivatives of montmorillonite, *Clay Minerals Bull.,* **1,** 214–220.
118. Uytterhoeven, J. (1960), Organic derivatives of silicates and aluminosilicates, *Silicates Inds.,* **25,** 403–409.
119. Uytterhoeven, J. (1962), Determination of the surface hydroxyl groups of kaolinite by organometallic compounds (CH_3MgI and CH_3Li), *Bull. Groupe Franç. Argiles,* **13,** 69–76.

SURFACE ACIDITY

120. Benesi, H. A. (1956), Acidity of catalyst surfaces—I. Acid strength from colors of adsorbed indicators, *J. Am. Chem. Soc.*, **78,** 5490–5494.
121. Benesi, H. A. (1957), Acidity of catalyst surfaces—II. Amine titration using Hammett indicators, *J. Phys. Chem.*, **61,** 970–973.
122. Fowkes, F. M., Benesi, H. A., Ryland, L. B., Sawyer, W. M., Detling, K. D., Loeffler, E. S., Folckemer, F. B., Johnson, M. R., Sun, Y. F. (1960), Clay-catalyzed decomposition of insecticides, *Agricultural and Food Chemistry*, **8,** 203–210.
123. Fripiat, J. J. (1964), Surface properties of alumino-silicates, *Clays, Clay Minerals*, **12,** 327–358.
124. Mortland, M. M., and Raman, K. V. (1968), Surface acidity of smectites in relation to hydration, exchangeable cation and structure, *Clays, Clay Minerals*, **16,** 393–398.
125. Solomon, D. H. (1968), Clay minerals as electron acceptors and/or electron donors in organic reactions, *Clays, Clay Minerals*, **16,** 31–40.
126. Lloyd, M. K., and Conley, R. F. (1970), Adsorption studies on kaolinites, *Clays, Clay Minerals*, **18,** 37–46.
127. Conley, R. F., and Lloyd, M. K. (1971), Adsorption studies on kaolinites—II: Adsorption of amines, *Clays, Clay Minerals*, **19,** 273–282.
128. Fripiat, J. J. (1972), Time concepts in surface chemistry, *Proc. Intern. Clay Conf.*, Madrid, **2,** 219–232.
129. Hawthorne, D. G., and Solomon, D. H. (1972), Catalytic activity of sodium kaolinites, *Clays, Clay Minerals*, **20,** 75–78.
130. Solomon, D. H., and Murray, H. H. (1972), Acid-base interactions and the properties of kaolinite in non-aqueous media, *Clays, Clay Minerals*, **20,** 135–141.
131. Bailey, G. W., and Karickhoff, S. W. (1973), An ultraviolet spectroscopic method for monitoring surface acidity of clay minerals under varying water content, *Clays, Clay Minerals*, **21,** 471–478.
132. Frenkel, M. (1974), Surface acidity of montmorillonite, *Clays, Clay Minerals*, **22,** 435–442.
133. Helsen, J. A., Drieskens, R., and Chaussidon, J. (1975), Position of exchangeable cations in montmorillonite, *Clays, Clay Minerals*, **23,** 334–335.

ORIGIN OF LIFE

134. Bernal, J. D. (1951), *The Physical Basis of Life*, Routledge and Kegan Paul, London.
135. Oparin, A. I. (1957), *The Origin of Life on Earth*, Academic, New York.
136. Cairns-Smith, A. G. (1965), The origin of life and the nature of the primitive gene, *J. Theor. Biol.*, **10,** 53–88.
See also Appendix V, reference B-36.
137. Calvin, M. (1969), *Chemical Evolution*, Clarendon Press, London.
138. Friedlander, H. Z., and Frink, C. R. (1964), Organized polymerization. III. Monomers in montmorillonite, *Polymer Letters*, **2,** 475–479.
139. Blumstein, A. (1965), Polymerization of adsorbed monolayers. I. Preparation of the clay-polymer complex, *J. Polymer Sci.*, **A3,** 2653–2664; II. Thermal degradation of the inserted polymer, *J. Polymer Sci.*, **A3,** 2665–2672.
140. Blumstein, A., and Billmeyer, F. W., Jr. (1966), Polymerization of adsorbed monolayers. III. Preliminary structure studies in dilute solution of the inserted polymers, *J. Polymer Sci.*, **A4,** 465–474.

141. Blumstein, A., and Blumstein, R. (1967), Association in two-dimensionally cross-linked poly(methylmethacrylate), *Polymer Letters,* **5,** 691–696.
142. Blumstein, A., Blumstein, R., and Vanderspurt, T. H. (1969), Polymerization of adsorbed monolayers. IV. The two-dimensional structure of insertion polymers, *J. Colloid Interface Sci.,* **31,** 236–247.
143. Solomon, D. H., and Rosser, M. J. (1965), Reactions catalyzed by minerals. I. Polymerization of styrene, *J. Appl. Polymer Sci.,* **9,** 1261–1271.
144. Solomon, D. H., and Swift, J. D. (1967), Reactions catalyzed by minerals. II. Chain termination in free radical polymerization, *J. Appl. Polymer Sci.,* **11,** 2567–2575.
145. Solomon, D. H., and Loft, B. C. (1968), Reactions catalyzed by minerals. III. The mechanisms of spontaneous interlamellar polymerizations in aluminosilicates, *J. Appl. Polymer Sci.,* **12,** 1253–1262.
146. Paecht-Horowitz, M., Berger, J., and Katchalsky, A. (1970), Prebiotic synthesis of polypeptides by heterogeneous polycondensation of amino acid adenylates, *Nature,* **228,** 636–639.
147. Theng, B. K. G. (1970), Formation of two-dimensional organic polymers on a mineral surface, *Nature,* **228,** 853–854.
148. Pezerat, H., and Mantin, I. (1967), Polymerisation cationique du styrène entre les feuillets d'une montmorillonite acide, *C. R. Acad. Sci., Paris,* **265,** 941–944.
149. Pezerat, H., and Vallet, M. (1972), Formation de polimère inséré dans les couches interlamellaires de phyllites gonflantes, *Proc. Intern. Clay Conf.*, Madrid, **2,** 419–430.
150. Pinck, L. A. (1962), Adsorption of proteins, enzymes, and antibiotics by montmorillonite, *Clays, Clay Minerals,* **9,** 520–529.
151. Degens, E. T., Mathejar, J., and Jackson, T. A. (1970), Template catalysis: asymmetric polymerization of amino-acids on clay minerals, *Nature,* **227,** 492–493.
152. Cloos, P., Calicis, B., Fripiat, J. J., and Makay, K. (1966), Adsorption of amino-acids and peptides by montmorillonite I: Chemical and X-ray studies; II: Identification of adsorbed species and decay products by infrared spectroscopy, *Proc. Intern. Clay Conf.,* Jerusalem, **1,** 223–232; 233–246.
153. Fripiat, J. J., Poncelet, G., van Assche, A. T., and Mayaudon, J. (1972), Zeolite as catalysts for the synthesis of amino acids and purines, *Clays, Clay Minerals,* **20,** 331–340.
154. Gatineau, L., and Mering, J. (1966), Relation ordre-désordre dans les substitutions isomorphes des micas, *Bull. Groupe Franç. Arg.,* **18,** 67–74.
155. Alcover, J. F., Gatineau, L., and Mering, J. (1973), Exchangeable cation distribution in Nickel- and Magnesium-Vermiculites, *Clays, Clay Minerals,* **21,** 131–136.
156. Besson, G., Mifsud, A., Tchoubar, C., and Mering, J. (1974), Order and disorder relations in the distribution of the substitutions in smectites, illites and vermiculites, *Clays, Clay Minerals,* **22,** 379–384.
157. Thompson, T. D., and Tsunashina, A. (1973), The alteration of some aromatic amino acids and polyhydric phenols by clay minerals, *Clays, Clay Minerals* **21,** 351–362.

CHAPTER TWELVE

Electrokinetic and Electrochemical Properties of Clay-Water Systems

Knowledge of the electrokinetic and electrochemical properties of clay-water systems is important in several fields of clay technology—for example, in the interpretation of electric resistivity and potential logs in boreholes, of electrical effects accompanying ground-water flow, or of acidity measurements in soils with the pH meter.

Electric properties of clay-water systems are determined principally by the structure of the electric double layers on the clay surfaces. Although the negative-face double layer probably plays the major part, possible contributions of the edge double layer should not be overlooked in the interpretation of electrochemical and electrokinetic behavior of the clay systems.

The electrochemistry and electrokinetics of dispersed systems in general have been subjects of considerable confusion and controversy in the literature. As mentioned earlier, the significance of the zeta potential as a quantitative stability criterion has been overestimated in the classical literature, and its computation from electrokinetic data has been oversimplified. Additional uncertainties in the interpretation of these data must be resolved in dealing with the irregular capillary systems found in shaly sands or shales, which are of interest in electric-logging techniques. The geometrical effects in such capillary systems are very difficult to evaluate.

Considerable difference of opinion has prevailed in the interpretation of electrochemical-potential measurements on clay suspensions and on clay "membranes" or shales, but in recent years a clearer understanding has been reached.

Evidently, electrokinetics and electrochemistry of dispersed systems are involved subjects which do not readily lend themselves to an elementary treatment. Nevertheless, some of the factual knowledge will be presented in this chapter, and an attempt will be made to indicate the pitfalls in interpretations. Some selected references will be given as entries to the recent literature.

I. ELECTROKINETIC PHENOMENA (12)

Electrokinetic phenomena occur when two phases move with respect to each other while the interface is the seat of an electric double layer. In this motion, for example, between a solid and a liquid phase, a thin layer of liquid adheres to the solid surface, and the shearing plane between liquid and solid is located in the liquid at some unknown distance from the solid surface. Part of the counter-ion atmosphere therefore moves with the solid, and part moves with the liquid. The electric double-layer potential at the shearing plane is called the electrokinetic or zeta potential.

The following electrokinetic phenomena are distinguished:

Electrophoresis. A suspended charged particle moves in an applied electric field.

Dorn potentials or *sedimentation potentials* and *centrifugation potentials.* An electric potential is created when charged particles are forced to move through a liquid by gravity or by centrifugal forces, respectively.

Electrosmosis (*electro-osmosis* or *electroendosmosis*). A liquid flows along a charged surface when an electric field is applied parallel to the surface in the liquid.

Streaming potential. An electric potential is created when a liquid is forced to flow along a charged surface.

In the theoretical treatment of electrokinetic phenomena, electric currents must be evaluated, and hence the conductance of the systems must be considered. In the liquid close to the surface, the conductance will be abnormally high because of the high ion concentration in the counter-ion atmosphere. This so-called *double-layer conductance* (or more commonly *surface conductance*) will be discussed first.

A. Surface Conductance (1–11)

Consider a capillary of length L and diameter d, filled with a fluid with a specific conductance of K_w ohm^{-1} cm^{-1}. The contribution of the bulk fluid conductance to the total conductance of the capillary is then $k_w = [\frac{1}{4} \pi d^2/L]K_w$. The difference between the conductance of the capillary and that of the bulk fluid is defined as the surface conductance, which is parallel with the bulk conductance. The surface conductance is $k_s = (\pi d/L)K_s$, in which K_s is the specific surface conductance, measured in ohm^{-1}.

For a sufficiently accurate measurement, the surface conductance should be of the same order as the bulk conductance in the capillary. It is desirable therefore that $\frac{1}{4}\, dK_w \lessapprox K_s$. Or, since the surface conductance is often only about 10^{-9} ohm^{-1}, a capillary with a radius of about 20 μm would be required for nearly pure water with a conductivity of 10^{-6} ohm^{-1} cm^{-1}. At higher electrolyte concentrations, proportionally smaller capillaries will be needed for accurate results. The experimental difficulties inherent in the work with very narrow, well made capillaries has seriously hampered the collection of accurate data on surface conductance, especially at higher electrolyte concentrations. In many applications, however, the contribution of surface conductance can be neglected with respect to the bulk conductance, unless very low electrolyte concentrations and very narrow capillaries are involved.*

In order to measure surface conductance, the geometry of the system must be known, and most data in the literature refer to glass from which capillaries or other systems of known geometry can be made. Values of the order of 10^{-9} ohm^{-1} have been obtained for glass and extremely dilute electrolyte solutions.

Surface conductance data on clays have been derived for montmorillonites from the conductance of homogeneous bentonite gels in pure water where it is assumed that the geometry of the gel network can be represented by a cubic card house of plates (1). The values obtained were somewhat higher, but still of the same order of magnitude as those for the surface conductance of glass. The surface conductance measures the product of the number of counter-ions per cm^2 of the surface and the average mobility of these ions. Since the charge density of the bentonite surfaces is known the average mobility of the counter-ions can be computed from the surface conductance. For sodium bentonite an average mobility of 0.55 times the infinite dilution mobility of sodium ions was found; for calcium bentonite it was 0.19 times the infinite dilution mobility of calcium ions. Since the average is taken over the ions on both the external and the interlayer surfaces it is still possible that the mobility of the ions on the exterior surfaces is higher than that of the interlayer cations.

The effect of electrolytes on the surface conductance of clays cannot be evaluated from conductance measurements of gels. The geometry of the gel is considerably affected by the addition of small amounts of electrolytes, as demonstrated by the sharp drop of the yield stress, and can no longer be represented by the cubic card house. Moreover, anion exchange may occur at the edges and would result in a change in composition and conductivity of the bulk solution.

* In dc and in low-frequency ac measurements an electrosmotic contribution should be added.

In the absence of experimental data on the effect of electrolytes on surface conductance, the following theoretical notes are of interest:

For clay surfaces, the density of charge is independent of the electrolyte concentration; therefore, the specific surface conductance changes only if the average mobility of the ions changes. Possibly it becomes lower because of the compression of the double layer, which might result in a greater retarding effect by the oppositely charged surface on the counter-ions. For surfaces of constant potential (glass, sand), the density of charge increases with increasing electrolyte concentration. Hence an increase in surface conductance with increasing electrolyte concentration can be expected, although the increase may be partly offset by a decreased average mobility.

Surface conductance may play an important role in the conductance of capillary systems such as porous rocks, in which very narrow capillaries often occur. Knowledge of the conductivity of rocks as a function of electrolyte concentration, which is obtained empirically for a variety of rock types of varying porosity, is used in the interpretation of resistivity logs of boreholes. On the basis of empirical correlations, the porosity and type of rock encountered in the borehole can be derived from the resistivity log, particularly when combined with the information gained from the "self-potential" logs which will be discussed later.

A theoretical interpretation of the conductivity data for porous formations is extremely difficult. In the first place, little is known about the magnitude of the surface-conductance contribution and its change with increasing electrolyte concentration. The pore geometry, which affects the total conductance of the porous formation, is unknown, and it may change with changing electrolyte concentration—for example, owing to the flocculation of clays in the pores. In addition, cation and anion exchange reactions with the added salt would significantly alter the composition of the bulk fluid in the pores. The problem of the interpretation of the conductance of capillary systems is therefore still in the empirical stage.

B. Electrophoresis (13, 14)

When an electric field is applied to a suspension of particles carrying a negative double-layer charge, the particles move toward the positive electrode, and the counter-ions move to the negative electrode, except those which are located between the particle surface and the "slipping plane" within the layers of water which adhere to the moving particle.

In the stationary state, which is reached a very short time after the field is applied, the particles move with a constant velocity, and the total force on the particles is zero. Therefore, in first approximation, the *electric force* on the charged particle, f_1, is equal to the *hydrodynamic frictional force* on the particle by the liquid, f_2. A detailed analysis of the problem shows that two additional forces oppose the electric force. An additional frictional force, f_3, results from the movement of water with the counter-ions, which move in a direction opposite to that of the particle. A second retarding force, f_4, is caused by a distortion of the diffuse atmosphere of counter-ions around the particle. Since the counter-ions move in an opposite direction, the double layer in front of the particle must be constantly restored as it is broken off behind the particle. The restoration of the double layer takes a finite time, the relaxation time; hence, on the average, the double layer will be shifted somewhat with respect to the particle and will thus exert an electric retardation force. The force f_3 is called the *electrophoretic retardation force;* f_4 is called the *relaxation force.*

The complete formula relating the electrophoretic mobility and the zeta potential is derived by the theoretical evaluation of the four forces acting on the particles. Since such an evaluation is possible only for limited cases, no formula is available which covers every situation.

Additional complications arise when the particles are nonspherical and when they carry two different double layers, as they do in clays. In general, therefore, it is advisable to report electrophoretic results in terms of the observed mobility of the particles [e.g., in cm/s per volt/cm] rather than in terms of a zeta potential computed from some of the simpler formulas.

Electrophoretic mobilities of clay particles are usually in the range of 1 to 3×10^{-4} cm/s per volt/cm.

The classical equation for the computation of the zeta potential from the electrophoretic velocity is

$$u_e = \frac{\epsilon \zeta E}{4\pi\eta} \tag{1}$$

in which E is the applied field strength, and ϵ and η are the dielectric constant and the viscosity of the medium, respectively. Correction terms containing the zeta potential appear in the corrected equation; hence the electrophoretic velocity is in general not proportional to the zeta potential, as suggested by the classical approximate equation (1).

C. Electrosmosis

When an electric double layer exists on the wall of a capillary filled with a liquid, the application of an electric field between the ends of the capillary causes flow of the liquid. When the counter-ion charge in the liquid is positive, the liquid moves toward the negative electrode; when the charge is negative, the liquid moves toward the positive electrode. This effect, the electrosmotic effect, can be treated as follows:

Consider the liquid divided in thin concentric layers parallel with the axis of the capillary. The force exerted on the ions comprised in each thin layer by the electric field is transmitted to the whole fluid layer. In the stationary state, each layer moves with a constant velocity; therefore, the force from the electric field is compensated for by the frictional force of the liquid on the layer. If the thickness of the double layer is small compared with the radius of the capillary, the linear velocity of the layers, which is zero at the slipping plane, increases rapidly across the thickness of the double layer and becomes constant throughout practically the whole diameter of the capillary. Integration of the differential equation for the two equal forces on the cylindrical layers of infinitesimal thickness yields the following equation for the linear electrosmotic velocity of the liquid plug:

$$u_{\text{e.o.}} = \frac{\epsilon \zeta E}{4\pi\eta} \tag{2}$$

in which the symbols have the same meaning as those in equation (1). In addition to the assumption of a relatively small double-layer thickness, the values of ϵ and η are assumed to have the bulk value up to the slipping plane in the double layer.

In practice, the volume rate of electrosmotic flow is often measured instead of the linear velocity. If A is the cross sectional area of the capillary, the volume rate of flow is $J_v = Au_{\text{e.o.}}$. According to Ohm's law, $E = i/AK_w$, in which i is the electric current in the capillary, and K_w is the specific conductance of the liquid. The volume rate of flow can therefore be found from

$$\frac{J_v}{i} = \frac{\epsilon\zeta}{4\pi\eta K_w} \tag{3a}$$

When the surface conductance cannot be neglected with respect to the bulk conductance of the liquid in the capillary, equation (3a) is modified to

$$\frac{J_v}{i} = \frac{\epsilon\zeta}{4\pi\eta(K_w + SK_s/A)} \tag{3b}$$

in which K_s is the specific surface conductance, and S is the circumference of a cross section of the capillary.

D. Streaming Potentials

When a pressure difference is applied between the ends of a capillary, the counter-ions are forced to flow with the flowing liquid, and a potential difference is generated between the ends of the capillary by the convection current. The potential difference creates a conductance current in the opposite direction, and a stationary state is obtained in which the convection current is equal to the conductance current. The evaluation of these currents leads to the following expression for the streaming potential (when the same assumptions are made as were made in the derivation of the formula for electrosmosis, and when only the case of laminar flow is considered):

$$\frac{E}{P} = \frac{\epsilon\zeta}{4\pi\eta K_w} \tag{4a}$$

or

$$\frac{E}{P} = \frac{\epsilon\zeta}{4\pi\eta(K_w + SK_s/A)} \tag{4b}$$

in which P is the applied pressure drop per unit length.

When the equations for electrosmosis and for streaming potentials are compared, it appears that $J_v/i = E/P$. This relation is known as the *Saxén relation* (1892).

E. Electrokinetic Phenomena in Porous Media (15–18)

The dimensions of the capillaries do not enter equations (3*a*) and (4*a*) for electrosmosis and streaming potentials. It can be shown that these equations are equally valid for porous media with intricate networks of capillaries, as long as the same conditions are fulfilled for which equations (3*a*) and (4*a*) have been derived—that is, small thickness of the double layer with respect to the diameter of the capillaries, and absence of surface conductance of any significance with respect to the bulk liquid conductance in the capillaries.

In many porous systems of practical interest, these conditions are often not fulfilled. Usually, the capillaries are very narrow, and surface conductance is not always negligible. We shall discuss the direction in which the magnitude of streaming potentials and electrosmotic fluid transport can be expected to change when the various assumptions made in the derivation of equations (3*a*) and (4*a*) are eliminated.

First, a single capillary will be considered. The assumption of constancy of the dielectric constant and of the viscosity of the medium up to the slipping plane in the double layer may not be warranted. In the region of the double layer, these constants may change owing to the strong electric field in the neighborhood of the surface. Since any change may be expected to go in the direction of a decrease of the dielectric constant and an increase of the viscosity, such a change would amount of a reduction of the streaming potential or of the electrosmotic fluid transport.

Second, if the capillary is so narrow that the double layer is appreciably distorted by the curvature of the wall, it can be shown that the streaming potential and the electrosmotic velocity become smaller than expected. For example, a decrease of about 20 percent is predicted in a capillary with a radius of 0.1 μm and 0.001N univalent electrolyte.

Finally, if surface conductance may not be neglected, the streaming potentials and electrosmotic velocities decrease according to modified equations (3*b*) and (4*b*).

Apparently, the elimination of every assumption leads to smaller streaming potentials and electrosmotic velocities than those predicted by equations (3*a*) and (4*a*). This fact applies equally to single capillaries and to porous media. In the porous media, however, some additional factors must be considered.

In the first place, if surface conductance is not negligible, it can be shown quite generally that for most types of capillary networks the streaming potentials and electrosmotic velocities are even more reduced than in single capillaries.

Such an effect can be demonstrated in a simple way by considering a narrow and a wide capillary of equal length in series. The applied pressure differential is distributed over the two capillaries in the ratio $P_1 : P_2 = 1/r_1^4 : 1/r_2^4$; therefore, the largest pressure differential is that over the narrowest capillary, in which surface conductance is more important. Hence surface conductance carries more weight in the reduction of the streaming potential than would be expected from the measured average conductance of the two capillaries in series and from the total pressure differential.

In the measurement of streaming potentials in compressible porous media, such as filter cakes, the average pore size will be reduced when the differential pressure is increased. The corrections related to small pore sizes therefore become of increasing importance with increasing pressure. The result is that the increase in streaming potential is not proportional to the pressure differential.

Since each of the corrections tends to reduce the streaming potential, it is not surprising that streaming potentials in porous formations are comparatively low. For example, the streaming potentials in filter cakes of drilling fluids are much lower than would be expected by substituting the electrophoretically determined zeta potential of the clay particles in the uncorrected streaming potential equation (4*a*).

II. ELECTROCHEMISTRY OF DISPERSED SYSTEMS (19–28)

A. Ion Activities and pH Measurements

Some confusion is apparent in the literature concerning the interpretation of electrochemical measurements in dispersions in terms of "degree of dissociation" of the double layer and the "activity" of the ions in a suspension. Since these terms are chosen in analogy with the same terms in the treatment of electrolyte solutions, their meaning in the latter systems will be discussed first.

1. Degree of Dissociation and Ion Activity in Electrolyte Solutions. Consider a cell consisting of two reversible electrodes, one of which is reversible to the cation and the other to the anion of the electrolyte contained in the cell. The emf of this cell can be computed from the electrolyte concentration by applying the thermodynamically derived Nernst equation and assuming that the electrolyte is completely dissociated into cations and anions. The measured emf appears to be smaller than the computed emf in many electrolyte solutions.

There may be two reasons for the lower apparent ion concentration: (*1*) the molecules of the electrolyte may be only partly dissociated into ions, and (*2*) electrostatic interactions between the ions may curtail their effectiveness in the electrochemical experiments.

If the fraction of the molecules which are dissociated into ions is called α (the *degree of ionization* or *dissociation*), the actual concentration of the ions of a univalent electrolyte is equal to αC, in which C is the electrolyte concentration. The effective concentration or the *activity* of the ionized portion of the molecules may amount to only a fraction f of their actual concentration. The fraction f is called the *activity coefficient*. This coefficient is a measure of the nonideality of the solution which is primarily due to electrostatic ion interaction in solution.

In general, therefore, the total effective concentration of the ions is equal to $f\alpha C$.

Hence emf measurements in solutions of known concentration C yield values for the product $f\alpha$, but such measurements cannot give information on the two constants f and α separately.

The problem of the separation of α and f has been approached by an evaluation of the magnitude of f from a theoretical treatment of electrostatic interactions of ions in solution. The Debye-Hückel theory deals with this problem. Values for f were derived from an analysis of the electric potential distribution around an ion where a "cloud" of ions of opposite sign accumulates, analogous to the diffuse layer of counter-ions around a charged sol particle. Approximate expressions were obtained for f, which could be expected to approach closest to the real values in dilute solutions. Indeed, good agreement between theory and experiment was obtained for dilute solutions of so-called *strong electrolytes* when it was assumed that under such conditions these electrolytes were completely dissociated. For another group of electrolytes, the *weak electrolytes,* limited dissociation of the molecules had to be assumed to explain the low electrochemically effective concentrations of ions in their solutions.

This analysis, with some proper refinements of the treatment, is supported by other observations and by considerations of the structure and binding energies between the atoms of the compounds forming strong or weak electrolytes. The analysis has therefore been widely accepted.

2. Determination of Ion Activities in Solutions. Although the Debye-Hückel theory gives expressions for the activities of single ions in solution, no method is known with which single-ion activities can be measured. One cannot measure the potential of a single reversible electrode which would depend on the activity of the ion to which the electrode is reversible. A cell consisting of two electrodes must be used. If the two electrodes are reversible to the anion and to the cation in the system respectively, only the product of the cation and anion activity can be derived from the measured emf. Combining a reversible electrode with a nonspecific calomel electrode with salt bridge as the reference electrode does not supply an answer either, since the potential at the liquid junction with the reference electrode cannot be evaluated without making assumptions about the single-ion activities. Also, osmotic measurements give activity products only.

Agreement between theory and experiment mentioned above has therefore been obtained for ion-activity products only and not for single-ion activities.

A splitting of experimentally determined ion-activity products into single-ion activities would be entirely arbitrary. Nevertheless, in the description and the interpretation of many experiments, the use of single-ion activities is very convenient. Although such single-ion activities would not be thermodynamically determined, they do have a physical meaning, and they can be computed in principle from electrostatic theory. Certain arbitrary agreements have therefore been adopted regarding the evaluation of single-ion activities, and the adopted values have been chosen to agree as closely as possible with the results of the Debye-Hückel theory and its refinements. In this convention, the activity is defined as previously stated.

3. The pH-Scale and Measurements of pH in Electrolyte Solutions. On the basis of the Debye-Hückel theory and our knowledge of the dissociation of weak acids, the hydrogen-ion activity, a_H, of a number of acid solutions has been calculated as accurately as possible. Such solutions have been made standards with which the hydrogen activity of other solutions might be compared. In this way, a so-called *pH scale* was originally devised, defining pH as equal to $-\log a_H$.

The determination of the pH of an unknown solution is carried out by measuring the emf of a cell consisting of a reversible hydrogen electrode (usually the glass electrode) and of a calomel electrode with a salt bridge as the nonspecific reference electrode. This electrode combination is calibrated by measuring the emf of the cell when it contains various standard solutions of known pH.

With the calomel electrode with the salt bridge, a liquid junction is introduced between the KCl solution in the salt bridge and the unknown solution. At this liquid junction, a diffusion potential is created which depends on the difference between the cation and anion mobilities and on the change of the ion activities across the liquid junction. Since these factors are likely to be different for the junction with the unknown solution and with the standard solution used for calibration, the magnitude of the liquid-junction potential contribution to the emf of the cell will be different in the two measurements. Consequently, one cannot expect to determine pH values in the unknown solution which agree exactly with the standard pH scale. The modern operating definition of pH is therefore given in terms of the value obtained by comparing the emf of a given cell containing the unknown solution with that of the same cell containing the standard solution; the possible difference in the liquid-junction potential is disregarded. In reporting pH values, therefore, it is important to describe the electrode combination used.

Possible variations in the liquid-junction potentials in the unknown solutions and the standard solutions are kept at a minimum by the use of rather concentrated KCl solutions in the salt bridge. The current in the liquid junction is then mainly carried by the potassium and chloride ions, and since their mobilities are very nearly equal, the liquid-junction potential is small. As long as the ionic strength of unknown and standard solution is small with respect to that of the concentrated KCl solution, the liquid-junction potentials will vary very little.

4. MEASUREMENT OF pH IN SOLS AND SUSPENSIONS. The pH in acid sols or suspensions is commonly measured with a pH meter, consisting of an electrode which is reversible to the hydrogen ion (e.g., a glass electrode), and a calomel electrode with salt bridge as the reference electrode. The cell is calibrated in the usual way with standard solutions of known pH. It has been mentioned earlier that this electrode combination gives pH values closely comparable with those of the calibrating solutions as long as the liquid-junction potentials in the unknown solution and in the calibrating solutions are not too different. In a sol or a suspension, the liquid-junction potential is often significantly different from that in the calibrating solutions. These differences between junction potentials for true solutions and suspensions can be demonstrated as follows:

Figures 38*a–f* show beakers containing a suspension in contact with a supernatant liquid which is supposed to be in thermodynamic equilibrium with the suspension. A reversible hydrogen electrode (glass electrode) and a

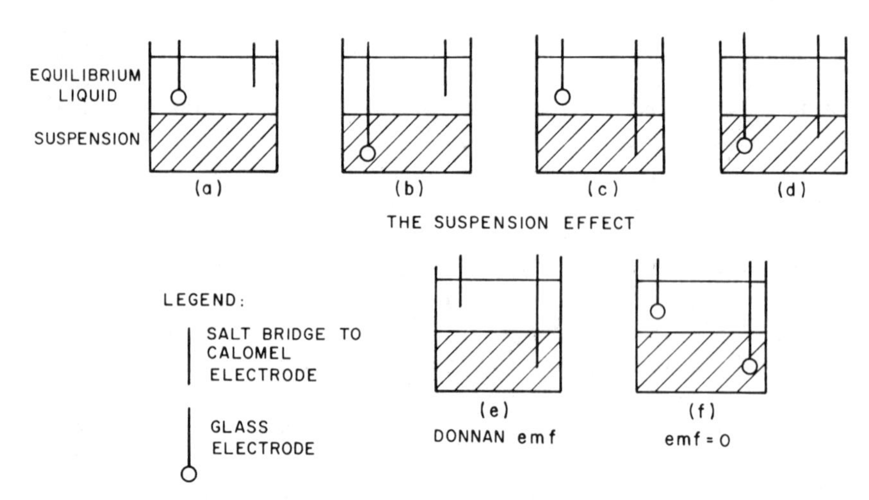

Figure 38. Suspension effect and Donnan emf.

calomel reference electrode with salt bridge are placed in the beakers in six different arrangements. In Figure 38*f*, a glass electrode in the supernatant equilibrium liquid is combined with a glass electrode in the suspension. Since equilibrium is supposed to exist between the two phases, the potential of the two reversible electrodes must be the same in both phases; otherwise work could be gained from a system in thermodynamic equilibrium, which is contrary to the Second Law. The emf of cell *f* must therefore be zero, as it is indeed observed to be.

Since the potential of the reversible hydrogen electrode is independent of its position, the emf's of cells *a* and *b* are equal and those of *c* and *d* are equal. The emf's of cells *a* and *b* on one hand appear to be substantially different from those of cells *c* and *d* on the other hand, which indicates that the potential at the calomel electrode with the salt bridge depends on whether the salt bridge is in contact with the suspension or in contact with the equilibrium liquid.

The difference between the emf's of the cells *a* or *b* with the salt bridge in the equilibrium liquid, and the emf's of the cells *c* or *d* with the salt bridge in the suspension, is called the *suspension effect*. It can easily be seen that the suspension effect is identical with the *Donnan emf,* which is defined as the emf of cell *e* consisting of two calomel electrodes with salt bridge, one immersed in the suspension and one in the equilibrium liquid.

With increasing suspension concentration, the suspension effect increases, hence the measured pH decreases. This effect is shown in Figure 39 for an acid clay suspension. The pH measured with a normal pH meter is plotted versus the suspension concentration.

Overbeek (27) has made a theoretical evaluation of the suspension effect or Donnan emf. He showed that the main contributing factor to the effect is the reduced mobility of the counter-ions owing to a retarding effect by the oppositely charged particle surfaces. As far as the theory has been tested, at least qualitative agreement between theory and experiment has been obtained (28).

Evidently, pH values of acid suspensions obtained with the pH meter cannot be compared directly with those of true solutions because of the considerable uncertainty about the magnitude of the liquid-junction potential between suspension and salt bridge.

The same applies to analogous measurements of pNa, pCa, and so on, in sodium clays, calcium clays, and so on, carried out with cells consisting of an electrode which is reversible to the specific cation, and a calomel electrode with salt bridge as the reference electrode. Therefore, "cation activities" in suspensions as derived from these measurements may be seriously in error. In the next section we shall discuss potentiometric measurements in suspension in which no liquid junction between suspension and reference electrode is involved, and we shall see what exactly can be learned from such measurements.

Before going into this question, we shall discuss briefly some types of

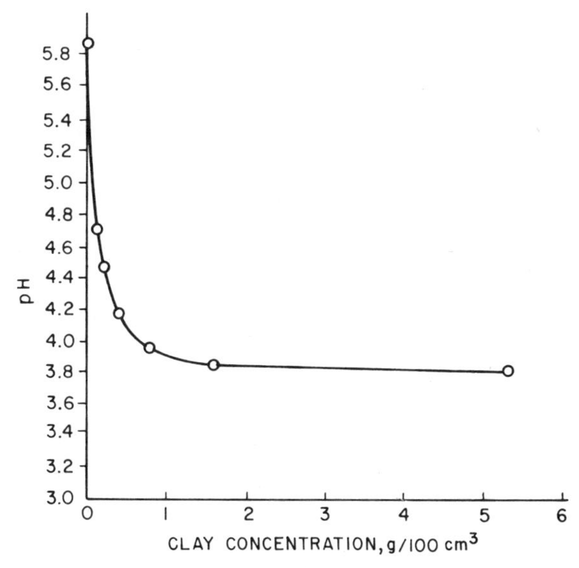

Figure 39. Values of pH for acid bentonite suspensions as a function of clay concentration. (Data from H. Pallman, *Kolloidchem. Beih.*, **30**, 334 (1930).)

electrodes which are reversible to specific cations other than hydrogen, which occur frequently as counter-ions in clay suspensions.

Marshall (22, 23, 32) has developed so-called *clay-membrane electrodes* which act as reversible electrodes for a variety of cations depending on their composition. These electrodes are made by attaching to the end of a glass tube a thin but strong clay membrane obtained by heating a thin flake of clay. The tube is filled with a salt solution of known cation activity, and the clay membrane is equilibrated with this solution to bring it in the corresponding cation form. The electrode is placed in the liquid in which the cation activity must be determined, together with a reference electrode. At the membrane of the electrode, the solutions of known and unknown cation activity are separated by the clay membrane and a membrane potential is created which is a function of the ratio of the cation activities on either side of the membrane, as will be discussed in Section B. The choice of the cation species in the solution in the electrode determines the cation for which the electrode acts as a reversible electrode. Owing to possible imperfections of the membrane, the electrode requires calibration with solutions of known cation activity.

In addition to clay-membrane electrodes, membrane electrodes of organic ion exchange materials have been developed, as well as special glass electrodes which are reversible to cations other than hydrogen.

5. "Degree of Dissociation" and "Cation Activities" in Clay Suspensions. In the literature, potentiometric measurements in clay suspensions have been reported in terms of the cation activities in these suspensions, and from these data certain conclusions have been drawn regarding the bonding energies between specific cations and the clay surface.

As pointed out in the previous section, the presence of a liquid junction between suspension and salt bridge in the measuring cell introduces an element of uncertainty. However, the liquid junction can be eliminated simply by using an anion reversible electrode as the reference electrode, although this arrangement will sometimes require the special addition of an anion to the suspension, for example, chloride ion when an Ag/AgCl electrode is chosen as the reference electrode.

For a further discussion of the meaning of such potentiometric measurements in suspensions, we shall consider a suspension in the state in which it is equilibrated with a solution, for example, a sedimented suspension in contact with its supernatant liquid, or a suspension separated from its equilibrium liquid by a neutral semipermeable membrane.

As discussed before, electrodes which are reversible to some cation or anion in the system, will have the same potentials in the suspension and in the equilibrium liquid. Therefore, the emf of the cell with the anion and cation reversible electrode combination is the same when containing the suspension and when containing the equilibrium liquid. The ion activity product may be derived from the emf value for either phase.

For the equilibrium liquid, single-ion activities may be computed from the ion-activity product according to the conventions which are based on the Debye-Hückel theory for strong electrolytes.

Hence from an emf measurement with a pair of reversible electrodes in a given suspension, one can derive the ion activities in the liquid which would be in equilibrium with that suspension. Essentially the same information may be obtained by separating and analyzing the equilibrium liquid. One may completely analyze this solution chemically, or one may be satisfied with a pH measurement in the equilibrium solution using a normal pH meter, from which reasonable values of the hydrogen-ion activity in this solution may be derived. Systems in which more than one cation is present will be discussed shortly.

Since the methods of isolating the equilibrium liquid from a suspension are not always completely satisfactory (26), the emf measurement in the suspension has some advantage over the equilibrium liquid analysis. On the other hand, the emf measurement with a pair of reversible electrodes in the suspension may require the special addition of certain anions by which the suspension will be changed. As emphasized earlier, the use of a nonspecific electrode with salt bridge is not recommended.

What can be learned about the suspension properties from the composition of the equilibrium liquid derived from direct emf measurements or from the analysis of the equilibrium liquid? In other words, how should the ion-activity product in the suspension be interpreted? In answering this question, the Debye-Hückel theory, which is applicable to the equilibrium liquid, must obviously be substituted by the double-layer theory which applies to the suspension.

The ion distribution in a diffuse double layer of the Gouy type is sketched in Figure 10*b*. The cation concentration increases from a certain minimum value far from the surface in the equilibrium liquid, in the direction of the clay surfaces in the suspension. The anion concentration far from the surface in the equilibrium liquid is maximal, and decreases in the direction of the clay surfaces in the suspension. In the equilibrium liquid, the concentrations of cations and anions are equal, whereas in the suspension, the total cation charge compensates the total anion charge and the total negative charge on the clay surfaces.

When we know the ion concentrations far from the surface in the equilibrium liquid from emf measurements as described above, and if the charge density of the surface as well as the clay concentration are known, the ion distribution in the suspension may be computed in principle on the basis of the double-layer theory, allowing for a possible overlapping of the double layers at high clay concentration. Also, the electric potential distribution (see Figure 11*b*) can be computed.

However, the results of these computations will depend on the choice of the model for the double-layer structure. One may choose the Gouy model in which the ions are considered as point charges, or one may choose a more refined model, taking into account ion sizes as well as interactions between the ions and between the ions and the solvent molecules, or between the charged surface and the solvent molecules. Another alternative is to assume that specific interactions between the clay surface and the counter-ions occur. Then the computations may be based on the Stern model, including a specific adsorption energy of the cations at the clay surface.

Using these different models as a basis for the computations, different ion distributions and different potential distributions will be calculated. There is, however, no way to check by emf measurements which model applies to the real situation since a reversible electrode does not register any potential-fluctuations in the suspension, or between the suspension and the equilibrium liquid.

It is quite possible that an electric potential difference exists between the suspension and the equilibrium liquid, owing to the special distribution of charge and potential in the suspension. However, such a potential difference, the so called *Donnan potential,* is not accessible to experimental determination (27). Only the previously defined *Donnan emf* can be measured, but this emf cannot be separated into the electrode potentials and the Donnan potential without making arbitrary assumptions or accepting models.

Any potential difference between the suspension and the equilibrium liquid will be compensated by differences in single-ion activities* in both phases. Hence the average single-ion activities in the suspension are not necessarily the same as those in the equilibrium liquid, but such differences are not accessible to determination. If one wishes to report emf measurements with reversible electrodes in suspensions in terms of single-ion activities computed on the basis of the conventions which actually apply to electrolyte solutions, this should be done with the understanding that these are the single-ion activities in the equilibrium liquid of the suspension from which the properties of the suspension may be derived on the basis of some choice of a double-layer model.

* The single-ion activities are here defined by the chemical part of the electrochemical potential of the ions, according to common practice.

Since both a double-layer model with counter-ion adsorption and a model without counter-ion adsorption could fit the emf data obtained for the suspension, no information can be obtained from such data regarding specific adsorption of counter-ions, or, as this is sometimes loosely called, a limited degree of dissociation of the clay particle.

Potentiometric measurements in suspensions of mixed ion forms of a clay have been performed using sets of an anion reversible electrode and different cation reversible electrodes (24). Again, such measurements supply essentially the same information as is obtained by analyzing the equilibrium liquid in cation-exchange equilibrium experiments. The Gouy theory predicts that the concentration ratio of two cation species of the same valence in suspension and equilibrium solution are equal. In practice, according to emf measurements in the suspension or the analysis of suspension and equilibrium liquid, these ratios are usually found to be different. Thus, the existence of specific ion effects is indicated, but from these measurements alone it is impossible to determine whether such specificity is due to differences in specific adsorption energies of the two cation species at the clay surface, or merely to differences in ion-to-ion or ion-to-solvent interactions. One has to resort to different types of experiments, that is, those in which double-layer interaction effects can be studied, to make a case for specific counter-ion adsorption (See Appendix III–E.)

Potentiometric "ion-activity" measurements in soils, or in other words, the determination of ion activities in the equilibrium liquid, are considered important for the study of the uptake of cations by plant roots (20).

B. Membrane Potentials (29–32)

Membrane potentials are created when two electrolyte solutions of different concentration* are separated by a porous medium or gel (the membrane) in which the pore walls carry an electric double layer. These potentials can be measured between two identical electrodes placed in the two solutions separated by the membrane. *Membrane potentials,* together with streaming potentials, are the components of the so-called *self-potentials* observed between an electrode located in the mud pit and another electrode suspended in the drilling fluid in a borehole opposite certain porous formations.

The creation of the membrane potential has been explained as follows: In the absence of the membrane, a liquid-junction diffusion potential occurs, the magnitude of which depends on the differences between the transport numbers of the cations and anions and on the ratio of the ion activities on

* When the membrane separates two solutions of different composition, so-called bi-ionic potentials are created.

each side of the liquid junction. The insertion of the membrane between the two liquids results in the creation of a much larger potential difference and possibly in the change of sign of the potential differences. The increase of the diffusion potential is attributed to the increase of the differences between the cation and anion transport numbers in the membrane. This explanation is supported by measurements of the transport numbers.

The cause of the changes in transport numbers is explained as follows: In a core containing clay or shale, for example, the pore walls are negatively charged, and the passage of anions through the core is seriously hampered, particularly when the pores are narrow, although those anions which are able to enter the membrane may travel faster. The cations can freely enter the capillaries, although they may move somewhat slower through the capillary, since they may be retarded by the negative charges on the wall. Consequently, a membrane potential is created, with the concentrated solution negative with respect to the dilute solution. Certain membranes are able to block the passage of one kind of ion completely, and charge transport is entirely due to the other ion. Such an *ion selective* membrane is called a *perfectly selective* or *100 percent efficient* membrane. The maximum possible potential difference between the electrolyte solutions occurs across a perfect membrane, and this maximum potential is determined by the ratio of the cation activities in the two solutions (for a negative membrane) according to the Nernst formula. The previously mentioned determination of cation activities by means of clay membranes is based on the applicability of the Nernst formula.

However, when some passage of anions through a negative membrane occurs, the membrane is less efficient and the membrane potential decreases. If such membranes with an *anion leak* are used to determine ion activities, the Nernst formula must contain a correction which is assessed by calibration of the membrane electrodes with solutions of "known" ion activities.

Recently, it was recognized that another correction must be applied to the observed membrane potentials in the calculation of ion activities because some *transport of water* takes place through the membrane. Actually, the above interpretation of the processes taking place in a membrane is somewhat naive; the creation of the membrane diffusion potential as a result of a modification of the cation and anion transport numbers is only part of the story. Since an electric potential occurs across the membrane, electrosmosis must occur. Furthermore, the separation of two solutions of different concentration will lead to the occurrence of normal osmosis. Finally, if the experiment is carried out under such conditions that a pressure difference is built up between the separated solutions, streaming potentials will be generated.

The analysis of these interrelated phenomena is extremely important for the understanding of the motion of salt and water in geological formations and in soils.

According to Staverman (29), three driving forces must in general be considered in a membrane concentration cell—the electric potential difference E, the pressure difference P, and the difference in chemical potential of all the components in the two solutions which are separated by the membrane. Owing to these forces, three simultaneous currents occur in the membrane cell—an electric current, fluid flow, and transport of components. The membrane cell is not an equilibrium system; it strives to the ultimate equilibrium in which all the forces have become zero. By the application of nonequilibrium thermodynamics to the membrane system, general relations between all forces and currents can be derived. This treatment enables one to define the minimum number of independent measurements that must be performed to describe a membrane system completely. Several investigators have attacked the problem of membrane behavior on this fundamental basis (30).

The general theory describes the separate processes by keeping one of the forces zero. When $P = 0$, relations are derived for membrane diffusion potentials; when $E = 0$, normal osmosis occurs; when the chemical-potential difference of the components is kept zero by use of the same solution on either side of the membrane, the general relation between electrosmosis and streaming potential, known as the Saxén relation, follows from the theory ($J_v/i = E/P$). (See reference 29).

The derived general relations are strictly phenomenological; they give no information about the mechanism of transport of charge, fluid and components in the cell as a function of membrane geometry, or charge density and structure of the double layer on the pore walls. Several models for the mechanism of the creation of membrane potentials, such as the one discussed in the beginning of this section, have been proposed, and ready-to-apply formulas have been derived on the basis of such models. We shall refrain from reproducing any of these formulas. Their handling and application should be left to the specialist who can judge their limitations.

References

SURFACE CONDUCTANCE AND CATION DIFFUSION

1. van Olphen, H., and Waxman, M. H. (1958), Surface conductance of sodium bentonite in water, *Clays, Clay Minerals*, **5**, 61–80.
2. van Olphen, H. (1957), Surface conductance of various ion forms of bentonite in water, *J. Phys. Chem.*, **61**, 1276–1280.
3. Gast, R. G. (1963), Relative effects of tortuosity, electrostatic attraction and increased

viscosity of water on self-diffusion rates of cations in bentonite-water systems, *Proc. Intern. Clay Conf.*, Stockholm, **1**, 251–260.

4. Gast, R. G., and East, P. J. (1964), Potentiometric, electrical conductance and self diffusion measurements in clay-water systems, *Clays, Clay Minerals*, **12**, 297–310.
5. Gast, R. G., and Spalding, G. E. (1966), Demonstration of a quantitative relationship between activity and diffusion coefficients of Na ion in bentonite-water systems, *Proc. Intern. Clay Conf.*, Jerusalem, **1**, 331–340.
6. Shainberg, I., and Kemper, W. D. (1966), Electrostatic forces between clay and cations as calculated and inferred from electrical conductivity, *Clays, Clay Minerals*, **14**, 117–132.
7. Cremers, A., van Loon, J., and Laudelout, H. (1966), Geometry effects for specific electrical conductance in clays and soils, *Clays, Clay Minerals*, **14**, 149–162.
8. Weiler, R. A., and Chaussidon, J. (1968), Surface conductivity and dielectrical properties of montmorillonite gels, *Clays, Clay Minerals*, **16**, 147–164.
9. Calvet, R., and Chaussidon, J. (1969), Diffusion des cations compensateurs dans la montmorillonite aux faibles hydratations, *Proc. Intern. Clay Conf.*, Tokyo, **1**, 635–647.
10. Lorenz, P. B. (1969), Surface conductance and electrokinetic properties of kaolinite beds, *Clays, Clay Minerals*, **17**, 223–232.
11. Shainberg, I., and Levy, Rachel (1975), Electrical conductivity of Na-montmorillonite suspensions, *Clays, Clay Minerals*, **23**, 205–210.

ELECTROKINETICS

12. Rutgers, A. J., de Smet, M., and Rigole, W. (1975), Surface conductance, see Appendix V, reference B-12, 309–341.
13. Street, N. (1956), The zeta-potential of kaolinite particles, *Australian J. Chem.*, **9**, 450–466.
14. Olivier, J. P., and Sennet, P. (1967), Electrokinetic effects in kaolinite-water systems. I. The measurement of electrophoretic mobility, *Clays, Clay Minerals*, **15**, 345–356.

ELECTROKINETIC PHENOMENA IN CAPILLARY SYSTEMS

15. Overbeek, J. Th. G., and Wijga, P. W. O. (1946), On electro-osmosis and streaming potentials in diaphragms, I, *Rec. Trav. Chim.*, **65**, 556–563.
16. Mazur, P., and Overbeek, J. Th. G. (1951), General quantitative relationships between electrokinetic effects, II, *Rec. Trav. Chim.*, **70**, 83–91.
17. Overbeek, J. Th. G. (1953), Thermodynamics of electrokinetic phenomena, *J. Colloid Sci.*, **8**, 420–427.
18. van Est, W. T., and Overbeek, J. Th. G. (1952), Electrokinetic effects in networks of capillaries, I, *Koninkl. Ned. Akad. Wetenschap., Proc.*, Ser. A, **55**, 347–362; III, *Rec. Trav. Chim.*, **72**, 97–104 (1953).

ELECTROCHEMISTRY

19. Loosjes, R. (1942), *pH Metingen in suspensies*, Thesis, Utrecht. See also Kruyt, H. R. (1952), *Irreversible Systems*, Volume I of *Colloid Science*, Amsterdam, Houston, New York, London, 185–187.
20. Schuffelen, A. C., and Loosjes, R. (1946), Importance of the ion activity of the medium and the root potential for the cation adsorption by the plant, *Koninkl. Ned. Akad. Wetenschap., Proc.*, Ser. A, **49**, 80–86.

21. Peech, M., Olsen, R. A., and Bolt, G. H. (1953), The significance of potentiometric measurements involving liquid junction in clay and soil suspensions, *Soil Sci. Soc. Am. Proc.*, **17,** 214–222.
22. Marshall, C. E. (1954), Multifunctional ionization as illustrated by the clay minerals, *Clays Clay Minerals,* **2,** 364–385.
23. Marshall, C. E. (1956), Thermodynamic, quasithermodynamic, and non-thermodynamic methods as applied to the electrochemistry of clays, *Clays, Clay Minerals,* **4,** 288–300.
24. Davis, L. E. (1955), Ion pair activities in bentonite suspensions, *Clays, Clay Minerals,* **3,** 290–295.
25. Schofield, R. K., and Taylor, A. W. (1955), The measurement of soil pH, *Soil Sci. Soc. Am. Proc.,* **19,** 164–167.
26. Bolt, G. H. (1961), The pressure filtrate of colloidal suspensions, I—Theoretical considerations; II—Experimental data on homoionic clays, *Kolloid-Z.,* **175,** 33–39 (I), 144–150 (II).
27. Overbeek, J. Th. G. (1953), Donnan EMF and suspension effect, *J. Colloid Sci.,* **8,** 593–605.
28. Bloksma, A. H. (1957), An experimental test of Overbeek's treatment of the suspension effect, *J. Colloid Sci.,* **12,** 135–143.
See also Appendix III, References.

MEMBRANE POTENTIALS

29. Staverman, A. J. (1952), Nonequilibrium thermodynamics of membrane processes, *Trans Faraday Soc.,* **48,** 176–185.
30. Lorimer, J. W., Boterenbrood, E. I., and Hermans, J. J. (1956), Transport processes in ion selective membranes, conductivities, transport numbers, and electromotive forces, *Discussions Faraday Soc.,* **21** (Membrane Phenomena), 141–149, 198–200.
31. Staverman, A. J., and Smit, J. A. M. (1975), Thermodynamics of irreversible processes. Membrane theory: Osmosis, electrokinetics, membrane potentials, see Appendix V, reference B-12, 343–385.

PREPARATION OF CLAY MEMBRANE ELECTRODES

32. Marshall, C. E., and Bergman, W. E. (1941), The electrochemical properties of mineral membranes, I—The estimation of potassium ion activities, *J. Am. Chem. Soc.,* **63,** 1911–1916.

APPLICATION

33. Wyllie, M. R. J. (1955), Role of clay in well-log interpretation, *Clays, Clay Minerals,* **1,** 282–305.

DIELECTRIC PROPERTIES

34. Dukhin, S. S. (1971), Dielectric properties of disperse systems, *Surface and Colloid Science,* Volume 3, pp. 83–166, Wiley-Interscience, New York.
35. Hasted, J. B. (1973), *Aqueous Dielectrics,* Chapman and Hall, London.

Synopsis

CHAPTER ONE—CLAY SUSPENSIONS AND COLLOIDAL SYSTEMS IN GENERAL

Clay Technology and Colloid Chemistry

When the mud chemist at the well site sees that the drilling mud is stiffening and that the mud pumps can no longer circulate the mud at the desired rate, he adds a few sacks of special chemicals to the several thousand barrels of mud, and a striking thinning effect is achieved.

The farmer who finds that after excessive rains his wet clayey soil is compacted and drainage has become poor restores the proper soil structure by working lime or gypsum into the wet soil,

Such treatments of systems of finely divided solids in a liquid are examples of applied colloid chemistry. It is typical for these procedures that the bulk physical properties of the dispersions are strongly affected by relatively small changes in the composition of the liquid phase. In colloid chemistry, such effects are explained by subtle changes in the system on a microscopic scale.

It is the theme of this book to show that knowledge and understanding of the rules and theory of colloid chemistry can serve as a most useful guide in handling practical clay problems with optimum efficiency.

The Clay Suspension

When a dilute clay suspension is observed in the field of a microscope, very few particles can usually be discerned. These belong to the largest size fraction of the clay. The majority of the particles are too small to be resolved in the microscope. They are of *submicroscopic* size. However, their presence can be detected by the use of side illumination in the microscope. In this *ultramicroscopic* arrangement, the light beam does not enter the objective lens, and the field of observation is dark. Particles in the path of the light beam scatter the light in all directions and appear as light specks on the dark background in the ultramicroscope. The light specks are, however, not images of the particles (Figure 1).

The particles appear to move rapidly in a random fashion, changing direction frequently. This motion, known as the *Brownian motion* of the particles, is caused by random collisions of the surrounding vibrating water molecules with the particle. Owing to the Brownian motion, the particles collide frequently, but after the collision they separate again.

The picture changes completely when a very small amount (a few tenths of 1 percent) of salt is dissolved in the suspension. Upon collision, the particles stick together, and agglomerates grow in the suspension (Figure 6).

The particle agglomeration can be observed in the suspension with the naked eye. The agglomerates appear as "flocs" which settle rapidly. Finally, the suspension separates into a bottom sediment and a clear, particle-free supernatant liquid. This phenomenon is called *flocculation* or *coagulation*. The suspension has become *colloidally unstable*. The original fresh-water suspension, which remains homogeneous for a long time, is called *colloidally stable, peptized,* or *deflocculated*.

Particle Interaction

In the following, we shall digress for a while from clay suspensions and deal with dispersions of small solid particles in a liquid. Such systems are called *hydrophobic colloids*. Their properties depend largely on the properties of the extensive interface between the particles and the liquid.

The dispersions are called either *sols* or *suspensions,* depending on the particle size. Customarily, the term sol is used when the particle size is below about 1 μm (see Table 1). However, the borderline between sol and suspension is arbitrary, and there is no difference in principle between a soil and a suspension, although the coarser particles in the latter settle more rapidly than the small particles in the sol.

The microscopic observations mentioned in the previous paragraph pose the question of why particle collision leads to sticking in salt solutions but not in fresh-water suspensions. Particle attraction apparently prevails in a salt solution but not in fresh water. In the modern theory of the stability and flocculation of hydrophobic colloids, the following picture of particle interaction has been developed:

Particle association in salt solutions is caused by general van der Waals attraction forces which operate between the atoms of the two particles. The total attractive force between the particles is obtained by the summation of all attractive forces between all atom pairs. Since this force is practically independent of the salt content of the liquid, it is equally effective in a fresh-water system. However, particle association in a fresh-water solution is prevented by repulsive forces between the particles. These forces are a result of the electric charges on the surfaces of the particles. The presence of such charges is evident from the well-known movement of the sol particles in an electric field, called *electrophoresis*.

The effectiveness of the repulsive forces can be shown to decrease with increasing ion concentration of the liquid phase of the sol. Therefore, these forces will prevent particle association in fresh-water solutions by counteracting the van der Waals attraction, but in salt solutions they are no longer powerful enough to prevent flocculation.

Colloid Systems

Colloidal systems are two-phase systems in which large kinetic units are dispersed in a homogeneous phase. The term "colloid" should be applied to the whole system and not to the dispersed substance.

The two principal classes of colloidal systems are the *hydrophobic* and the *hydrophilic* colloids. Although these terms literally mean "water repelling" and "water attracting," this terminology of the colloidal systems is somewhat misleading since the particles in a hydrophobic sol are certainly well wetted by water. Hydrophobic colloids are dispersions of solid particles in a liquid, featuring a large interfacial area. The common representatives are dispersions of small inorganic particles in water, such as gold sols, sulfur sols, and clay sols. Hydrophilic colloids consist of macromolecular compounds and water. In modern colloid chemistry they are considered as true solutions of macromolecules and macro-ions in water and at present they are more frequently called *macromolecular colloids*. Their colloid chemical properties are due to the large size of the molecules and ions. Representatives are solutions of natural gums and synthetic polyelectrolytes.

Whereas hydrophobic sols flocculate with rather small amounts of salt, hydrophilic sols are very insensitive to salt. Sometimes they are precipitated by large amounts of salt owing to a reduction of the solubility of the macromolecular compound ("salting out").

CHAPTER TWO—PROPERTIES OF HYDROPHOBIC SOLS

The Origin of the Particle Charge

A suspended particle may obtain its charge in two different ways, either by specific adsorption of certain ions or from interior defects of the crystals.

ION ADSORPTION. In most hydrophobic sols, the particles obtain their charge by the specific adsorption of certain ions which are available in solution. As a whole, the suspension is not charged; therefore, the particle charge must be internally compensated. The compensation of the particle charge is effected by an accumulation of an equivalent amount of ions of opposite sign in the solution near the particle surface. The particle charge and the accumulated compensating-ion charge together form a so-called *electric double layer*. The ions accumulated in the solution are called the *counter-ions* or *gegenions* (Figure 7).

Since the electric double layer is responsible for the repulsion between the particles and the stability of the suspension, the presence of a certain amount of a special electrolyte is required to create a stable sol. This requirement amends the previous statement that a suspension would be stable in fresh water. However, the necessary amount of *peptizing electrolyte* is usually extremely small, and in many cases, the dissociation products of the slightly soluble colloidal particles or of water act as peptizing ions.

The preferential adsorption of certain kinds of ions on the particle surface may be due to chemical bonds ("chemisorption"), hydrogen bonds, or van der Waals attraction. Often, the same ions as those constituting the particles act as peptizing ions. At the surface of the particle, where the crystal structure is disrupted, those vacant spots which the next ion would have occupied if the crystal had an opportunity to grow further are preferred adsorption sites for such ions.

For example, in an AgI sol, an excess of silver or of iodide ions, depending on their relative availability in solution, is adsorbed on the silver-iodide surface, and either a positive or a negative sol is obtained. The AgI particle acts as a reversible electrode, and the electric surface potential is determined by the concentration of Ag and I ions in solution. Therefore, these ions are called *potential-determining ions*. At a certain concentration of iodide ions and of silver ions, the product of which is the solubility product of AgI, the surface is uncharged (point of zero charge). With an increase of the silver ion concentration (which is accompanied by a decrease of the iodide ion concentration), the surface becomes positively charged, having a slight excess of silver ions in the surface layer of the crystal. A negative particle is obtained by increasing the iodide ion concentration (which is accompanied by a decrease of the silver ion concentration). (See Figure 7.)

As another example, potential-determining ions for an aluminum-hydroxide sol are both aluminum ions and hydroxyl ions (or possibly aluminate ions) for a positive and a negative sol, respectively. In this example the availability of the two ions in solution depends on the pH of the solution, and the sign of the particle charge is positive in acid solution and negative in alkaline solution. At some intermediate pH, the sol is not charged.

For many sols, ions of only one type act as potential-determining ions. For example, a silicic acid sol is naturally negatively charged, with silicate anions probably acting as potential-determining ions.

INTERIOR CRYSTAL CHARGE. Another possible cause of particle charge is the occurrence of imperfections in the interior of the crystal, involving a certain unbalance of charge. This charge is compensated by counter-ions which are accumulated near the particle surface in the solution (Figure 7). A well-known example is a clay double layer. These counter-ions remain associated with the negative clay crystal upon drying and resuspending.

Flocculation Rules

Different electrolytes have a different "flocculating power" for a given sol, which means that the *critical coagulation concentration* (c.c.c.) or the *flocculation value* varies. The flocculation value of a sol is determined by the addition of increasing amounts of electrolyte to the sol in a series of test tubes; after an arbitrarily chosen time, one determines by visual inspection which sols are flocculated and which are still stable. In this way, the flocculation value is enclosed between a lower and an upper limit (Figure 8).

Numerous flocculation series experiments have led to the formulation of certain empirical rules, the most important of which is the *Schulze-Hardy rule* formulated in 1900. The rule states that the flocculating power of *indifferent* or *inert electrolytes,* that is, electrolytes which do not specifically react with the sol particles, is primarily determined by the valence of the ion of opposite charge of the particle charge. The higher the ion valence, the greater its flocculating power. For example, in order to flocculate a clay suspension in which the particles are negatively charged, a lower concentration of $CaCl_2$ than of NaCl is required.

Obviously, any proposed theory of colloidal stability should be able to explain this universal flocculation rule.

Special Effects

REVERSAL OF CHARGE—IRREGULAR SERIES. Deviations from the Schulze-Hardy rule are observed when specific interactions occur between the added electrolyte and the particle. An abnormally low flocculation value is observed if the added electrolyte neutralizes the particle charge by reaction with the charging ions. However, when, for example, an excess of negative ions react with the positive surface charge, forming complex anions, reversal of the particle charge occurs.

Charge reversal also occurs when the relative concentrations of positive and negative potential-determining ions in such sols as AgI or $Al(OH)_3$ are varied, as discussed earlier.

A third method of effecting charge reversal is by the addition of soluble aluminum salts to a negative sol. First the sol is flocculated by minute amounts of Al salt, but, upon further addition of Al salt, a stable positive sol is created. Finally, an excess of Al salt flocculates the positive sol. When a flocculation series experiment is conducted with an Al salt, the following sequence is observed: stable negative—flocculated—stable positive—flocculated. Such a series is called an *irregular series.*

MUTUAL FLOCCULATION. When a positive and a negative sol are mixed, *mutual flocculation* occurs, owing to the attraction between the positive and negative particles.

FLOCCULATION BY ALCOHOLS, AND OTHER. The addition of water-miscible organic solvents such as alcohol or acetone to a sol often induces flocculation.

GELATION—A SPECIAL CASE OF FLOCCULATION. The addition of electrolytes to certain sols results in *gelation* instead of flocculation. In the gel the particles are agglomerated to form a single "floc" extending throughout the available volume. In this way, a homogeneously appearing rigid, and often quite elastic, mass is obtained. Spherical particles may link to strings of beads, platelike particles to a card-house structure, and rods to a scaffolding. Gelation occurs in certain silica, alumina, and clay sols.

The concentrations of different electrolytes which are required to cause gelation are governed by the Schulze-Hardy rule, indicating that gelation is indeed a special case of flocculation.

COUNTER-ION EXCHANGE. The ions in the counter-ion atmosphere may be exchanged for other species of ions of the same sign if they are available in the solution. If there are no specific interactions between particle and counter-ion or between different counter-ions, the ratios of two ion species of the same valence in the counter-ion atmosphere and in the bulk solution are equal. When two ion species of different valence are present, there is a preference of the double layer for the ions of the highest valence.

The above summary of the most important phenomena occurring in hydrophobic sols demonstrates that the simple adjustment of the composition of the liquid phase by the addition of rather small amounts of chemicals offers many interesting possibilities for the adjustment of the degree of stability and flocculation, hence for the tailoring of the bulk physical properties of colloidal systems in practice, which are very sensitive to changes in particle interactions.

CHAPTER THREE—THE THEORY OF THE STABILITY OF HYDROPHOBIC SOLS

The Electric Double Layer

Counter-ions accumulate near the particle surface owing to the electrostatic attraction by the surface charge. Simultaneously, they have a tendency to diffuse away from the surface toward the bulk of the solution, where their concentration is low. This situation is analogous to that in the earth's atmosphere, in which the gravitational forces on the gas molecules are counteracted by diffusion. Consequently, a diffuse atmosphere of counter-ions is created around the particle. A model of such a diffuse electric double layer is sketeched in Figure 9. Honoring the Frenchman who first treated this model quantitatively it is also called a *Gouy double layer.*

The most important conclusions concerning the structure of the double layer as a function of the electrolyte concentration in the solution are as follows:

The extension of the double layer in the solution decreases with increasing electrolyte concentration. At very low electrolyte concentration, the counter-ion concentration, which decreases asymptotically with increasing distance from the surface, is still appreciable at a distance which is comparable with the diameter of the particle. However, with increasing electrolyte concentration, a considerable compression of the diffuse atmosphere toward the surface occurs.

The degree of double-layer compression is primarily determined by the concentration and valence of the ions of opposite sign of the particle charge. The compression is more pronounced if the valence of these ions is higher. This feature indicates the importance of the ions of opposite sign which are known to govern flocculation according to the Schulze-Hardy rule.

Double-Layer Repulsion

When two particles carrying a diffuse double layer approach each other, what happens as a result of their Brownian motion? It can be shown that as soon as the diffuse counter-ion atmospheres begin to interfere, work must be performed to bring the particles closer together. In order words, there will be a repulsion ("double-layer repulsion") between the particles. The repulsive energy V_R at a certain particle distance is the work which must be performed. In the upper part of Figure 12, the repulsive energy is plotted versus the particle distance for three different electrolyte concentrations. With increasing electrolyte concentration, the particles will be able to come closer before the repulsion becomes significant because of the greater compression of the double layer.

van der Waals Attraction

As pointed out before, in order to explain the phenomenon of flocculation, an attractive force between the particles must exist. Obviously, in order to be able to compete with the double-layer repulsion, the attractive forces must be of comparable magnitude and range of action. These requirements are indeed met by the total van der Waals attraction between the particles, which is obtained by the summation of the attractive forces between all the atom pairs.

In the lower part of the diagrams of Figure 12, the van der Waals attraction energy V_A, derived from current information about the magnitude of the van der Waals forces, is plotted as a function of particle distance, with

V_A considered negative. These curves are practically the same at different electrolyte concentrations.

The Net Energy of Particle Interaction

Finally, the repulsive and attractive energies must be combined. If we consider the repulsion to be positive and the attraction to be negative, their summation gives the net interaction energy as a function of particle separation.

In order to complete the picture, however, two other particle-interaction forces should be considered. Contrary to the double-layer repulsion and the van der Waals attraction, these two forces are of a short-range character. At close approach of the particles, the *Born repulsion* prevents the interpenetration of the crystal lattices of the particles. Another cause of short-range repulsion is the presence of one or two molecular layers of water which are more or less tightly adsorbed on the particle surface. These water layers must be desorbed when the particles approach each other so closely that there is no longer room for the adsorbed water layers. Since the desorption costs work, the dehydration of the surface manifests itself by a short-range repulsion (*solvation repulsion*).

With these additional short-range repulsive energies taken into account, the net interaction energy as a function of the distance is represented by the curves of Figure 13 for three different electrolyte concentrations.

The steep rise of the curve at close approach is dominated by the two above mentioned short-range repulsive energies. The deep minimum at rather small particle separations indicates that van der Waals attraction predominates in this region. At larger separations, the double-layer repulsion dominates at the lowest electrolyte concentrations. The long-range repulsion maximum in the curve becomes lower with increasing electrolyte concentration. At the highest electrolyte concentration, such a long-range maximum is absent, and particle attraction predominates at any particle separation.

The Net-Interaction-Energy Curve and Colloidal Stability

The net-interaction-energy curves, or "potential curves of interaction," can be translated in terms of sol stability and flocculation by the following considerations:

Owing to their Brownian motion, two particles may approach each other in a relative position at which the deep minimum in the potential energy occurs and become associated because of the prevailing attraction. The rate at which this process occurs may be computed from the theory of diffusion. If repulsion between the particles exists in a certain range of particle distances between the position of the minimum and infinite separation, the mutual approach of the particles is counteracted, and the rate of association is reduced.

At a high electolyte concentration, at which the potential curve shows no repulsion at any particle distance (Figure 13*c*), particle agglomeration occurs at a maximum rate (*rapid coagulation*). At intermediate electrolyte concentrations (Figure 13*b*), the coagulation process is slowed down by the long-range epulsion (*slow coagulation*). At very low electrolyte concentrations (Figure 13*a*), the coagulation process is so much retarded by the appreciable long-range repulsion that it may take weeks or months before flocculation becomes perceptible in the sol. Under these conditions, the sol is, for all practical purposes, considered "stable."

Figure 13 is a concise representation of stability. A large section of the potential curve above the distance coordinate, or a large *energy barrier,* reduces the rate at which the sol particles associate by "jumping over the barrier." When the barrier becomes smaller, the rate of coagulation increases, and in the absence of a barrier, the rate is maximal. The magnitude of the barrier is determined by the magnitude and range of the double-layer repulsion energy. The addition of electrolytes causes a compression of the double layer and therefore a reduction of the range of repulsion and a reduction of the magnitude of the energy barrier in the interaction curve. Since the compression of the double layer is ruled by the concentration and valence of ions of opposite sign, the stability theory is able to explain the Schulze-Hardy rule.

CHAPTER FOUR—SUCCESSES OF THE THEORY OF STABILITY—FURTHER THEORIES AND REFINEMENTS

Stability, Flocculation, and the Schulze-Hardy Rule

According to the preceding chapter, colloidal stability and instability are only arbitrary terms—stable and unstable sols differ only in the rate of coagulation. Therefore, the rate of coagulation would be an appropriate measure of the stability. However, the velocity of agglomeration is not constant, but decreases during the coagulation process owing to the continuous reduction of the number of particles. According to kinetic studies of the process, the time in which the number of particles is halved appears to be a suitable constant to describe the course of the process. This so-called *flocculation time* should be an appropriate measure of the degree of stability.

In routine practice, stability and flocculation are studied by means of the simple flocculation series test described previously. Since the electrolyte concentration is determined at which the sol flocculates in a specified time, the flocculation value represents the amount of electrolyte required to obtain a specified average coagulation rate, that is, a specified degree of instability. Hence the flocculation test is a fundamental test, despite its arbitrary character.

The flocculation value is an appropriate relative measure of the original stability of a sol—the higher the stability, the greater will be the amount of electrolyte to reduce the stability to the specified level prevailing at the flocculation value. For example, the effectiveness of a peptizing agent in improving the stability of a sol can be measured by the increase of the flocculation value.

On the basis of the understanding of the fundamental nature of the flocculation value, the Schulze-Hardy rule may be derived from the stability theory. The general agreement between the predicted and observed ratios of flocculating power of different types of electrolytes is a valuable support of the stability theory.

Limits of Particle Size

The repulsive energy between small particles is smaller than that between large particles carrying an identical double layer because of the larger interaction surface area of the latter. It can be shown from the stability theory that this difference is the principal reason for the observed instability of sols containing small particles.

Flocculation by Water-Miscible Organic Solvents

According to the Gouy theory, the double layer is compressed if the dielectric constant of the medium is decreased by the addition of water-miscible organic solvents. This compression results in a reduction of the size of the energy barrier in the potential curve of interaction and thus in a decreased stability of the sol. Although instability is usually induced by the addition of organic solvents to a sol, there are some exceptions. These must be explained by considering secondary effects of the organic solvent on the structure of the double layer.

Counter-Ion Exchange

The counter-ion exchange equilibrium can be quantitatively predicted by the Gouy theory, provided that specific effects can be neglected. In the absence of such effects, the ratios of the concentration of two species of ions of the same valence in the double layer and in the equilibrium liquid are predicted to be the same, whereas a preference of the counterion atmosphere should exist for the ion of the highest valence in the presence of two ion species of different valences. In certain sols in which the condition of nonspecifity is sufficiently fulfilled, the observed exchange equilibrium can be quantitatively predicted from the theory. Therefore, deviations from the predicted ion distribution will be indicative of specific effects, requiring a refinement of the picture.

Refinements of the Gouy Theory

Certain limitations of the applicability of the Gouy theory may be expected, since the theory does not take into account the size of the ions, mutual interactions between the ions, and interactions between ions and solvent, ions and the surface, and solvent and the surface. Several refinements of the Gouy theory have been considered, and different models for the structure of the double layer have been proposed, allowing an appropriate analysis of those systems in which specific interactions are important. A well-known alternative double-layer model is Stern's model, in which the limitation of the approach of a counter-ion to the surface to within one ion radius is accounted for. The diffuse Gouy atmosphere is separated from the surface by a molecular condenser in which the electric potential drops linearly with the distance. Hence the Gouy atmosphere is no longer determined by the electric surface potential but by the electric potential at the borderline of the molecular condenser (the *Stern layer*) and the Gouy layer (Figure 14). In this model, specific interactions between the surface, the solvent, and the counter-ions can be easily introduced.

Previous Stability Theories

The theory that particle *hydration forces* are responsible for sol stability, the rigidity of a gel, and swelling cannot be upheld, since such forces are likely to be effective only at a very small distance from the particle surface. They affect the potential curves of interaction only in the separation range of two to four water layers.

Contrary to earlier concepts, the *zeta potential* of a particle, which is derived from the electrophoretic mobility, is only an approximate criterion for sol stability, since it is an ill-defined parameter for the structure of the electric double layer.

"Entropy" Stabilization

In some dispersed systems, electric double-layer repulsion does not exist, because there are no ions present in the system. Nevertheless, such systems can be colloidally stable. Particle repulsion in those dispersions has been analyzed to result from a decrease in entropy when the particles approach each other. For example, such a situation exists when long-chain molecules are attached with one point at the particle surface, but are otherwise free to move around this point in the solution. Upon approach of the particles, at a distance which is smaller than twice the length of the chains, the freedom of motion of the chains becomes restricted, resulting in a decrease of entropy. This decrease in entropy is manifested as a repulsion.

The repulsive energy as a function of distance is obtained from a statistical evaluation of the change in entropy. This repulsion potential curve may be combined with the van der Waals attraction potential curve to obtain the net potential curve of particle interaction.

The stability of certain sols in hydrocarbons can be explained on this basis, although in some hydrocarbon systems ionized compounds may be formed and such compounds may create a double layer which is sufficiently well developed to stabilize the sol.

CHAPTER FIVE—CLAY MINERALOGY

AND

CHAPTER SIX—PARTICLE SIZE AND SHAPE, SURFACE AREA, AND CHARGE DENSITY

Shape and Structure of Clay Particles

Concerning the various techniques of observing particles in a clay suspension, another step in magnifying power is realized by use of the electron microscope. With this instrument, which operates at high vacuum, a picture of the dry clay particles is obtained. The particles appear to be large thin plates and sometimes lath-, rod-, or needle-shaped.

The final step in magnification is crystal structure analysis from X-ray diffraction patterns, which supplies a picture of the crystalline regularities in the arrangement of the atoms.

The knowledge concerning the crystal structure of different clays has contributed greatly to an understanding of their surface properties, which are of primary importance in the colloid chemical behavior of clay suspensions. Therefore, the main structural principles will be described in some detail.

According to Pauling (Chapter Five, reference 2), who was the first to elucidate the structure of clay minerals, each platelike clay particle consists of a stack of parallel layers. Each layer is a combination of tetrahedrically arranged *silica sheets* (T) and octahdrically arranged *alumina* or *magnesia sheets* (O). 2:1 *layer clays* (montmorillonites and illites) are built of layers composed of two tetrahedral sheets with one octahedral sheet in between (T-O-T); 1:1 *layer clays* (kaolinites) are built of layers consisting of one tetrahedral and one octahedral sheet (T-O).

Montmorillonite Clays (2 : 1 Layer Clays)

The atom arrangement in a 2:1 layer clay is shown in the schematic drawing in Figure 18, representing a unit of structure which repeats itself laterally in the layer. All valences of the atoms are satisfied in the sketched structure. Of the three possible octahedral positions, two are occupied by trivalent Al (*dioctahedral arrangement*). Alternatively, all three positions may be filled by divalent Mg (*trioctahedral arrangement*). These electroneutral structures represent the compositions of the micaceous minerals pyrophyllite and talc, respectively.

Montmorillonite clays are derived from these two prototype minerals by substitution of some of the elements by others; for example, tetravalent Si is replaced partly by trivalent Al, or Al in the octahedral sheet is partly replaced by divalent Mg without the third vacant position being filled. Such substitutions of electropositive elements by those of lower valence result in an excess of negative charge of the layers. The net negative charge is compensated by the adsorption of cations on the layer surfaces, both on the interior and on the exterior surfaces of the stack.

The most typical property of montmorillonites is the phenomenon of *interlayer swelling* (or *intracrystalline swelling*) with water. In the presence of water or water vapor, the dry clay takes up water which penetrates between the layers and pushes them apart a distance equivalent to one to four monomolecular layers of water. Owing to this effect, the layer-repetition distance in the stack is increased from about 10 A for the dry clay to approximately 12.5–20 A for the wet clay.

The compensating cations which are adsorbed on the layer surfaces may be exchanged for other cations; therefore, they are called the exchangeable cations of the clay. The amount per unit weight of clay is the *cation exchange capacity* (CEC) of the clay, which is usually of the order of 80–100 meq per 100 g for montmorillonites. The CEC is a measure of the amount of substitution in the crystal.

Illites (2 : 1 Layer Clays)

Illites have the same basic structure as montmorillonites, but they do not show interlayer swelling. The total amount of isomorphous substitution is usually larger than that for montmorillonites. Substitution of Si by Al in the tetrahedral sheet is predominant. The compensating cations are mainly potassium ions.

The lack of interlayer swelling is attributed to the strong electrostatic attraction between the potassium ions and the two charged layers on each side, owing to a favorable geometric arrangement and the large number of potassium ions.

Because of the lack of interlayer swelling, the cations between the layers are not available for exchange, as in montmorillonites; therefore, the CEC involves only the ions on the exterior surfaces of the stack. Hence, the CEC is lower than for montmorillonites (about 20–40 meq per 100 g) despite the greater charge deficiencies in the crystal.

Kaolinites (1:1 Layer Clays)

The arrangement of atoms in the layers of the 1:1 layer minerals, the kaolinites, is shown in Figure 19. Kaolinites conform very closely with the sketched electroneutral structure. The small amount of isomorphous substitution, which usually escapes precise analysis, is compensated by exchangeable cations located exclusively on the exterior surfaces of the layer stacks. The CEC is of the order of 2–10 meq per 100 g. Alternatively, the charge may originate from adsorbed silica-alumina compounds.

The species in any group of minerals differ in the degree and type of substitution and in the geometry of stacking of the layers in the particle. In many clays, layers of different types of clay are stacked in the same particle. These are called *mixed-layer clays*.

The number of layers stacked in a single particle is usually small in montmorillonites (say, two to five) and is larger in illites. The thickest particles occur in kaolinite suspensions.

CHAPTER SEVEN—ELECTRIC DOUBLE-LAYER STRUCTURE AND STABILITY OF CLAY SUSPENSIONS

Electric Double-Layer Structure of Clays

Electrophoresis shows that the suspended clay particles are negatively charged. The negative charge of the particle is obviously a consequence of the net negative crystal charge. When the dry clay is contacted with water, the adsorbed compensating cations spontaneously form a diffuse counter-ion atmosphere, which, together with the crystal surface charge, constitutes a negative double layer. This double layer is an example of a double layer of fixed charge which is entirely determined by the crystal substitutions. There are indications that this double layer is more adequately described by using the Stern model instead of the Gouy model and by assigning a certain specific adsorption energy to the counter-ions at the surface.

Although the layer surface is the largest part of the particle surface, the relatively small surface of the edges of the clay plates should also be considered. Here the silica and alumina sheets are broken, and the situation at the edge surfaces is therefore analogous to that at the surface of silica and alumina sol particles. On such surfaces a double layer is created by the adsorption of potential-determining ions. An alumina particle is either positively or negatively charged, depending on the pH of the solution. A silica particle is usually negatively charged, but in the presence of small amounts of Al ions in solution, reversal of charge takes place. By analogy, the edge surface of a clay particle may also be either negatively or positively charged, and there are strong indications that the charge is positive (see, e.g., the electron micrograph in Figure 22, showing the preferential adsorption of negative gold sol particles at the edges of kaolinite plates). Whatever the sign of the edge double layer may be, it is certainly a different type double layer than that existing on the flat layer surfaces.

Flocculation and Gelation

In suspensions of platelike particles, three different modes of particle association must be considered in the flocculating system: edge-to-edge, edge-to-face, and face-to-face. In clay suspensions, these associations will be governed by different potential curves, since three different combinations of the two double layers are involved, and since the total van der Waals attraction energies are different for the three modes of association.

The consequences of the three modes of association for the physical and thus also for the technological properties of the clay suspensions will be quite different. Edge-to-face and edge-to-edge association will lead to voluminous card-house structures and, therefore, at moderate clay concentrations, to gelation. Face-to face association, on the other hand, merely leads to thicker and possibly larger particles, which will not appear as "flocs" or cause gelation of the system. Edge-to-edge and edge-to-face association and dissociation are sometimes described as *flocculation* and *deflocculation* processes, whereas face-to-face association and dissociation are called *aggregation* and *dispersion,* although the latter are just different kinds of flocculation and deflocculation (Figure 23).

In the usual flocculation series experiments with clay suspensions, it appears that the flocculation values for different salts obey the Schulze-Hardy rule for a negative sol, provided that corrections are applied for ion exchange, which takes place simultaneously between the flocculating cations and the exchangeable cations present on the clay.

Indications of the occurrence of different modes of particle association upon the addition of increasing amounts of electrolyte are obtained when the changes in physical properties of the suspensions are studied. Particularly the flow behavior ("rheological" behavior) is very sensitive toward changes in particle association.

Particle Association and Flow Properties of Clay Suspensions

The viscosity of a dilute clay suspension and the rigidity of a concentrated clay suspension as measured by its "yield stress" increase when edge-to-edge and edge-to-face association (flocculation) occur because of the formation of voluminous aggregates in dilute suspensions and interlinked card-house structures in concentrated suspensions. When face-to-face association occurs ("aggregation"), the viscosity decreases, since the particles become less asymmetrical, and the yield stress decreases, because the number of links in the rigid card-house decreases owing to the smaller number of particles.

The change in the viscosity of a dilute sodium montmorillonite suspension with increasing salt concentration is shown in Figure 24*a*. The change in the yield stress in a more concentrated system is represented by the curve marked "Na-montmorillonite" in Figure 24*b*.

When the hypothesis is adopted that the edge surfaces indeed carry a positive double layer, these changes may be interpreted as follows:

In the salt-free system, both double layers of the clay particle are well enough developed to prevent particle association by van der Waals attraction. However, owing to the opposite charges of the edge and the face double layers, edge-to-face association takes place (internal mutual flocculation), causing a relatively high viscosity and yield stress. The particles are associated in a double-T fashion, in which the edge-to-face attraction outweighs the face-to-face repulsion.

In the presence of a few milliequivalents of NaCl, both double layers are compressed, and their effective charge will be reduced. Consequently, both the edge-to-face attraction and the face-to-face repulsion will diminish. Apparently, under these conditions, the attraction becomes less than the repulsion in the double-T arrangement, and the particles are disengaged. Hence the card-house structure breaks down, and the yield stress of the concentrated suspension is dramatically reduced. In dilute solutions, the particles are disengaged and the viscosity decreases.

When the amount of NaCl in the suspension is increased, both double layers are further compressed. The van der Waals attraction between edges and faces enhances the opposite-charge attraction, and once more the subtle balance between edge-to-face attraction and face-to-face repulsion becomes favorable for the formation of the card-house structure. Simultaneously, edge-to-edge association by van der Waals attraction may occur. Consequently, the viscosity and the yield stress increase. At very high salt concentrations, the yield stress decreases somewhat (not shown in Figure 24*b*). This observation may be explained by the simultaneous occurrence of face-to-face association by which the number of particles in the card-house is reduced, and hence its strength decreases.

Since different clays may be expected to have different double-layer properties, they do not necessarily react in the same way to the addition of electrolytes. The flow behavior at different salt concentrations depends on which of the three types of association predominates at a certain salt concentration, and the relative degrees of such associations are dependent on the double-layer properties. For example, in kaolinites, the rise of the yield stress after the initial decrease does not occur, indicating that in these systems face-to-face association ("aggregation") prevails at the higher salt concentrations (see Figure 24*b*, curve marked kaolinite).

CHAPTER EIGHT—PEPTIZATION OF CLAY SUSPENSIONS

Peptization (Deflocculation) by Special Inorganic Salts

The gelation of suspensions of sodium montmorillonite, both in the presence and in the absence of salt, shows that particle association occurs under these conditions. Hence the system may not be called colloidally stable. This is a direct consequence of the presence of two double layers of opposite sign. Obviously, deflocculation can be achieved and gelation can be prevented by reversing either charge.

The most efficient peptization method is to reverse the positive edge charge into a negative one. One means of achieving such a reversal of charge is by adsorption of an excess of anions at the edge surfaces. In clay technology, it is common practice to peptize clay suspensions with certain phosphates, oxalates, and other salts. Small amounts of such salts disperse the flocs in dilute suspensions and convert stiff, flocculated, concentrated suspensions into thin, freely flowing liquids. The anions of these peptizing salts have in common that they react with the type of cations which are exposed at the particle edges, such as Al. Consequently, they will be chemisorbed at the edges. When an excess of anions is involved, for example, when complex anions with Al are formed, a powerful negative double layer is created at the edges. Even in the presence of some salt, this double layer prevents edge-to-face association as well as edge-to-edge association, the card-house structure is broken down, and a stiff suspension becomes thin.

This analysis of the peptization mechanism is supported by the observation that an alumina sol, which is a model for the edge surface structure, reacts in an analogous manner to the same peptizers as clay.

Since the edge surface area is comparatively small, the adsorption capacity of the clay for the anions is small, and only relatively small amounts of the peptizing chemicals are required. This fact makes the method economically attractive.

Relative Peptizing Power of Chemicals and Other Contributing Factors

A measure of the relative peptizing power of various chemicals on a clay suspension is the relative increase of salt tolerance of the suspension as reflected by an increased salt-flocculation value. Figure 25*a* shows the effect of the peptizing salt sodium polymetaphosphate on the NaCl flocculation value of a sodium montmorillonite sol (curve 1).

In this particular example, a peptizer concentration of 20 meq/dm^3 (about 0.2 percent) causes an increase of the NaCl flocculation value from 20 meq/dm^3 (about 0.1 percent salt) to about 340 meq/dm^3 (about 2 percent salt). Optimum tolerance for salt is reached at about 0.4 percent of the peptizer. At higher peptizer concentration, the flocculating effect of the simultaneously added sodium ions makes itself felt, and finally the sol flocculates by the peptizer alone at a concentration of 600 meq/dm^3, or about 6 percent. The addition of too much peptizer is said to result in "overtreatment."

The peptizing power of various chemicals for a given clay varies considerably. Also, the sequence of increasing peptizing power of a number of chemicals varies with the type of clay. One reason for this variation is that in different clays different cations are exposed at the edge surfaces, which react differently with the peptizing anions in the chemisorption process.

In addition to the charging effect by the anions, there are two additional contributions to the improved salt tolerance of the clays. One contribution may result from the exchange of calcium ions present in the natural clays by the sodium ions of the peptizer, resulting in conversion of the usually rather unstable calcium clay into a more stable sodium clay. From curve 2 in Figure 25 it can be read that this conversion accounts for an increase of the salt-flocculation value from a few meq/dm^3 for the calcium clay to 20 meq/dm^3 for the sodium clay, which is a small effect as compared with the total peptizing effect due to anion adsorption.

A second contribution is the result of a reduction of the activity of the cations and therefore of their flocculating power owing to the presence of polyvalent anions in solution. In the absence of such a contribution, for example, when the peptizer is KF, which forms complex AlF_6^{3-} ions at the edges, the total stabilizing effect is much smaller than in the case of a peptizer with a higher anion valency.

Peptization by Alkali

In analogy with reversing the charge of a positive alumina sol by raising the pH, an alternative method of peptization by charge reversal of the clay particle edges is to add alkali to the suspension. Such an effect of alkali is indeed observed in many clay suspensions, although the effect is usually comparatively small, possibly because of the absence of any contribution of the activity reduction of the cations by the anion, which is only monovalent in this case.

Sometimes, rather large effects of alkali are observed in impure clays, particularly when they show an acid reaction. In this case, the contribution resulting from conversion of the rather unstable hydrogen clay into the sodium clay will be important. In addition, such clays often contain organic acids, which, upon neutralization by the alkali, may act as efficient peptizers, as will be discussed later.

CHAPTER NINE—TECHNOLOGICAL APPLICATIONS OF STABILITY CONTROL: SEDIMENTATION, FILTRATION, AND FLOW BEHAVIOR

Sedimentation and Stability

The sedimentation rate of a flocculated suspension is usually much faster than that of a stable suspension, although the difference is not pronounced in coarse suspensions, which settle rather rapidly in the stable condition. In practice, solid-liquid separations by settling under gravity or by centrifugation are often accelerated by the addition of flocculants.

Of great technological importance is the difference of the sediment properties in stable and unstable suspensions. A much denser sediment is obtained from a stable suspension than from a flocculated suspension. This difference can be explained as follows: In a stable suspension the particles settle individually. When crowding into the lower part of the vessel, they will slide and roll along each other, since they repel each other. In this way, they are able to reach the lowest possible position and become arranged in a close-packed fashion. In a flocculated suspension, on the other hand, the haphazardly formed loose flocs will settle as such and will pile up at the bottom as a voluminous sediment, as long as the particles are not so big that the gravity forces can break the links between the particles (see Figure 27).

Filtration and Stability

Analogous effects are observed when suspensions are filtered. A stable suspension yields a rather compact filter cake of low porosity and often of low permeability also, so the filtration process is slowed down considerably when the first layers of the filter cake are established. The filter cake of a flocculated suspension is voluminous and porous, allowing filtration to proceed at a fast rate.

Applications

Soils. Root growth and proper drainage require a porous soil. Therefore, the clay in the soil should be in a flocculated condition. If rain water leaches out the flocculating electrolyte in the soil, the resulting peptization of the clay leads to compaction of the wet soil and the formation of impermeable filter cakes of clay within the pores between the sand grains. In this condition, root growth is hampered, and drainage is poor. Therefore, a flocculating concentration of electrolyte should be maintained in the soil. Lime or gypsum is commonly used for this purpose because of their moderate solubility and the powerful flocculating effect of the divalent calcium ions on the negative clay, according to the Schulze-Hardy rule.

Drilling Fluids. Drilling fluids build up a filter cake on porous formations owing to the larger hydrostatic pressure of the clay suspension with respect to the pressure in the fluids present in the formation. In order to keep filtrate loss in the formation low and the filter cakes thin, the clay suspension should be in a deflocculated condition. Such a suspension has adequate "plastering" qualities. Montmorillonite clays, such as bentonitic clays, are particularly effective permeability reducers for the filter cakes, probably because the particles are very thin and somewhat flexible. In addition, the finer particles penetrate into the formation and effectively clog the pores by the formation of microscopic filter cakes. However, the latter is undesirable when the producing zone is being drilled, since the subsequent oil production flow is also adversely affected by the clay particles in the pores. Therefore, in well completion, muds with coarser particles are preferred, and either flocculated or "aggregated" muds are more adequately used. In such muds, other means for preventing excessive fluid losses must be devised, as will be discussed later.

Rheological Properties and Stability

The relations between various modes of particle association and flow properties (rheological properties) of clay suspensions have already been discussed. In practice, these properties are extremely important, and it appears that adjusting the state of stability of the suspensions by slight modifications in the composition of the water phase is a very useful tool in tailoring the flow properties to suit the particular purpose.

It must be emphasized that when pilot tests on the effects of certain chemical additions are performed in the laboratory, the flow properties should be measured under the conditions prevailing in the field. Therefore, an analysis of the type of rheological information required must precede the choice of the viscometer to be used in pilot tests.

Drilling Fluids. More effort is required of the mud pumps to circulate a drilling fluid when it is in the flocculated state, and a certain yield stress is developed. Peptization causes a reduction of the yield stress, and less pumping pressure is needed to maintain a certain rate of circulation which is sufficient to cool the bit and to carry up the drill cuttings. Therefore, drilling muds should be in the colloidally stable condition to ensure adequate pumping characteristics. At the same time, their plastering properties will be adquate under those conditions. On the other hand, a certain rigidity of the mud is desirable to prevent cuttings from falling to the bottom of the hole during an interruption of circulation. An elegant way to reconcile the opposite requirements is to make the mud somewhat thixotropic. In a thixotropic system, which is a somewhat flocculated system, the particle links are temporarily broken by stirring and are only slowly restored during rest. The circulation pressure is hardly affected, but during an interruption, the mud stiffens sufficiently to keep the cuttings suspended. Therefore, a certain degree of flocculation in a mud has some advantage over a well-peptized mud.

Ceramics. Ceramic objects are often made by pouring a clay suspension in a porous mold (slip casting). A certain degree of peptization will assure good pouring properties. On the surface of the mold, a concentrated layer of clay is deposited, since the water is removed by suction of the capillaries of the mold. This process is comparable with the formation of a filter cake. The properties of the final object will largely depend on the structure of the "filter cake," which can be adjusted at will by regulation of the stability properties of the original suspension.

Paints. A certain degree of flocculation of the pigment in a paint, particularly when accompanied by thixotropy, is desirable for the following reasons:

During brushing or spraying, the paint will be sufficiently thin for proper application and for the brush marks or drops to spread out evenly. A gradual thixotropic stiffening will prevent the paint from dripping off. At the same time, the thixotropic stiffening of the paint in the container will retard settling of the pigment. Finally, during long storage the sediment will have such a loosely built structure that homogenization will be relatively easy. If, on the other hand, the paint were in the peptized state, a compact sediment would be formed which is homogenized only with great difficulty.

CHAPTER TEN—INTERLAMELLAR AND OSMOTIC SWELLING—APPLICATIONS

The swelling of clays can be discussed in terms of particle interaction. Two stages of swelling should be distinguished.

In the first stage, up to four monolayers of water penetrate between the layers of an expanding clay, or between flat surfaces of adjoining particles. In the parting of the surfaces, three forces are operative—the van der Waals attraction, the electrostatic interaction of charged surfaces and cations, and the adsorption energy of the water (hydration energy). The last term appears to be the predominant term. The net potential curve of interaction of two plates as a function of the distance in the range considered can be derived from water-vapor-adsorption isotherms of expanding clays and simultaneous basal spacing measurements. From these net interaction energies, the forces operating in this stage of swelling are derived to be of the order of a few thousand atmospheres for the entrance of the first monolayer of water. The forces decline rapidly with the entrance of successive water layers. These computations show that the complete removal of the last layers of adsorbed water in a clay sediment by overburden pressure is not likely to occur in nature.

In the second stage of the swelling process in which the plates are separated to distances larger than equivalent with the thickness of four layers of water, the surface hydration energy is no longer important, and the swelling is now governed by double-layer repulsion, which is identical with the "osmotic pressure" of the system. The "swelling pressure" at a given plate distance is defined and measured by the confining pressure which must be applied to the system to prevent further swelling. These pressures, which vary between several times 10 atm to 0.1 atm or less with increasing particle distance, are of importance in foundation engineering and in the first stages of compaction of sediments in geology.

Since forces other than the double-layer repulsion (or osmotic pressure) are operative, the swelling pressure can only be identified with the osmotic pressure if the other forces are negligible.

Experiments on the free swelling of oriented flakes of montmorillonite clays have shown that an equilibrium layer distance is reached, indicating that the swelling is limited by an attractive force. This attractive force is probably not the van der Waals attractive force, but a cross-linking force supplied by occasional nonparallel particles, due to positive edge to negative face attraction. Hence even in well-oriented flakes, limited swelling due to cross linking occurs.

Consequently, the swelling pressure as defined by the confining pressure in compaction experiments is not identical with the osmotic pressure. Since the clay matrix is able to support pressure (grain pressure), the measured swelling pressure is probably larger than the osmotic pressure, and cannot be used to evaluate the double-layer repulsion. Irreversibility of swelling and compaction can be explained by irreversible changes in the geometry of the cross-linked clay matrix.

CHAPTER ELEVEN—INTERACTION OF CLAYS AND ORGANIC COMPOUNDS

Low-Molecular-Weight Compounds

ORGANIC ANIONS, SPECIFICALLY TANNATES. Tannates, particularly those of the red quebracho wood extract, are commonly used in drilling fluids. These tannates are water-soluble polyphenolates. They peptize clay suspensions when present in small amounts, that is, a few tenths of one percent. The mechanism of peptization is essentially the same as for inorganic peptizing chemicals. The tannate anion is chemisorbed at the particle edges by complexing with Al ions. In this way, a stabilizing negative double layer is built up.

Since the quebracho is a condensation product of a monomer with three phenolic groups, the quebracho anion is polyvalent. Hence its edge-charging effect as well as the activity reduction of the flocculating cations in the system will contribute greatly to the improvement of the salt tolerance of the clay suspension. It is possible, therefore, to treat the clay suspension with as much as 1–2 percent of quebracho and even an excess of alkali without decreasing the salt tolerance. Such heavily treated muds, the so-called "red muds," have the advantage that formation clays and drill cuttings are not as easily disintegrated as by muds of low peptizer content. The red mud is said to have "formation-conserving" properties. This fact seems paradoxical in view of the deflocculating effect of the tannate anions. It can be explained as follows: Before a peptizing anion can exercise its peptizing action, the edges of the clay particles must be accessible. The first step in the disintegration of a lump of clay is osmotic swelling, which is due to the face double-layer repulsion. Contrary to a mud with low peptizer concentrations, the high electrolyte concentration in the red mud represses the osmotic swelling because of a compression of the face double layers. Therefore, the edge surfaces do not become readily accessible for the peptizing anions, and the disintegration is considerably retarded.

A defect of the red muds is that they tend to stiffen at elevated temperatures. Under these conditions, the alkali probably reacts with the octahedral part of the clay which is exposed at the edges, and the reaction products, such as sodium aluminate, may well have a flocculating effect on the suspension. It has been found that the stiffening at high temperatures can be prevented by the addition of solid lime to the red mud. The reaction products are probably made harmless by precipitation with the lime. The calcium ions do not exert their usual strong flocculating effect on the clay suspension, probably because of complexing with the tannate anions. It is likely that these complex calcium-tannate anions are adsorbed on the clay-particle edges and thus peptize the clay.

At still higher temperatures, the lime red muds are no longer applicable, since they tend to solidify under such conditions. Cementlike silicates are probably formed by reaction of lime and alkali with the clay. The high-temperature solidification can be largely prevented or retarded by substituting lime with barium hydroxide.

The above interpretation of the red-mud and lime-red-mud behavior has been based on studies of model systems in which the tannate was substituted by simple polyphenols such as pyrogallol and the clay was substituted by the model sols alumina and silica.

Organic Cations. When water-soluble amine salts or quaternary ammonium salts are added to a clay suspension, the clay is flocculated or precipitated, but the clay can often be redispersed by the addition of more organic salt. Then the particles in the repeptized sol appear to have acquired a positive charge.

Initially, the organic cations are adsorbed by exchange with the exchangeable cations on the clay. The amine group becomes attached to the surface, and simultaneously the hydrocarbon chains displace the water from the clay surface and also become attached to the surface (Figure 34*a*). The exchange adsorption is often almost quantitative, indicating a strong preference of the surface for the organic cation. When all the exchange positions are occupied by the organic cations, the clay surface has become preferentially wet by oils and has lost its hydrophilic character. In this condition the clay is an effective emulsifier and also may be dispersed in oil. The loss of the water-wetting properties of the clay particles under these conditions may explain their flocculation or precipitation from the colloidal solution.

Upon further addition of organic salt, additional organic cations are adsorbed by association of the hydrocarbon part of their molecule with that of the previously adsorbed ions (Figure 34*b*). The ionic group points toward the water phase, and, together with the anions of the amine salt, a positive double layer is created, resulting in repeptization of the suspension.

In montmorillonites the organic cations are adsorbed on both the exterior and interior layer surfaces. From the measurement of the c spacing of the complex, the arrangement of the adsorbed molecules may be derived.

Polar Organic Compounds. Many organic compounds containing polar groups, such as alcohols, amines, and ketones, are strongly adsorbed on the layer surfaces of a clay displacing the adsorbed water. Such organo-clay complexes may be prepared either by mixing the clay-water suspension with the organic liquid or by contacting the dry clay directly with the organic liquid.

As discussed before, the clay is usually flocculated by a water-miscible organic compound because of a lowering of the dielectric constant of the medium, according to the Gouy theory. However, exceptions to this rule are known, and the refinement of the Stern model of the double layer is required to explain these exceptions on the basis of changes in the Stern layer due to adsorption of the organic compound.

Macromolecular Compounds

Polyelectrolytes. A very important method of stability control of clay suspensions is treatment with synthetic or natural polyelectrolytes. These compounds consist of long-chain macromolecules with ionized groups along the chain. Examples of polyanions are oxidized starch, carboxymethylcellulose, gum arabic, or synthetic polymethacrylates. Examples of polycations are synthetic polyamines. Proteins, which are amino acids, act as polycations in acid and as polyanions in an alkaline environment. In colloid chemistry, polyelectrolyte-water systems are referred to as "hydrophilic colloids." However, it has been recognized that polyelectrolyte solutions are true solutions in water, and that the colloidal properties of these solutions are due to the large size of the macromolecules, or ions. The hydrophilic colloids do not "flocculate" by the addition of salt, but they are sometimes "salted-out" at very high salt concentration owing to a reduction of their solubility.

The effect of polyelectrolytes on the stability of clay suspensions depends in a remarkable way on the concentration of the macro-ions. When present in a concentration of a few tenths of one percent and higher, they have a tremendous peptizing action. The salt tolerance of the clay suspension becomes so high that the clay remains suspended in concentrated salt solutions. Therefore, hydrophilic colloids are called *protective colloids.* On the other hand, when polyelectrolytes are present in a concentration of a few thousandths of 1 percent, they *sensitize* the clay suspensions toward flocculation by salt, as shown by a decrease of the salt-flocculation value. The two effects are demonstrated by the data in Figure 35.

At low polyelectrolyte concentrations, the long-chain ions are adsorbed on more than one clay particle simultaneously, thus providing a link between the particles, which promotes flocculation. The bridging will be easier in the presence of some salt, which allows the particles to come closer together, and which promotes the adsorption of the polyion at the clay surface.

At relatively high concentrations of polyelectrolyte, the polyions may envelop the individual particles and lend them their solubility properties and insensitivity to salt.

In technology, both the protective and the sensitizing action of polyelectrolytes is used to advantage. The *protective action* of larger amounts of polyelectrolytes enables the maintenance of stability of drilling fluids when drilling is conducted under marine conditions or in salt domes.

The following applications are based on the *sensitizing action* of very small amounts of polyelectrolytes:

Flocculation aids. When tenaceous suspensions are separated by settling or centrifugation, small amounts of polyelectrolytes aid the flocculation by normal flocculating salts, yielding flocs which are usually easier to handle.

Soil conditioners. Small amounts of polyelectrolytes give to the soil the desired flocculated condition at a low electrolyte level. They resist leaching better than the normal flocculating electrolytes owing to their adsorption on the clay surface.

Mud extenders. Polyelectrolytes improve the "yield" of a mud-clay without seriously affecting the plastering properties.

NONIONIC POLYMERS. Little work has been done thus far on the analysis of the interaction between clays and nonionic polymers, which are macromolecules containing polar groups instead of ionized groups along the chains. Examples of such compounds are natural starches or synthetic polyoxyethylene compounds. Like the low-molecular-weight polar organic compounds, they are adsorbed on the flat layer surfaces of the clays. Their effect on the stability of clay suspensions is rather unpredictable. Some compounds flocculate a clay suspension, and others do not. The nonionics may have both a sensitizing and a protective action, depending on their concentration.

Intercalation and Intersalation in Kaolinites

Species of the kaolinite family expand by interlayer adsorption of certain salts such as potassium acetate ("intersalation"). They also intercalate urea and other molecules which have a strong hydrogen-bonding ability, for example, dimethylsulfoxide, hydrazine, and others. Once the crystals have expanded, other compounds can be intercalated by displacement, and some species form regular interlayer hydrates when treated with water.

Clay-organic Reactions. Clays and the Origin of Life

A variety of reactions between clays and organic molecules has been observed, indicating the presence of acid sites and oxidation sites on the clay surfaces, as well as hydroxyl groups. Such reactions are very sensitive to the state of hydration of the clay. They are mostly occurring in the dry or nearly dry state.

Adsorption processes and catalytic activity of clays may be important in understanding the generation of prebiotic molecules of some complexity. Stereospecificity of reactions, imposed by surface and interlayer constraints, have been observed, and perhaps crystal imperfection patterns in clays may induce replication and provide information transfer.

CHAPTER TWELVE—ELECTROKINETIC AND ELECTROCHEMICAL PROPERTIES OF CLAY-WATER SYSTEMS

The electric properties of clay-water systems are primarily determined by the electric double layers on the clay surfaces. The face double layer is the most important contributor, although secondary effects of the edge double layer should not be overlooked. In clay technology, such properties are important in the study of electric logging procedures in boreholes and their interpretations, of pH and other ion-activity determinations in soils, and of the movement of water and electrolytes in soils and geological formations.

Reference is made to the main text of Chapter Twelve, which presents a rather brief summary of the present state of knowledge of the following phenomena:

Surface conductance or double-layer conductance.
Electrophoresis in suspensions.
Dorn effect and centrifugation potentials in suspensions.
Electrosmosis or electroendosmosis in capillaries and in porous media.
Streaming potentials in capillaries and in porous media.
pH measurements in suspensions, suspension effect, and Donnan potentials.
Membrane potentials.

APPENDIX I

Note on the Preparation of Clay Suspensions

In connection with certain technological problems, it is often desirable to study the behavior of different kinds of clays or of certain clays in different ion forms. The following note may be helpful in preparing well-defined clay suspensions.

The selection of a good clay sample which represents one of the main groups of clay minerals should be guided by X-ray analysis amplified by differential thermal analysis and by chemical analysis, including determination of the cation exchange capacity. Valuable information about many representative clays can be obtained from reports issued under the auspices of the American Petroleum Institute Project 49.

The purification of the clay and the conversion to a single ion form may be carried out as follows:

If the sample is suspected of containing organic material, the dry powder should be digested with concentrated hydrogen peroxide and left in contact for a day or so. The excess hydrogen peroxide may then be decomposed by gentle heating on a water bath.

The clay is suspended in distilled water (or deionized water) by vigorous stirring in some type of high-shear-dispersing equipment. The clay concentration should be kept as high as can be handled by the stirring equipment.

The concentrated suspension is diluted with distilled water until it is sufficiently dilute to allow settling of the clay by gravity. Usually, the raw clay is contaminated with a flocculating amount of electrolytes, so settling is rather rapid. The clear supernatant liquid is siphoned off, and the sediment is redispersed in distilled water. This procedure is repeated until the clay remains suspended owing to peptization, which occurs as soon as the salt content is lowered to below the flocculation value. Sometimes, peptization may be promoted by the addition of ammonia. At this point the coarse mineral impurities in the clay may be disposed of, since they still settle rapidly, and the peptized clay suspension can be siphoned off from the bottom sediment. Mineral impurities with the same range of particle sizes as that of the clays are difficult to remove. Their presence can be detected with the electron microscope.

The next step is to convert the suspension into the desired ion form—for example, the sodium form—by ion exchange. This conversion may be carried out in two ways:

A concentrated NaCl solution—for example, 1–2*M*—is added to the suspension. The clay flocculates and settles comparatively rapidly. The supernatant, which contains the majority of the ions which were originally present on the clay in exchange position, is discarded, and the NaCl treatment is repeated until the exchange is complete.

Alternatively, the clay may be converted into the sodium form with the aid of exchange resins in the sodium form. A batch procedure may be followed in which the clay suspension is shaken with beads of the resin. The clay suspension is then separated from the beads by decanting, the resin is regenerated by NaCl treatment and washing, and the contact with the clay suspension is repeated. The conversion with the aid of exchange resins can be carried out also by passing the dilute clay suspension through a column packed with beads of the resin in the desired ion form. Between repeated passes, the resin is regenerated by passing NaCl solution through the column and then washing. Often, a small amount of a salt of the exchanging ion is added to the suspension to accelerate exchange. This amount should be smaller than the flocculation value.

In all exchange procedures, it is desirable to use a large excess of exchanging ions with respect to the total exchange capacity of the clay in the system, particularly when replacing divalent ions on the clay with monovalent ions, because of the preference of the clay for the divalent cations.

The completion of the exchange process is indicated when, upon repetition of the procedure, those ions which were originally in exchange position on the clay are no longer released according to an analysis of the equilibrium liquid.

Alternatively, completion of the reaction can be determined by measuring the amount of exchanging cations taken up by the clay until this amount is constant. This check requires samples of the clay to be washed free of electrolyte.

After conversion of the clay, the suspension must be washed free of electrolyte, for example, by repeated decanting of the supernatant salt solution after settling and replenishment with conductivity water. In the first stages, when the clay is still flocculated, settling under gravity is usually rapid enough. In the later stages, when the clay becomes peptized, the super- or ultracentrifuge should be used. If desired, the clean clay suspension may be fractionated according to particle size by choosing appropriate conditions of centrifugation.

The best method by which completion of the washing procedure can be determined is to measure the conductivity of the supernatant liquid after each centrifugation until the conductivity is constant. Because of the slight solubility of the clay, the conductivity will not reach the low level of the conductivity of the water used in the washing.

A clay may be prepared in many different ion forms according to the procedure described above. Certain limitations inherent in the system often make it impossible, however, to prepare a clay in which 100 percent of the exchange positions are occupied by a single cation species. Owing to the slight solubility of the clay, the equilibrium liquid always contains a small amount of the cations which are present in the octahedral sheet of the clay, such as aluminum or magnesium ions. To a slight extent, these ions will occupy some of the exchange positions in the equilibrium condition. For example, when a clay with magnesium ions in both octahedral and in exchange positions is converted into a sodium clay, it is observed that the concentration of magnesium ions in the equilibrium solution decreases rapidly during the first few treatments with the sodium resin or with a sodium salt and reaches a constant value which is determined by the solubility of the octahedral magnesium ions. It can be expected that after washing the final product will contain some magnesium in exchange position.

Large effects of the solution of clays on the course of the exchange process are encountered when the hydrogen form of a clay is prepared either by HCl treatment or by conversion with a resin in the hydrogen form. The acid attacks the clay plates at the edges, and aluminum ions are dissolved and are subsequently transferred to the exchange sites, where they are held in preference to the hydrogen ions (4–18). The amount of Al in exchange position in a "hydrogen" clay can be measured according to a procedure given by Low (9). During the conductometric titration of the clay with NaOH, a curve consisting of two intersecting straight lines with different slopes is obtained before the equivalence point is reached. During the first part of the titration, hydrogen ions are exchanged; during the second part, aluminum ions are replaced. For example, the conversion of a Wyoming bentonite suspension with $0.01N$ HCl led to the occupation of 60 percent of the exchange sites by Al. In the case of conversion with a hydrogen resin of the same clay, 40 percent of the exchange sites became occupied with Al. Because of the attack of the lattice by acids, it is, in general, not recommended to prepare various ion forms of a clay via the hydrogen form followed by the addition of an equivalent amount of the base of the desired metal to the hydrogen clay.

An alternative procedure for the removal of the excess electrolyte after conversion is electrodialysis. This method is not recommended for clays, since electrodialysis appears to accelerate hydrolysis, by which, for example, a sodium clay will be partly converted into a hydrogen clay.

Hydrolysis of the clay often takes place to some extent merely when the clay is washed with distilled water. Upon prolonged washing, the pH of the suspension decreases. This process is faster for one type of clay than for another. Therefore, the pH should be checked regularly during washing. The check should be made in the suspension rather than in the filtrate or supernatant liquid because of the suspension effect (see Figure 38). Hydrolysis of the clay is usually prevented by washing with alcohol.

In research, it is often desirable to study a certain particle-size fraction of the clay in different ion forms. In that case, the various ion forms should be prepared from the same fraction of the clay, preferably from a fraction separated after conversion to the Na form. If one were to prepare the various ion forms of the raw clay first and then separate a certain centrifuge fraction of each ion form, these fractions would not represent the same fraction of the raw clay, since the particle size is very much dependent on the ion form of the clay.

References

PREPARATORY TECHNIQUES

1. Lewis, D. R. (1953), Replacement of cations of clay by ion exchange resins, *Ind. Eng. Chem.*, **45,** 1782.
2. Bolt, G. H., and Frissel, M. J. (1960), The preparation of clay suspensions with specified ionic composition by means of exchange resins, *Soil Sci. Soc. Am. Proc.*, **24,** 172–177.
3. *Clay Minerals Bull.*, **3** (1956), 1–47 (13 papers).

ACID TREATMENT OF CLAYS

4. Coleman, N. T., and Harward, M. E. (1953), The heats of neutralization of acid clays and cation exchange resins, *J. Am. Chem. Soc.*, **75,** 6045–6946.
5. Harward, M. E., and Coleman, N. T. (1954), Some properties of H^+ and Al^{3+} clays and exchange resins, *Soil Sci.*, **78,** 181–188.
6. Wiegner, G. (1931), Some physicochemical properties of clay, II—Hydrogen clay, *J. Soc. Chem. Ind. (London)*, **50,** 1035.
7. Paver, H., and Marshall, E. C. (1934), The role of aluminum in the reactions of the clays, *J. Soc. Chem. Ind. (London)*, **53,** 750.
8. Mukherjee, J. N., Chatterjee, B., and Banerjee, B. M. (1947), Liberation of H^+, Al^{+++}, and Fe^{+++} ions from hydrogen clays by neutral salts, *J. Colloid Sci.*, **2,** 247.
9. Low, P. F. (1955), The role of aluminum in the titration of bentonite, *Soil Sci. Soc. Am. Proc.*, **19,** 135–139.
10. Laudelout, H., and Eeckman, J. P. (1958), La stabilité chimique des suspensions d'argile saturée par l'ion hydrogène, *Vortragsveröffentlichungen der zweiten und vierten Kommission der Internationalen Bodenkundigen Gesellschaft, Hamburg,* **II,** 193–199.
11. Eeckman, J. P., Massart, Y., and Laudelout, H. (1959), Stabilité chimique et propriété catalytique des suspensions de montmorillonites acides, *Bull. Groupe Franç. Argiles,* **11,** 3–8.
12. Eeckman, J. P., and Laudelout, H. (1961), Chemical stability of hydrogen-montmorillonite suspensions, *Kolloid-Z,* **178,** 99–107.

13. Heyding, R. D., Ironside, R., Norris, A. R., and Prysiazniuk, R. Y. (1960), Acid activation of montmorillonite, *Can. J. Chem.*, **38**, 1003–1016.
14. Granquist, W. T., and Sumner, C. G. (1959), Acid dissolution of a Texas bentonite, *Clays, Clay Minerals*, **6**, 292–308.
15. Davis, L. E. (1961), The instability of neutralized H-Al bentonites, *Soil Sci. Soc. Am. Proc.*, **25**, 25–27.
16. Spain, J. M., and White, J. L. (1961), Titration of H-Al bentonite with tetra-alkyl ammonium bases, *Soil Sci. Soc. Am. Proc.*, **25**, 480–483.
17. Barshad, I. (1960), The effect of the total chemical composition and crystal structure of soil minerals on the nature of the exchangeable cations in acidified clays and in naturally occurring acid soils, *Trans. Intern. Congr. Soil Sci., 7th, Madison, Wis.*, Vol. II, 435–444.
18. Schwertmann, U., and Jackson, M. L. (1963), Hydrogen-aluminum clays: A third buffer range appearing in potentiometric titration, *Science*, **139**, 1052–1054.

APPENDIX II

Miscellaneous Computed Data for Montmorillonites

A. TOTAL LAYER SURFACE AREA AND SURFACE CHARGE DENSITY

As an example, consider a montmorillonite clay in the sodium form having the following unit-cell formula:

$$\underset{\vdots \atop Na_{0.67}}{(Si_8)(Al_{3.33}Mg_{0.67})O_{20}(OH)_4}$$

The unit-cell weight for this formula is 734. Therefore, 734 g of this clay contain 6.02×10^{23} unit cells (Avogadro's number).

Each cell has a surface area of 5.15×8.9 A^2 on each side. (The dimensions of the unit cell are slightly dependent on the type of isomorphous substitutions in the lattice.)

Thus, the total surface area of 1 g of clay is

$$\left(\frac{1}{734}\right) \times 6.02 \times 10^{23} \times 2 \times 5.15 \times 8.9 \text{ A}^2/\text{g} = 752 \text{ m}^2/\text{g}$$

The CEC of the above material is 91.5 meq per 100 g, or 0.915 meq/g,* which is equivalent to $0.915 \times 6.02 \times 10^{20}$ monovalent cations per gram. This number of cations covers a surface area of 750×10^{20} A^2. Thus, the surface area which is available per monovalent exchangeable cation is

$$\frac{(750 \times 10^{20})}{(0.915 \times 6.02 \times 10^{20})} = 136 \text{ A}^2/\text{ion}$$

Alternatively, the surface area per cation follows immediately from the

* CEC = (0.67/unit-cell wt.) × 1000 × 100 meq per 100 g.

unit-cell formula and the unit-cell dimensions:

$$\frac{(2 \times 5.15 \times 8.9)}{0.67} = 136 \text{ A}^2/\text{ion}$$

The interlayer area per cation is $\frac{1}{2} \times 136 = 68$ A²/ion.

In general, there is about $\frac{2}{3}$ of a monovalent cation per unit cell.

For divalent cations, the surface area per ion is twice as large as for monovalent cations.

The surface charge density can be computed as follows: On each sq cm of surface area, there are $10^{16}/136$ elementary charges. This is equivalent to a charge density of 3.5×10^4 esu/cm², or 11.7 μC/cm².

The data for the CEC and the surface area per gram for a given clay lattice vary somewhat with the type of exchangeable ion, since the unit-cell weight comprises the weight of the exchangeable ion. For example, such a variation amounts of 13 percent between a hydrogen and a cesium clay (Table II-1).

B. AVERAGE PARTICLE DISTANCE IN SUSPENSIONS OF CLAYS FOR PARALLEL AND CUBIC STACKING OF THE PLATES

The following computations give an indication of particle distances in clay suspensions at various clay concentrations, assuming that the plates are *parallel* to each other.

Table II-1

Unit Cell Weight and Cation Exchange Capacity (CEC) and Total Surface Area per Unit Weight of a Montmorillonite Clay as a Function of the Species of Exchangeable Cation.

Ion form	Area m²/g	Unit-cell wt.	CEC, meq/100 g
Li	762.9	723.3	92.5
Na	751.8	734.0	91.5
K	740.9	744.8	90.0
Rb	711.3	775.8	86.5
Cs	675.7	816.6	82.0
Mg	759.2	726.8	92.0
Ca	753.5	732.3	91.5
Sr	737.8	747.9	89.5
Ba	721.7	764.6	87.5
H	767.2	719.3	93.0

1. Bentonite

Every particle consists of n layers 10 A thick and 10 A apart, since they are fully hydrated. The average particle distance is b A.

The particle thickness is $(n - 1) \times 20 + 10$ A. The distance between the center planes of two particles is $b + (n - 1) \times 20 + 10$ A. Let the total volume occupied by 1 g of clay be $a \times a \times a$ A^3. Then, there are

$$\frac{a}{b + (n - 1) \times 20 + 10} \text{ squares of } a \times a \text{ A}^2$$

Since 1 g of dry bentonite represents about 375 m^2 of layers,

$$\frac{375 \times 10^{20}}{n} = \frac{a^3}{b + (n - 1) \times 20 + 10}$$

Then, the relation between the clay concentration, c, in grams per 100 cm^3 of suspension and the particle distance, b, becomes

$$c = \frac{100}{10^{-24} \times a^3} = 2670 \times \frac{n}{b + (n - 1) \times 20 + 10} \text{ g/1000 cm}^3 \text{ of suspension.}$$

Some figures for the average distance, b, and the corresponding clay concentration for various values of n, the average degree of layer stacking in a particle, are given in Table II-2.

2. Illite

For a clay such as illite, which does not contain interlayer water, the relation between c and b is

$$c = 2670 \times \frac{n}{(b + 10n)}$$

Some figures are given in Table II-2.

For *cubic arrangement* of the clay plates, the values for c at a given value of b should be multiplied by 3. Some figures for bentonite are given in Table II-2. The table shows, for example, that plates of a diameter of 3000 A and 2 layers thick can build a continuous cubic network throughout the total volume available if the clay concentration is 5.28 g per 100 cm^3 of suspension. For plates of 10,000 A diameter, which are quite common in bentonite suspensions, and 2 layers thick, only 1.6 g of clay per 100 cm^3 of suspension is required to build a coherent cubic network throughout the whole volume.

A good illustration of the expanse of the plates in a clay suspension is given by the computation of the length of the ribbon obtained by lining up the particles in a sodium bentonite sol. For example, the length of the rib-

Table II-2

Particle Distances and Corresponding Clay Concentrations in Clay Suspensions and Gels

Average particle distance (b), A	Clay concentration (c), g/100 cm³				
	$n^a = 1$	$n = 2$	$n = 4$	$n = 7$	$n = 10$
	Bentonite, Parallel Stacking				
100	24.3	41.0	62.8	81.0	—
200	12.7	23.2	39.5	56.5	68.3
300	8.6	16.2	28.8	43.5	54.3
500	5.2	10.0	18.7	29.5	38.6
750	3.5	6.8	13.0	21.2	28.3
1000	2.65	5.2	10.0	16.5	22.5
1500	1.77	3.47	6.8	11.4	16.0
2000	1.33	2.62	5.15	9.75	12.2
3000	0.88	1.76	3.48	6.0	8.4
5000[b]	0.53	1.06	2.10	3.65	5.4
	Bentonite, Cubic Stacking				
100	72.9				
200	38.1	69.6			
300	25.8	48.6			
500	15.6	30.0	56.1		
750	10.5	20.4	39.0	63.6	
1000	7.95	15.6	30.0	49.5	67.5
1500	5.91	10.4	20.4	34.2	48.0
2000	3.99	7.86	15.5	26.2	36.6
3000	2.65	5.28	10.4	18.0	25.2
5000[c]	1.59	3.18	6.3	10.9	16.2
	Illite, Parallel Stacking				
100	24.3	44.5	76.0		
500	5.2	10.2	19.8	32.5	44.3
1000	2.65	5.25	10.3	17.4	24.3
5000	0.53	1.06	2.10	3.70	5.5

[a] n = the average degree of layer stacking in a particle.
[b] For larger particle distances, c is roughly inversely proportional to b.
[c] For larger particle diameters, c is roughly inversely proportional to b.

bon of bentonite particles considered as squares of 10,000 × 10,000 A² and consisting of 2 layers per particle in 100 cm³ of a 0.01 percent suspension (or 0.01 of clay) is

$$\left(0.01 \times \frac{375}{2} \times 10^{20}\right) \Big/ 10^4 \text{ A} = 188 \times 10^{14} \text{ A}$$

$$= 1880 \text{ km, or more than 1000 miles}$$

C. FORMULA COMPUTATION FROM CHEMICAL ANALYSIS

The general formula for the members of the montmorillonite group is

$$\underbrace{(Si_{4-y}Al_y)}_{\text{tetrahedral}}\underbrace{(Al_{a-y}Fe_b{}^3Fe_c{}^2Mg_dCr_e{}^3Mn_f{}^3Mn_g{}^2Li_h)}_{\text{octahedral}}O_{10}(OH,F)_2\text{— —}\underbrace{X_x}_{\text{exchange ions}}$$

In computation of the formula from the chemical analysis, TiO_2 and P_2O_5 may be considered as mineral impurities.

The following arbitrary assumptions are made:

1. There are 20 oxygen atoms and 4 OH groups (sometimes substituted by *F*) per unit cell, or 10 and 2, respectively, in the chemical formula representing half a unit cell.
2. All the Si present is assigned to the tetrahedral sheet.
3. The remainder of the tetrahedral positions are filled exclusively by Al. Any additional Al is assigned to the octahedral sheet, together with Mg, Fe, and others, except, of course, those cations which are in exchange position.

The analytical values for the percentage of the oxides of the crystal elements and the exchangeable ions must be reduced to atomic proportions. The percentage of the oxides of the crystal elements is divided by the molecular weight of the oxides and multiplied by the number of atoms of the positive element in the oxide. The percentages of the oxides of the exchangeable ions are divided by the equivalent weight of the oxides. Alternatively, the cation exchange capacity of the clay, expressed in equivalents of exchangeable ions per 100 g of clay, may be chosen (see Table II-3).

The atomic proportions thus obtained are represented by the same letters as those given in the general formula, but the letters are capitalized. These proportions must be multiplied by a factor K to obtain the actual amounts in the formula, for example, $K \times D = d$. The atomic proportion of Si, computed from the analysis, will be represented by Z. Therefore, $K \times Z = 4 - y$.

The total number of positive charges must be equal to the total number of negative charges. The latter is given by the formula as equal to 22. Therefore, according to assumption 1,

$$K(4Z + 3Y + 3A - 3Y + 3B + 2C + 2D + 3E + 3F + 2G + H + X) = 22$$

In this equation, K is the only unknown, since Y is eliminated, so K can be computed.

Table II-3

Conversion Factors for the Computation of Atomic Proportions in the Unit Cell from the Chemical Analysis and Cation Exchange Capacity

Crystal elements			Elements in exchange position		
Oxide	Factor[a]	Symbol	Oxide	Factor[b]	Symbol
SiO_2	1/60.06	Z	Na_2O	1/31.10	X
Al_2O_3	2/102.0	A	K_2O	1/47.09	X
Fe_2O_3	2/159.8	B	CaO	1/28.00	X
FeO	1/71.85	C	MgO	1/20.15	X
MgO	1/40.32	D			
Cr_2O_3	2/152.0	E			
Mn_2O_3	2/157.9	F			
MnO	1/70.93	G			
Li_2O	2/29.88	H			

[a] Percent oxide × factor = symbol.
[b] Percent oxide × factor = X. (Alternatively take X equal to the cation exchange capacity expressed in equivalents per 100 g.)

The value for Y follows from the requirement that the sum of the Si and Al atoms in tetrahedral coordination equal 4. Thus, according to assumptions 2 and 3, $K(Z + Y) = 4$.

Since K is known from the first equation, Y can be computed.

References

1. Marshal, C. E. (1935), Layer lattices and the base exchange clays, *Z. Krist.*, **91A,** 433.
2. Ross, C. S., and Hendricks, S. B. (1945), *Minerals of the montmorillonite group, their origin, and relation to soils and clays,* Professional Paper 205-B, U.S. Dept. of the Interior, Geological Survey.
3. Osthaus, B. B. (1954), Chemical determination of tetrahedral ions in nontronite and montmorillonite, *Clays, Clay Minerals,* **2,** 404–417.
4. Foster, M. D. (1951), The importance of the exchangeable magnesium and cation exchange capacity in the study of montmorillonite clays, *Am. Mineralogist,* **36,** 717–730.

APPENDIX III

Electric Double-Layer Computations (2–4)

This Appendix is intended to acquaint the reader with formulas which describe the properties of electric double layers and their interaction. Although these formulas are usually rather long, they are not difficult to handle numerically in certain specific problems. It is often more difficult, however, to decide which formula is applicable to the specific problem at hand. In the derivation of certain formulas approximations have been introduced; therefore, one must know whether these approximations are also justified in the particular application.

Furthermore, one must be aware of the boundary conditions which apply to the specific problem. Many double-layer computations in textbooks and in the literature are valid for surfaces of constant potential which are most common in colloid chemistry in general, since the surface potential is usually determined by the concentration of specific "potential-determining ions." In the case of the flat-layer surface of clays, however, we deal with a constant-charge surface, since the charge is determined by the isomorphous substitution in the lattice interior.

We shall give a brief derivation of the formulas for the potential and charge distribution in a single flat double layer of the Gouy type and of the Stern type. Next, the same characteristics will be derived for interacting double layers. Finally, the energy of double-layer interaction will be discussed. In several examples, the numerical handling of the formulas will be illustrated. In general, the formulations and notations of Verwey and Overbeek (Appendix V, reference B-6) will be followed.

Of special interest for the quantitative discussion of the electrical double-layer theory is the choice of the system of electrical units. In the classical literature, and still to a large extent in the contemporary literature, the three quantity electrostatic system is used rather than the rationalized four quantity system which is part of the SI system.

In the electrostatic system, the Coulomb equation is written $F = Q_1Q_2/\epsilon_r r^2$, and the Poisson equation is then $\nabla^2 \phi = -4\pi\rho/\epsilon_r$, in which ϵ_r is the static dielectric constant or the *relative* static permittivity.

In the rationalized four quantity system (part of the SI system) Coulomb's equation is $F = Q_1Q_2/4\pi\epsilon r^2$, and the Poisson equation becomes $\nabla^2 \phi = -\rho/\epsilon$, in which ϵ is now the static permittivity, which equals the product of the relative static permittivity and the permittivity in vacuum: $\epsilon = \epsilon_r\epsilon_0$.

Hence in the older literature on the diffuse electrical double layer the factor 4π appears in the equations, and the symbol ϵ_r stands for the dielectric constant of the medium, which is of the order of 80 for water. When using the rationalized four quantity system, these factors 4π are eliminated in the double layer formulas, whereas the symbol ϵ stands for the static permittivity which for water at room temperature is of the order of $80 \times 8.85 \times 10^{-12}$ $J^{-1}C^{-2}m^{-1}$ or Fm^{-1}. Factors of π will only occur in the formulas of the rationalized system when spherical problems are involved.

In view of the still popular use of the electrostatic system, the following discussion and calculations will be in esu units. However, an example will be given of a calculation using the rationalized four-quantity system, as adopted in the SI system.

A. THE SINGLE FLAT DOUBLE LAYER

1. Potential and Charge Distribution in a Single Flat Double Layer According to the Gouy Theory

The counter-ions of the double layer are subject to two opposing tendencies. Electrostatic forces attract them to the charged surface, whereas diffusion tends to bring them away from the surface toward the equilibrium solution, where their concentration is smaller.

Simultaneously, ions of the same sign as the surface charge are repelled by the surface, and back-diffusion from the equilibrium solution toward the surface counteracts the electric repulsion.

When equilibrium is established in the double layer, the average local concentration of ions at a distance x from the surface can be expressed as a function of the average electric potential, Φ, at that distance according to Boltzmann's theorem:

$$n_- = n_-^* \exp\left(\frac{v_- e\Phi}{kT}\right)$$

$$n_+ = n_+^* \exp\left(\frac{-v_+ e\Phi}{kT}\right) \qquad \text{(III-1)}$$

$$\rho = v_+ e n_+ - v_- e n_-$$

in which n_+ and n_- are the local concentrations of the positive and negative ions, and n_+^* and n_-^* are their concentrations far away from the surface in the equilibrium liquid. (When Φ is negative, $n_+ > n_-$; when Φ is positive,

Table III-1

Molarity mol/dm³	1-1 Valent		2-2 Valent	
	$1/\kappa$	Φ_0	$1/\kappa$	Φ_0
10^{-5} *M*	10^{-5} cm	322 mV	0.5×10^{-5} cm	161 mV
10^{-3} *M*	10^{-6} cm	208 mV	0.5×10^{-6} cm	104 mV
10^{-1} *M*	10^{-7} cm	94 mV	0.5×10^{-7} cm	47 mV

$n_- > n_+$.) The concentrations are expressed as numbers of ions per cm³. The valences of the ions are v_+ and v_-, e is the elementary charge, k is the constant of Boltzmann, T is the absolute temperature, and ρ is the local density of charge.

The local density of charge and the local electric potential are also related by Poisson's equation:

$$\frac{d^2\Phi}{dx^2} = -\left(\frac{4\pi}{\epsilon}\right)\rho \tag{III-2}$$

in which $d^2\Phi/dx^2$ is the variation of the field strength, $-d\Phi/dx$, with distance, and ϵ is the dielectric constant of the medium.

For the sake of simplicity of the formulations, we shall assume that $v_+ = v_-$, since the valence of the ion of the same sign as the surface charge is not very important. Then, $n_-^* = n_+^* = n$.

Combining equations (III-1) and (III-2), we obtain the fundamental differential equation for the double layer:

$$\frac{d^2\Phi}{dx^2} = \left(\frac{8\pi nve}{\epsilon}\right)\sinh\left(\frac{ve\Phi}{kT}\right) \tag{III-3}$$

It is convenient to rewrite this equation in terms of the following dimensionless quantities:

$$y = \frac{ve\Phi}{kT} \qquad z = \frac{ve\Phi_0}{kT} \qquad \xi = \kappa x$$

in which

$$\kappa^2 = \frac{8\pi ne^2v^2}{\epsilon kT}\,\text{cm}^{-2}$$

Then, equation (III-3) becomes simply

$$\frac{d^2y}{d\xi^2} = \sinh y \tag{III-3a}$$

Integrating once with the boundary conditions that for $\xi = \infty$, $dy/d\xi = 0$ and $y = 0$, we obtain

$$\frac{dy}{d\xi} = -(2 \cosh y - 2)^{1/2} = -2 \sinh\left(\frac{y}{2}\right) \qquad \text{(III-4)}$$

The second integration with the boundary conditions that for $\xi = 0$, $\Phi = \Phi_0$ or $y = z$ yields

$$e^{y/2} = \frac{e^{z/2} + 1 + (e^{z/2} - 1)e^{-\xi}}{e^{z/2} + 1 - (e^{z/2} - 1)e^{-\xi}} \qquad \text{(III-5)}$$

This equation describes the decay of the potential as a function of the distance from the surface at a given surface potential ($\sim z$) and at a given

Table III-2

Pairs of Values of Surface Potential ($\sim -z$) and Midway Potential ($\sim -u$) for Three Different Values of the Surface Charge (σ) at Various Electrolyte Concentrations ($\sim\kappa$) and Plate Distances (d)[a]

$(dy/d\xi)_0 = 840$			$(dy/d\xi)_0 = 560$			$(dy/d\xi)_0 = 280$		
κd	$-u$	$-z$	κd	$-u$	$-z$	κd	$-u$	$-z$
0.2555	5.0	13.47	0.2543	5.0	12.66	0.6943	3.0	11.27
0.1985	5.5	13.47	0.1973	5.5	12.66	0.5389	3.5	11.27
0.1540	6.0	13.47	0.1528	6.0	12.66	0.4181	4.0	11.27
0.1194	6.5	13.47	0.1182	6.5	12.66	0.3240	4.5	11.27
0.09249	7.0	13.47	0.09130	7.0	12.66	0.2507	5.0	11.27
0.07150	7.5	13.47	0.07032	7.5	12.66	0.1937	5.5	11.27
0.05516	8.0	13.47	0.05400	8.0	12.67	0.1493	6.0	11.27
0.04244	8.5	13.47	0.04126	8.5	12.67	0.1147	6.5	11.28
0.03253	9.0	13.48	0.03136	9.0	12.68	0.08776	7.0	11.28
0.02481	9.5	13.49	0.02366	9.5	12.70	0.06679	7.5	11.29
0.01881	10.0	13.50	0.01767	10.0	12.72	0.05049	8.0	11.31
0.01414	10.5	13.52	0.01304	10.5	12.77	0.03781	8.5	11.33
0.01052	11.0	13.55	0.00947	11.0	12.83	0.02799	9.0	11.37
0.00772	11.5	13.60	0.00674	11.5	12.93	0.02041	9.5	11.43
0.00557	12.0	13.67	0.00469	12.0	13.07	0.01460	10.0	11.52
						0.01021	10.5	11.65
						0.00697	11.0	11.84
						0.00463	11.5	12.08
						0.00301	12.0	12.39
						0.00191	12.5	12.76
						0.00120	13.0	13.16

(continued)

Table III-2 (*continued*)

$(dy/d\xi)_0 = 187$			$(dy/d\xi)_0 = 94$			$(dy/d\xi)_0 = 84$		
κd	$-u$	$-z$	κd	$-u$	$-z$	κd	$-u$	$-z$
0.8909	2.5	10.46	0.8803	2.5	9.09	2.7112	0.5	8.86
0.6907	3.0	10.46	0.6802	3.0	9.09	1.9516	1.0	8.86
0.5354	3.5	10.46	0.5248	3.5	9.09	1.4792	1.5	8.86
0.4145	4.0	10.46	0.4040	4.0	9.09	1.1373	2.0	8.86
0.3204	4.5	10.46	0.3099	4.5	9.10	0.8778	2.5	8.86
0.2472	5.0	10.47	0.2367	5.0	9.10	0.6776	3.0	8.86
0.1902	5.5	10.47	0.1798	5.5	9.11	0.5223	3.5	8.87
0.1458	6.0	10.47	0.1354	6.0	9.13	0.4015	4.0	8.87
0.1112	6.5	10.48	0.1010	6.5	9.16	0.3074	4.5	8.87
0.08428	7.0	10.49	0.07441	7.0	9.20	0.2342	5.0	8.88
0.06337	7.5	10.51	0.05390	7.5	9.27	0.1773	5.5	8.90
0.04713	8.0	10.54	0.03826	8.0	9.38	0.1330	6.0	8.92
0.03458	8.5	10.59	0.02653	8.5	9.53	0.09871	6.5	8.95
0.02493	9.0	10.67	0.01793	9.0	9.74	0.07219	7.0	9.01
0.01760	9.5	10.79				0.05184	7.5	9.09
0.01213	10.0	10.95				0.03643	8.0	9.21
0.00814	10.5	11.17				0.02497	8.5	9.39
						0.01668	9.0	9.63
						0.01087	9.5	9.92
						0.06941	10.0	10.28
						0.04357	10.5	10.68
						0.02703	11.0	11.11

(*continued*)

electrolyte concentration ($\sim \kappa^2$). It represents roughly an exponential decay, as sketched in Figure 11.

It is of interest to consider an approximation which is valid for small surface potentials ($\Phi \ll 25$ mV, $z \ll 1$).

The fundamental differential equation becomes

$$\frac{d^2\Phi}{dx^2} = \kappa^2\Phi \tag{III-3'}$$

and

$$\Phi = \Phi_0 \exp(-\kappa x) \tag{III-5'}$$

The decay of the potential with distance from the surface is now purely exponential. The center of gravity of the space charge coincides with the plane $\kappa x = 1$ or $x = 1/\kappa$. Hence $1/\kappa$ may be called the "thickness" of the

Table III-2 (*continued*)

$(dy/d\xi)_0 = 56$			$(dy/d\xi)_0 = 28$			$(dy/d\xi)_0 = 18.7$		
κd	$-u$	$-z$	κd	$-u$	$-z$	κd	$-u$	$-z$
2.6993	0.5	8.05	8.9158	0.001	6.67	8.8804	0.001	5.86
1.9397	1.0	8.05	6.6132	0.01	6.67	6.5778	0.01	5.86
1.4673	1.5	8.05	5.0031	0.05	6.67	4.2728	0.1	5.86
1.1254	2.0	8.05	3.6091	0.2	6.67	3.0013	0.35	5.86
0.8659	2.5	8.05	3.0366	0.35	6.67	2.4337	0.6	5.86
0.6658	3.0	8.06	2.6636	0.50	6.67	2.0521	0.85	5.86
0.5105	3.5	8.06	2.3826	0.65	6.67	1.8687	1.0	5.87
0.3897	4.0	8.07	1.9040	1.0	6.67	1.3965	1.5	5.87
0.2958	4.5	8.08	1.4317	1.5	6.67	1.0549	2.0	5.87
0.2227	5.0	8.10	1.0899	2.0	6.67	0.7959	2.5	5.89
0.1660	5.5	8.13	0.8305	2.5	6.68	0.5965	3.0	5.91
0.1221	6.0	8.17	0.6306	3.0	6.69	0.4423	3.5	5.95
0.08834	6.5	8.24	0.4756	3.5	6.71	0.3234	4.0	6.00
0.06262	7.0	8.35	0.3554	4.0	6.73	0.2322	4.5	6.09
0.04334	7.5	8.51	0.2623	4.5	6.77	0.1631	5.0	6.21
0.02923	8.0	8.72	0.1905	5.0	6.84			
0.01923	8.5	8.99	0.1357	5.5	6.94			
0.01236	9.0	9.33	0.09445	6.0	7.08			
			0.06409	6.5	7.28			
			0.04329	7.0	7.54			
			0.02739	7.5	7.86			
			0.01736	8.0	8.23			
			0.01084	8.5	8.65			
			0.00670	9.0	9.09			
			0.002513	10.0	10.03			

(*continued*)

double layer; it is equal to the "characteristic length" in the Debye-Hückel theory of strong electrolytes.

Finally, the total double-layer charge can be computed from the potential function as follows:

$$\sigma = -\int_0^\infty \rho dx = \frac{\epsilon}{4\pi}\int_0^\infty \frac{d^2\Phi}{dx^2}dx = -\frac{\epsilon}{4\pi}\left[\frac{d\Phi}{dx}\right]_{x=0}$$

Apparently, the surface charge is determined by the initial slope of the potential function. The result is

$$\sigma = \left(\frac{n\epsilon kT}{2\pi}\right)^{1/2}(2\cosh z - 2)^{1/2} = \left(\frac{2n\epsilon kT}{\pi}\right)^{1/2}\sinh\left(\frac{z}{2}\right) \qquad \text{(III-6)}$$

Table III-2 (*continued*)

$(dy/d\xi)_0 = 9.4$			$(dy/d\xi)_0 = 8.4$			$(dy/d\xi)_0 = 5.6$		
κd	$-u$	$-z$	κd	$-u$	$-z$	κd	$-u$	$-z$
8.7760	0.001	4.50	15.659	10^{-6}	4.28	15.545	10^{-6}	3.51
6.4734	0.01	4.50	13.356	10^{-5}	4.28	13.242	10^{-5}	3.51
3.2424	0.25	4.51	11.054	10^{-4}	4.28	10.940	10^{-4}	3.51
2.5240	0.5	4.51	8.7513	10^{-3}	4.28	8.637	10^{-3}	3.51
2.0868	0.75	4.51	6.4487	0.01	4.28	6.335	0.01	3.51
1.7650	1.0	4.52	4.1436	0.1	4.28	4.030	0.1	3.51
1.5080	1.25	4.52	2.4993	0.5	4.29	2.386	0.5	3.51
1.2939	1.5	4.53	1.7406	1.0	4.30	1.6292	1.0	3.54
0.9540	2.0	4.56	1.2700	1.5	4.32	1.1621	1.5	3.58
0.6979	2.5	4.61	0.9309	2.0	4.36	0.8289	2.0	3.66
0.5029	3.0	4.69	0.6760	2.5	4.41	0.5824	2.5	3.78
0.3551	3.5	4.80	0.4827	3.0	4.51	0.4003	3.0	3.94
0.2449	4.0	4.96				0.2684	3.5	4.17
0.1646	4.5	5.18				0.1756	4.0	4.45
0.1079	5.0	5.47				0.1124	4.5	4.80

(*continued*)

For small surface potentials,

$$\sigma \approx \left(\frac{n\epsilon}{2\pi kT}\right)^{1/2}(ve\,\Phi_0) = \left(\frac{\epsilon\kappa}{4\pi}\right)\Phi_0 \qquad \text{(III-6}'\text{)}$$

so that $C = \epsilon\kappa/4\pi$ may be written as the capacity of the double layer.

For a flat double layer of the Gouy model in which ions are considered as point charges and there is no specific interaction between ions, surface and solvent, we find that the electric potential as a function of distance is given by equation (III-5) (equation III-5′ for small potentials), the double-layer charge is given by equation (III-6) (or equation III-6′), and the distribution of ions in the double layer is described by equation (III-1), taking into account that the potential changes with distance according to equation (III-5). The computation of the total excess of counter-ions and the total deficit of ions of the same sign as the surface charge by integrating equation (III-1) will be discussed later.

From the formulas, the following rules are derived for double layers of constant potential and of constant charge, respectively, regarding the effect of the addition of electrolytes.

A. Surfaces of Constant Potential (Constant Concentration of Potential-Determining Ions). The charge increases proportionally

Table III-2 (*continued*)

$(dy/d\xi)_0 = 2.8$			$(dy/d\xi)_0 = 1.87$			$(dy/d\xi)_0 = 0.94$		
κd	$-u$	$-z$	κd	$-u$	$-z$	κd	$-u$	$-z$
15.741	10^{-7}	2.28	15.476	10^{-7}	1.67	14.906	10^{-7}	0.91
15.231	10^{-6}	2.28	14.965	10^{-6}	1.67	14.396	10^{-6}	0.91
12.928	10^{-5}	2.28	12.663	10^{-5}	1.67	12.093	10^{-5}	0.91
10.625	10^{-4}	2.28	10.360	10^{-4}	1.67	9.790	10^{-4}	0.91
8.323	10^{-3}	2.28	8.058	10^{-3}	1.67	7.488	10^{-3}	0.91
6.020	10^{-2}	2.28	6.448	0.005	1.67	5.185	10^{-2}	0.91
4.410	0.05	2.28	5.755	0.01	1.67	3.576	0.05	0.91
3.715	0.1	2.28	4.145	0.05	1.67	2.883	0.1	0.91
2.077	0.5	2.30	3.450	0.1	1.67	1.303	0.5	1.02
1.3351	1.0	2.38	1.8213	0.5	1.72			
0.8936	1.5	2.52	1.1075	1.0	1.86			
0.5948	2.0	2.73						

[a] The data are computed as follows:

$$\frac{dy}{d\xi} = -(2\cosh y - 2\cosh u)^{1/2}$$

For $y = z$, we obtain

$$\left(\frac{dy}{d\xi}\right)_0 = \frac{4\pi ev}{\epsilon kT}\frac{\sigma}{\kappa} = 1872\frac{\sigma}{\kappa}$$

Furthermore,

$$\int_z^u (2\cosh y - 2\cosh u)^{-1/2}\,dy = -\kappa d$$

Pairs of values of u and z are evaluated for the following conditions:

Normality of 1-1 electrolyte	0.964×10^{-5}	0.867×10^{-4}	0.964×10^{-3}	0.867×10^{-2}	0.964×10^{-1}	0.867
κ, cm^{-1}	10^5	3×10^5	10^6	3×10^6	10^7	3×10^7
$(dy/d\xi)_0$ for						
$\sigma = 1.5 \times 10^4$	280	94	28	9.4	2.8	0.94
$\sigma = 3.0 \times 10^4$	560	187	56	18.7	5.6	1.87
$\sigma = 4.5 \times 10^4$ esu/cm^2	840	280	84	28	8.4	2.8

for series of values of d between 6 and 240 A.

with the square root of the electrolyte concentration [$\sigma \propto \sqrt{n}$ according to equation (III-6)]; therefore, the initial slope of the potential curve increases in the same way, as represented by the dotted line in Figure 11. In Figure 10 the areas CAD represent the total double-layer charge; therefore, $C'A'D' : CAD = \sqrt{n'} : \sqrt{n}$.

B. SURFACES OF CONSTANT CHARGE DENSITY (FOR EXAMPLE, IF THE CHARGE IS DETERMINED BY INTERIOR-LATTICE SUBSTITUTIONS). The surface potential decreases in a somewhat complicated manner with increasing electrolyte concentration: $\sinh \frac{1}{2} ve\Phi_0/kT$ is inversely proportional to the square root of the electrolyte concentration according to equation (III-6). The initial slope of the potential curve remains constant, and the total surface area CAD is constant.

Problem: A flat surface has a constant charge density of 11.7 $\mu C/cm^2$. Compute the surface potential at the following electrolyte concentrations, assuming that the Gouy theory is valid: 10^{-5}, 10^{-3}, and $10^{-1}M$ NaCl, and 10^{-5}, 10^{-3}, and $10^{-1}M$ $MgSO_4$. Calculate $1/\kappa$ for each electrolyte concentration.

Solution:

$$
\begin{aligned}
n &= (\text{molarity} \times 10^{-3} \times \text{Avogadro's number}) \text{ ions/cm}^3 \\
&= M \times 10^{-3} \times 6.02 \times 10^{23} \text{ ions/cm}^3 \\
kT &= 0.4 \times 10^{-13} \text{ ergs at room temperature} \\
e &= 4.80 \times 10^{-10} \text{ esu} \\
\epsilon &= 80 \\
\sigma &= 11.7\ \mu\text{C/cm}^2 = (11.7 \times 10^{-6}/3.33 \times 10^{-10}) \text{ esu/cm}^2 \\
&= 3.5 \times 10^4 \text{ esu/cm}^2
\end{aligned}
$$

For $10^{-3}M$ NaCl, we find

$$n = 10^{-3} \times 10^{-3} \times 6.02 \times 10^{23} = 6.02 \times 10^{17} \text{ ions/cm}^3$$

$$1/\kappa = \left(\frac{\epsilon kT}{8\pi n e^2 v^2}\right)^{1/2} = \left(\frac{80 \times 0.4 \times 10^{-13}}{8\pi \times 6.02 \times 10^{17}\,(4.80 \times 10^{-10})^2 \times 1^2}\right)^{1/2}$$

$$= (0.0093 \times 10^{-10}) \approx 10^{-6} \text{ cm} = 100 \text{ A}$$

and [equation (III-6)]

$$\sigma = \left(\frac{2n\epsilon kT}{\pi}\right)^{1/2} \sinh\left(\frac{z}{2}\right)$$

or

$$\sinh\left(\frac{z}{2}\right) = \sigma\left(\frac{\pi}{2n\epsilon kT}\right)^{1/2}$$

$$= 3.5 \times 10^4 \times \left(\frac{\pi}{2 \times 6.02 \times 10^{17} \times 80 \times 0.4 \times 10^{-13}}\right)^{1/2} = 31.7$$

$$(1/2)z = 4.15 \qquad z = 8.3 = \frac{ve\Phi_0}{kT} \qquad \text{or} \qquad \Phi_0 = 208 \text{ mV}$$

Note:

$$\frac{ve\Phi_0}{kT} = z \quad \text{or} \quad \Phi_0 = \frac{kT}{ve}z = \frac{0.4 \times 10^{-13}}{4.80 \times 10^{-10}} \times 300 \times 10^3 z \text{ mV}$$

$$\approx 25 \text{ mV for } z = 1.$$

For the other concentrations and the divalent electrolyte, the following data are easily computed (Table III-1) (for $M = 10^{-5}$, $\sinh(z/2) = 317$; and for $M = 10^{-1}$, $\sinh(z/2) = 3.17$):

Solution (using SI units and the rationized four quantity system):

n = (molarity $\times 10^3 \times$ Avogadro's number) $= M \times 10^3 \times 6.02 \times 10^{23}$ ions/m^3

$kT = 0.4 \times 10^{-20}$ J

$e = 1.602 \times 10^{-19}$ C

ϵ = (static permittivity) $= 80 \times 8.85 \times 10^{-12}$ J^{-1} C^{-2} m^{-1} (or F m^{-1})

$\sigma = 11.7 \times 10^{-2}$ C m^{-2}

For $10^{-3}M$ NaCl we find:

$$1/\kappa = \left(\frac{\epsilon kT}{2ne^2v^2}\right)^{1/2} = \left(\frac{80 \times 8.85 \times 10^{-12} \times 0.4 \times 10^{-20}}{2 \times 1 \times 6.02 \times 10^{23} \times (1.602 \times 10^{-19})^2 \times 1^2}\right)^{1/2}$$

$$= (0.917 \times 10^{-16})^{1/2} \text{ m or about 100 A}$$

$$\sinh\left(\frac{z}{2}\right) = \left(\frac{1}{8n\epsilon kT}\right)^{1/2}$$

$$= 11.7 \times 10^{-2}\left(\frac{1}{8 \times 1 \times 6.02 \times 10^{23} \times 80 \times 8.85 \times 10^{-12} \times 0.4 \times 10^{-20}}\right)^{1/2}$$

$$= 11.7 \times 10^{-2}(7.33 \times 10^4)^{1/2} = 31.7$$

2. Potential and Charge Distribution in a Flat Double Layer According to the Stern Model

The application of the Gouy model to the electric double layer often leads to absurdly high local counter-ion concentrations. Since the ions are considered as point charges, they are considered present at the wall where the electric potential is the highest and where their concentration would be $c_+ = c \exp(-ve\Phi_0/kT)$. If the surface potential is, for example, -300 mV, and $c = 10^{-3}M$, the local concentration at the wall would be $c_+ = 10^{-3} \times e^{12}N = 160M$, which is not possible.*

In Stern's model of the double layer (Figure 12), the crowding of the ions at the surface is avoided by not allowing the Gouy ions to reach $x = 0$, since the size of the ions limits the distance of approach to the surface to a few angstrom units. Between the plane in which the surface charge is located and the plane of the centers of the counter-ions which are closest to the surface, there is no charge. The part of the double layer between these planes may be considered a molecular condenser in which the potential decreases linearly with the distance from the surface.

The counter-ions are statistically distributed over the Stern layer charge closest to the surface and a diffuse layer outside the Stern layer in the solution, in analogy with a Langmuir type derivation of the adsorption isotherm. A specific attraction between the surface and the counter-ions may be introduced in the form of a specific adsorption potential of the counter-ions at the surface.

Usually, in the Stern model, a decrease of the dielectric constant of the medium in the high field strength of the molecular condenser is accounted for.

The potential and charge distribution in a Stern-Gouy double layer can be described by the following set of equations: By an approximate statistical treatment, considering the number of avilable positions of the counter-ions in the Stern layer and in solution, Stern derives that

$$\sigma_1 = \frac{N_1 ve}{1 \times (N_A/M_w n) \exp[-(ve\Phi_\delta + \psi)/KT]} \qquad \text{(III-7)}$$

in which N_1 is the number of adsorption spots on 1 cm^2 of the surface, N_A is Avogadro's number, M_w is the molecular weight of the solvent, Φ_δ is the potential on the border between Stern and Gouy layers (Stern potential), and ψ is the specific adsorption potential of the counter-ions at the surface.

For the molecular condenser, we find

$$\sigma = \left(\frac{\epsilon'}{4\pi\delta}\right)(\Phi_0 - \Phi_\delta) \qquad \text{(III-8)}$$

in which ϵ' is the dielectric constant of the medium in the field of the molecular condenser, and δ is the thickness of the Stern layer.

Furthermore, the charge of the Gouy layer is given by equation (III-6), if

* Such crowding conditions, however, are seldom anticipated in clay systems.

we substitute Φ_δ for Φ_0:

$$\sigma_2 = \left(\frac{2\epsilon nkT}{\pi}\right)^{1/2} \sinh\left(\frac{ve\Phi_\delta}{2kT}\right) \tag{III-9}$$

Finally,

$$\sigma = \sigma_1 + \sigma_2 \tag{III-10}$$

Problem: Compute the Stern layer and Gouy layer charge and the Stern potential for a flat double layer with a constant surface charge of 11.7 μC at a concentration of 10^{-1}, 10^{-3}, and $10^{-5}M$ NaCl, and $10^{-3}M$ $MgSO_4$, when $\psi = 0$ and when $\psi = 0.1$ eV.

The following values may be assumed: The thickness of the Stern layer is equivalent to two monomolecular layers of adsorbed water, or 5 A, the dielectric constant of which is of the order of 3–6, as suggested by measurement of the capacity of the double layer on mercury in concentrated electrolyte solutions where the charge will primarily be in the Stern layer. These assumptions may well be applicable to a clay surface.

The number of adsorption spots on the surface $N_1 \approx 10^{15}$ per cm^2, or one spot per 10A^2, assuming that each site occupied by a water molecule in a monolayer is a potential cation site.

Solution: Equation (III-8) gives

$$3.5 \times 10^4 = \frac{6}{4\pi \times 5 \times 10^{-8}}(\Phi_0 - \Phi_\delta)$$

or

$$(\Phi_0 - \Phi_\delta) = 3.67 \times 10^{-3} \text{ esu} = 3.67 \times 10^{-3} \times 300 \times 10^3 = 1100 \text{ mV}$$

Write

$$y_0 = \frac{ve\Phi_\delta}{kT} \qquad X = \left(\frac{2\epsilon nkT}{\pi}\right)^{1/2} \qquad B = \frac{N_1 veM_w n}{N_A}$$

Then, since 1 is small with respect to the exponential term in the denominator, equation (III-7) becomes $\sigma_1 = B \exp [(y_\delta + \psi/kT)]$. Assuming that $\psi = 0$, then for 10^{-3} NaCl

$$B = \frac{10^{15} \times 1 \times 4.80 \times 10^{-10} \times 18 \times 10^{-6} \times 6.02 \times 10^{23}}{6.02 \times 10^{23}} = 8.5$$

and equation (III-9) becomes

$$\sigma_2 = X \sinh\left(\frac{y_\delta}{2}\right)$$

in which

$$X = \left(\frac{2 \times 80 \times 0.4 \times 10^{-13} \times 10^{-6} \times 6.02 \times 10^{23}}{\pi} \right)^{1/2} = 1110$$

and equation (III-10) becomes

$$3.5 \times 10^4 = 1110 \sinh\left(\frac{y_\delta}{2}\right) + 8.5 \exp y_\delta$$

from which it is found that $y_\delta = 7.4$, or $\Phi_\delta = 185$ mV, and thus $\Phi_0 = 1285$ mV. The ratio of the charges in Stern and Gouy layers is

$$8.5 \times \frac{e^{7.4}}{1110} \times \sinh\left(\frac{7.4}{2}\right) \quad \text{or} \quad \approx 38:62$$

For $10^{-3}M$ 2-2 valent electrolyte, we obtain

$$3.5 \times 10^4 = 1110 \times \sinh\left(\frac{y_\delta}{2}\right) + 2 \times 8.5 \exp y_\delta$$

from which it is found that $y_\delta = 6.9$, or $\Phi_\delta = 86$ mV, and the ratio of charges in Stern and Gouy layers is ≈50:50.

In the same manner, the following values are obtained for three different concentrations of 1-1 valent and 2-2 valent electrolyte:

	B	B		ϕ_δ (mV)		Stern:Gouy	
	(1-1)	(2-2)	X	(1-1)	(2-2)	(1-1)	(2-2)
$10^{-5}M$	0.085	0.17	111	300	144	37:63	50:50
$10^{-3}M$	8.50	17.0	1,110	185	86	38:62	50:50
$10^{-1}M$	850.	1700.	11,100	70	30	40:60	53:47

Assuming a specific adsorption potential $\psi = 0.1$ eV, we obtain $\psi = 0.1 \times 1.6 \times 10^{-12}$ erg and $\psi/kT = 4$. Then, for $10^{-3}M$ NaCl,

$$3.5 \times 10^4 = 1110 \sinh\left(\frac{y_\delta}{2}\right) + 8.5 \exp(y_\delta + 4)$$

from which it is found that $y_\delta = 4.20$ or $\Phi_\delta = 105$ mV, and $\sigma_1/\sigma_2 = 88:12$.

3. Corrections of the Gouy Theory According to Bolt

Another approach which gives the Gouy theory more physical reality has been presented by Bolt (1). In writing the Boltzmann distribution function, he adds several secondary energy terms to the primary potential energy term for the ion in the electric field of the double layer. The following energy terms are considered: the Coulombic interaction energy between the ions, the energy of polarization of the ions in the electric field, and the repulsive energy caused by noncoulombic interaction between the ions.

The effect of the field strength on the dielectric constant of the medium, which is most important near the surface where the field strength is highest, is taken care of by substituting, in the Poisson equation, the dielectric constant by an empirical expression relating the dielectric constant and the local field strength.

Short-range interaction between the surface and the ions is taken care of by introducing a distance of closest approach, as in the Stern model. Any specific interaction between surface and ion (the specific adsorption potential in the Stern theory) is not introduced in the general treatment.

Bolt estimates the various correction terms and shows that they can be taken care of by applying a correction to the distance coordinate, amounting to several angstrom units, to be subtracted from the real distance from the surface. The correction of the distance must be reduced by the distance of closest approach of the counter-ions, which is of the order of a few angstrom units.

The result of the estimates if that the Gouy theory in the uncorrected form gives results almost identical with those of the corrected theory when applied to the problem of double-layer interaction. However, in the treatment of the exchange equilibrium between ions of the same valence but of different size, the correction terms themselves determine the ion ratios in the double layer as a function of their ratios in the equilibrium liquid. Therefore, the corrected Gouy theory is important in the exchange problem.

B. INTERACTING FLAT DOUBLE LAYERS

1. Potential and Charge Distribution between Interacting Flat Double Layers According to the Gouy Model

The following figure shows schematically the potential distribution between two interacting flat double layers of the Gouy type:

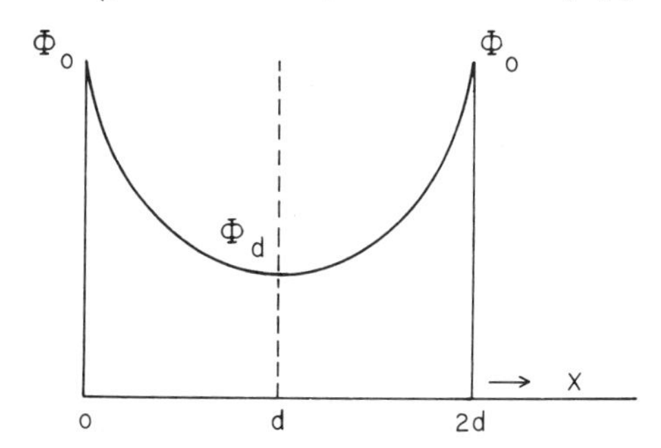

The distance between the plates is $2d$. In the midway plane, at a distance d from either surface, the field strength is zero: $(d\Phi/dx)_{x=d} = 0$. The potential midway between the plates is called Φ_d, and $ve\Phi_d/kT = u$. The fundamental differential equation for the potential function is

$$\frac{d^2y}{d\xi^2} = \sinh y$$

Integration with the boundary condition that for $x = d$, $y = u$, and $dy/d\xi = 0$ gives

$$\frac{dy}{d\xi} = -(2\cosh y - 2\cosh u)^{1/2} \tag{III-11}$$

When integrating a second time between the limits 0 and d for x (or between z and u for y), we obtain the following relation:

$$\int_z^u (2\cosh y - 2\cosh u)^{-1/2}\,dy = -\int_0^d d\xi = -\kappa d \tag{III-12}$$

Evaluating the integral using tables, we can obtain u, or the midway potential, for any chosen value of y (the surface potential), $2d$ (the plate distance), and κ (the electrolyte concentration). For most purposes, knowledge of the value of u suffices. The complete potential distribution can be computed, however, by computing the slope of the curve $dy/d\xi$ by means of equation (III-11) at any chosen value of y between z and u.

In Verwey and Overbeek's book (Appendix V, reference B-6, Table IX, p. 68, and Figure 10, p. 72) sets of values for κd, z, and u are tabulated. These sets are so chosen that, for a given surface potential ($\sim z$), the midway potential ($\sim u$) can be read for a series of values for κd. Thus, for a colloid of constant surface potential, the midway potential can be found at constant electrolyte concentration (constant κ) for varying plate distance ($2d$), or, at constant plate distance, the change of the midway potential with electrolyte concentration can be read. With interpolation, u can be computed from this table for most situations encountered in colloid chemistry dealing with colloids of constant surface potential. However, the table is not convenient for reading the midway potential as a function of distance or electrolyte concentration when the surface charge is a constant and the surface potential varies with distance and electrolyte concentration.

The computation of sets of z, u and κd values with the surface charge kept constant is somewhat more complicated. The surface charge is given by

$$\sigma = -\int_0^d \rho dx = -\frac{\epsilon}{4\pi}\left(\frac{d\Phi}{dx}\right)_0 = -\frac{\epsilon}{4\pi}\left(\frac{dy}{d\xi}\right)_0 \frac{\kappa kT}{ve}$$
$$= \left(\frac{\epsilon nkT}{2\pi}\right)^{1/2} (2\cosh z - 2\cosh u)^{1/2} \tag{III-13}$$

For a constant surface charge, therefore, $(2\cosh z - 2\cosh u)^{1/2}(n)^{1/2}$ is a constant independent of the plate distance. In order to compute the change of the midway potential ($\sim u$) with varying distance at a given electrolyte concentration (n, κ), sets of values for z and u are required for which

$$(2\cosh z - 2\cosh u)^{1/2} = \sigma\left(\frac{2\pi}{\epsilon nkT}\right)^{1/2} = \text{constant}$$

Table III-2 gives such sets of values for situations which are common in clay-water systems. The midway potential increases somewhat faster with decreasing distance than in the constant surface-potential case. For weak interaction (i.e., large d), the table shows z to be constant, so u changes in the same way with decreasing distance as in the constant–potential case. In this region, $\cosh u$ is small compared with $\cosh z$; therefore, according to equation (III-13), z must be practically constant.

A useful approximation in the computation of u is valid for small interaction or large values of κd. Under these conditions, the double layers are practically unperturbed, and the midway potential may be considered to be built additively from the separate double-layer potentials at the distance d. Hence $u = 2y_d$.

Equation (III-5) can be approximated for large values of κd by

$$y_d = 4\gamma e^{-\kappa d} \qquad \text{and} \qquad \gamma = (e^{z/2} - 1)/(e^{z/2} + 1)$$

Therefore,

$$u = 8\gamma e^{-\kappa d} \tag{III–14}$$

In principle, the ion concentration at any point between the plates is easily calculated according to equation (III-1) when the electric potential at any point in the double layer has been computed for the interacting double layers. Usually, the ion concentration midway between the plates is of primary interest. This concentration is found by substituting u in equation (III-1):

$$n_- = ne^u \qquad \text{and} \qquad n_+ = ne^{-u} \tag{III-1a}$$

Computation of the total excess of counter-ions and the total deficit of ions of the same sign as the surface charge will be discussed under Section C.

2. Potential and Charge Distribution between Interacting Surfaces with Stern-Type Double Layers

When two double layers of the Stern-Gouy model interact, the distribution of charges between the Stern layer and the Gouy layer varies with the particle separation. With decreasing distance between the surfaces, counter-ions shift from the Gouy layer to the Stern layer. During this process, the

Stern potential, $\Phi\delta$, increases. This increase is smaller for double layers with a constant surface potential than for those with constant surface charge, since, in the first case, the total double-layer charge decreases with decreasing distance.

The computation of the potential and charge distribution as a function of distance is rather laborious. Of the equations (III-7 through III-10) for the single double layer, equation (III-9) should be substituted by an equation which is analogous to equation (III-13):

$$\sigma_2 = \left(\frac{nKT\epsilon}{2\pi}\right)^{1/2} (2 \cosh z_\delta - 2 \cosh u)^{1/2} \qquad \text{(III-15)}$$

in which $z_\delta = ve\Phi\delta/kT$, $u = ve\Phi_d/kT$, and u is a function of d. For weak interaction, u will be small enough that cosh u can be replaced by 1.

Also, equation (III-7) should be revised, since the number of available positions of the ions outside the Stern layer will depend on the distance between the surfaces. In the case of the interaction of two flat surfaces of clay particles, the following treatment may give reasonable estimates, at least for weak interaction. Consider the Stern layer thickness δ to be equal to one or two monolayers of water in which the dielectric constant is of the order of 3. Then, the Gouy layer thickness is $d - \delta$. Put the statistical charge distribution between Stern and Gouy layers proportional to their respective volumes, that is, to their respective thicknesses. (This treatment is allowed, since it does not lead to excessive crowding of counter-ions at the surface.) When the Boltzmann distribution function is written, the potential of the ions in the Stern layer may be taken as $(ve\Phi_\delta + \psi)/kT$, and in the Gouy layer some weighted average potential should be considered. Neglecting this refinement, we can substitute equation (III-7) by

$$\frac{\sigma_1}{\sigma_2} = \left[\frac{\delta}{(d - \delta)}\right] \exp\left[\frac{(ve\Phi_\delta + \psi)}{kT}\right] \qquad \text{(III-16)}$$

C. FORCE AND ENERGY OF INTERACTION OF TWO FLAT DOUBLE LAYERS

1. Repulsive Energy of Double Layers of Constant Potential Derived from the Free-Energy Change

The repulsive energy, V_R, between opposite square centimeters of parallel plates resulting from the interaction of their double layers is equal to $V_R = 2(F_d - F_\infty)$, in which F_d is the free energy of the double layer per cm² when the plate distance is $2d$, and F_∞ is the free energy of the single noninteracting double layer per cm².

Computations of F_d and F_∞ for the case of double layers of constant potential have been carried out by Verwey and Overbeek. The resulting data for V_R, covering a wide range of situations of practical interest, have been presented in Table XI of their book and in several graphs. It should be stressed that the free-energy computations are valid for constant-potential surfaces only, for which the interaction involves a decrease of the double-layer charge by the desorption of potential-determining ions.

2. Direct Computation of the Interaction Force between Flat Double Layers

The interaction force between two flat double layers can be derived directly from the ionic concentration midway between the plates, which is determined by the value of u. Langmuir (Chapter Three, reference 13) has pointed out that the repulsive force is given by the osmotic pressure midway between the plates with respect to that of the equilibrium solution. Since these osmotic pressures are determined by the ion concentration, the force is found directly from the excess concentration midway between the plates. This excess concentration is equal to

$$n(e^{u} - 1) + n(e^{-\alpha} - 1) = 2n(\cosh u - 1)$$

Therefore, the force is

$$p = 2nkT(\cosh u - 1) \qquad \text{(III-17)}$$

This expression is valid for both constant-charge and constant-potential surfaces, as long as u is computed according to the appropriate procedures for either case, as discussed previously. Osmotic swelling pressures of clays can be computed by means of equation (III-17), based on the assumption that the Gouy theory is valid.

It is difficult, however, to evaluate the repulsive energy from the repulsive force, which varies in a complicated manner with the distance, as reflected in the change of u with the distance.

For small interaction, u has been shown to vary simply exponentially with the distance [see formula (III-14)]. In that region, the repulsive energy can be adequately derived from the repulsive force according to equation (III-17) by integrating the force with respect to the distance, substituting the approximate expression for u as a function of the distance. The result of this integration is

$$V_R = \left(\frac{64nkT}{\kappa}\right)\gamma^2 \exp(-2\kappa d) \qquad \text{(III-18)}$$

in which

$$\gamma = (e^{z/2} - 1)/(e^{z/2} + 1)$$

It is assumed that no changes in surface potential or double-layer charge occur. This relation is often a good approximation for values of $\kappa d > 1$.

3. Interaction Energy of Stern Double Layers

The change of the free energy of interacting double layers of the Stern-Gouy type with constant surface potential has been treated by Mackor (Chapter Five, reference 2). It appears that for small interaction ($\kappa d \gg 0.5$) the repulsive energy can be approximated by equation (III-18) if z is substituted by $z_\delta = ve\Phi_\delta/kT$, and Φ_δ has the constant value computed for the single double layer.

Problem: Compute the energy of interaction of two flat double layers of the Gouy type with a constant negative charge density of 3×10^4 esu/cm² at a plate distance of 80 A at a NaCl concentration of $0.1M$ and $0.01M$. Assume that the formula for weak interaction can be applied.

Compute the repulsive force under the same conditions.

Compute the local cation and anion concentration midway between the plates.

Solution: From Tables III-1 and III-2 the following values are read: $10^{-1}M$ NaCl: $\kappa = 10^7$; $(dy/d\xi)_0 = 5.6$; $\kappa d = 4.00$; $-u = 0.1$; $-z = 3.51$. For $10^{-2}M$ NaCl: $\kappa = 3.3 \times 10^6$; $(dy/d\xi)_0 = 17.0$; $\kappa d = 1.32$; $-u = 1.6$; $-z = 5.87$.

Equation (III-18) gives for $0.1M$ NaCl

$$V_R = \frac{64 \times 10^{-1} \times 10^{-3} \times 6.02 \times 10^{23} \times 0.4 \times 10^{-13}}{10^7} \gamma^2 e^{-8}$$

$$\gamma = \frac{e^{1.76} - 1}{e^{1.76} + 1} = 0.707; \ \gamma^2 = 0.50; \text{ and } V_R = 0.00254 \text{ erg/cm}^2 \text{ (mJ/m}^2)$$

[equation (III-17)]:

$$\begin{aligned} p &= 2 \times 10^{-1} \times 10^{-3} \times 6.02 \times 10^{23} \times 0.4 \times 10^{-13} (1.0050 - 1) \\ &= 2.4 \times 10^4 \text{ dyne/cm}^2 = 2.4 \times 10^3 \text{ N/m}^2 = 0.024 \text{ bar} \end{aligned}$$

[equation (III-1*a*)]:

$$c_d^+ = 10^{-1} \times e^{0.1} = 0.1105M; \ c_d^- = 10^{-1} \times e^{-0.1} = 0.0905M$$

For $0.01M$ NaCl, the results are respectively

$$\gamma = 0.899; \ V_R = 0.268 \text{ mJ/m}^2; \ p = 7.61 \times 10^4 \text{ N/m}^2 \text{ (0.761 bar)}; \text{ and } c_d^+ = 0.0495M \quad \text{and} \quad c_d^- = 0.00202M$$

Problem: Compute the interaction energy of two flat double layers of the Stern-Gouy type with a constant charge density of 3×10^4 esu/cm² at a distance of 125 A at a Na Cl concentration of $0.01N$. Assume that the formula for weak interaction can be applied and that the specific adsorption potential is zero.

Solution: The half-distance, d, between the plates is 62.5 A, and the thickness of the Stern layer, δ, is assumed to be 5 A. Therefore, the half-distance for Gouy layer interaction is 57.5 A $= d_{\text{Gouy}}$. Formula (III-16) gives

$$\frac{\sigma_1}{\sigma_2} = \left(\frac{5}{62.5} - 5\right) \exp z_\delta$$

and substituting

$$\sigma_1 = 3 \times 10^4 - \sigma_2$$

and equation (III-15):

$$\sigma_2 = \left(\frac{\epsilon nkT}{2\pi}\right)^{1/2} (2 \cosh z_\delta - 2)^{1/2} = X(2 \cosh z_\delta - 2)^{1/2}$$

we obtain

$$\exp z_\delta = \frac{3 \times 10^4 - X(2 \cosh z_\delta - 2)^{1/2}}{X(2 \cosh z_\delta - 2)^{1/2}} \quad \frac{62.5 - 5}{5}$$

$$X = (80 \times 10^{-2} \times 10^{-3} \times 10^{23} \times 6.02 \times 0.4 \times 10^{-13} \times 1/2\pi)^{1/2}$$
$$= 1.735 \times 10^3$$

The value of z_δ which fits the above equation is $z^\delta = 3.3$, or $\Phi_\delta = 82$ mV. The repulsive energy is found from equation (III-18). In this equation,

$$\kappa = \frac{8\pi \times 10^{-2} \times 10^{-3} \times 6.02 \times 10^{23}(4.80 \times 10^{-10})^2 \times 1^2)^{1/2}}{(80 \times 0.4 \times 10^{-13})^{1/2}}$$
$$= 3.28 \times 10^6 \text{ cm}^{-1}$$

Thus,

$$2\kappa d = 2 \times 3.28 \times 10^6 \times 57.5 \times 10^{-8} = 3.77$$

and

$$e^{-2\kappa d} = 0.023$$

$$\gamma = \frac{e^{1.85} - 1}{e^{1.85} + 1} = \frac{5.36}{7.36} = 0.728 \qquad \text{and} \qquad \gamma^2 = 0.53$$

Therefore,

$$V_R = \frac{64 \times 10^{-2} \times 10^{-3} \times 6.02 \times 10^{23} \times 0.4 \times 10^{-13}}{3.28 \times 10^{6}} 0.53 \times 0.023$$

$$= 0.058 \text{ erg/cm}^2 \text{ (mJ/m}^2\text{)}$$

D. CATION-EXCHANGE CAPACITY AND NEGATIVE ADSORPTION (10–20)

The cation-exchange capacity of a negative double layer may be defined as the excess of counter-ions in the double layer which can be exchanged for other cations (~ area *BAD* in Figure 10). The negative adsorption (also called Donnan exclusion) of anions due to the repulsion of these ions by the negative surface is defined as the total deficit of these ions in the double layer (~ area *BCD*). The sum of cation-exchange capacity, as defined above, and negative adsorption is equivalent to the double-layer charge (~ area *CAD*).

As pointed out previously, for a constant-potential surface, the total charge of the double layer increases proportionally with $\sqrt{n}$ and decreases with decreasing plate distance. For a constant-charge surface, the total charge is independent of both electrolyte concentration and plate distance.

The total excess of cations and the total deficit of anions can be computed by integrating the local concentrations minus the equilibrium concentration between $x = 0$ and $x = \infty$ for a single double layer and between $x = 0$ and $x = d$ for interacting double layers, taking into account that the local potential changes with distance according to the Poisson equation.

Such computations have been carried out by Klaarenbeek (10), Schofield (11) and Bolt *et al.* (12) for Gouy double layers, and have been successfully applied to special problems.

For a single double layer, the integration leads to the following simple expression for the total deficit of charge:

$$\sigma_- = \left(\frac{\epsilon n k T}{2\pi}\right)^{1/2} \left[\exp\left(\frac{v e \Phi_0}{2kT}\right) - 1\right] \qquad \text{(III-19)}$$

The total charge is given by equation (III-6), and the difference gives the total excess of positive charge.

The deficit of negative charge is the following fraction of the total charge:

$$\frac{\sigma_-}{\sigma} = -\frac{[\exp(ve\Phi_0/2kT) - 1]}{[\exp(-ve\Phi_0/2kT) - \exp(ve\Phi_0/2kT)]} \qquad \text{(III-19}a\text{)}$$

For a surface of constant potential, the total deficit of charge is easily computed from equation (III-19) for any given electrolyte concentration, n. From equation (III-19*a*) it follows that the deficit is a constant fraction of the total charge, independent of the electrolyte concentration. Therefore, the ratio between excess positive charge and deficit negative charge remains the same, as illustrated in figure 10a.

For a surface of constant charge, the surface potential must first be computed from equation (III-6) for the given electrolyte concentration. The deficit of negative charge then follows from equation (III-19). The deficit fraction of the total charge is no longer independent of the electrolyte concentration, since the absolute value of Φ_0 decreases with increasing electrolyte concentration; thus, the quotient in equation (III-19*a*) increases with decreasing Φ_0 (Φ_0 is negative). Therefore, for constant-charge surfaces, such as those in clays, the anion deficit increases with increasing electrolyte concentration, and the excess of cations, or the cation-exchange capacity as defined above, decreases. (Compare Figure 10*b*.)

The cation-exchange capacity of clays is often determined by saturating the clay with an excess of ammonium salt, separating the ammonium clay from the solution, washing, and determining the nitrogen content of the clay. This cation-exchange capacity is equivalent to the total charge of the clay, and it would be identical with the excess of cations in the double layer under those conditions at which the negative adsorption is negligible, that is, at very low electrolyte concentrations.

An alternative procedure for determining the cation-exchange capacity is to treat the clay with a concentrated ammonium salt solution and to measure the decrease of the ammonium ion concentration in the equilibrium liquid, assuming that the ammonium salt concentration is high enough to effect complete exchange of the counter-ions in the double layer. Obviously, the first method gives higher values for the cation-exchange capacity than the second, the difference being the deficit of anions in the double layer at the ammonium salt concentration used in the second method. Since the deficit of anions increases with increasing salt concentration, the difference between the values obtained with the two methods may be appreciable.

When dealing with clays, one generally assumes that the cation exchange capacity is equivalent to the total charge density, but this is apparently not correct. In keeping with established customs in the field of clays, the cation-exchange capacity of a clay may be specifically defined as the maximum exchange capacity, which is indeed equivalent to the surface-charge density. This maximum exchange capacity is determined directly with the first method and can be found by the second method by correcting for negative adsorption.*

Since the maximum cation exchange capacity is equivalent with the surface charge, the maximum CEC is a materials constant for clays. For particles with constant potential surfaces, however, the surface charge and therefore the cation exchange capacity increases with increasing electrolyte concentration according to the double-layer theory. For those particles, the CEC cannot be considered a materials constant, and it is perhaps advisable not to refer to a cation exchange "capacity" for such particles. This is of particular importance when discussing the exchange properties of amorphous clay materials, which have constant potential surfaces, hence giving a CEC value for such clay materials is meaningless unless the conditions of measurement are given. The same applies to such oxides as aluminum or ferric oxides.

The computation of the deficit of anions and the excess of cations between two surfaces with interacting double layers can be carried out along the same lines as for a single double layer. In this case, however, the integration leads to elliptic integrals which must be solved with the use of tables [Klaarenbeek (10), Bolt (12)].

Measurements of negative adsorption per unit weight of clay can be used to evaluate the surface area per unit weight of clay. Schofield (3) has proposed this method. Schofield's general mathematical treatment of the diffuse double layer, which has many attractive features, leads to the following explicit relation which is valid for a single or for weakly interacting double layers (small values of u):

$$\frac{\gamma_-}{n^*} = q(\rho\beta n^*)^{-1/2} - \frac{4}{\rho\beta\gamma} \quad \text{and} \quad \beta = \frac{8\pi F^2}{\epsilon RT} \qquad \text{(III-20)}$$

in which γ_- is the total deficit of repelled ions per unit surface area in meq/cm^2, γ is the total charge density of the surface in meq/cm^2, n^* is the normality of the bulk solution, ρ is the valence of the attracted ions, and ρ/p is the valence of the repelled ions.

q is a number which depends on the valency ratio of the ions p. For symmetric salts, $q = 2$. Other values are tabulated by Schofield (11). β has the value of 1.06×10^{15} cm/meq for water at 20°C.

* In the older clay literature, a correction for the increase of the anion concentration in the equilibrium liquid caused by negative adsorption of the anions was made on the assumption that the higher anion concentration was caused by a withdrawal of water from the liquid by the clay to form a hull of electrolyte-free adsorption water on the clay surfaces. However, this has no physical meaning; the low anion concentration near the surface is merely a result of the repulsion of the anions by the negative surface.

The surface area per gram is found as follows: The negative adsorption is measured per gram of solid (γ_-^*). When γ_-^*/n^* is plotted versus $q/\sqrt{\rho\beta n^*}$, the surface area per gram is determined from the slope of the straight line obtained. Although, theoretically, the charge density per unit surface area could be obtained from the intercept of the straight line with the $q/\sqrt{\rho\beta n^*}$ coordinate, the accuracy with which the intercept can be located is insufficient for the precise evaluation of the surface charge density. A reliable measure of the surface charge density can be derived, however, by combining the determined surface area per gram with the cation exchange capacity per gram.

In analyzing data for negative adsorption, one should consider the fact that low values are obtained when anion-exchange adsorption can occur at the positive edges of the clay particles. This effect can be avoided by reversal of the edge charge by phosphate addition. Bolt (12) has shown that at low electrolyte concentrations, at which the anion adsorption has relatively the largest effect, low values for the negative adsorption were indeed obtained and that previous addition of small amounts of phosphates gave higher values which were in line with the values expected according to formula (III-20).

E. CATION-EXCHANGE EQUILIBRIUM

1. Monovalent Cation Exchange

The uncorrected Gouy theory predicts that the ratio of counter-ions of the same valence in the double layer is the same as that in the equilibrium solution. Formalistically, the exchange equilibrium data can be presented in the mass-law form as follows:

$$K = \frac{(A)_i(B)_e}{(A)_e(B)_i} \tag{III-21}$$

in which A and B are two monovalent ions, and $(\ )_i$ and $(\ )_e$ denote their equilibrium concentrations in the double layer and in the equilibrium solution, respectively. According to the Gouy theory, the equilibrium constant, K, is unity. This conclusion is a direct consequence of considering the ions as point charges and neglecting any specific interactions.

As mentioned previously, Bolt (1) has corrected the Gouy theory, complying with the physical realities of ion size and various interaction energies. Bolt showed that the introduced corrections could, in principle, explain possible deviations of K from 1 to 4. Specific applications of the theory to experimental observations have not yet been presented.

Bolt points out that commonly observed preferences of the double layer for certain ions should be treated by means of the corrected Gouy theory before one resorts to assuming differences in specific adsorption forces between the surface and the competing counter-ions. Of course, the latter would merely amount to data-fitting when observed deviations from the uncorrected Gouy theory are attributed entirely to differences in adsorption energies, but the choice is arbitrary (see Chapter Twelve, Section II-A-5). However, if other observations demonstrate that the counter-ions are present mainly in the Stern layer, it would be more appropriate to treat ion exchange as an exchange between more or less strongly adsorbed Stern counter-ions. In that case, the difference in adsorption potentials ($\psi_A - \psi_B$) of the two competing ions can be derived from

$$K = \exp \frac{(\psi_A - \psi_B)}{kT} \qquad \text{(III-21}a\text{)}$$

Relatively small differences between the adsorption potentials would correspond to comparatively large deviations of K from 1.

In clays, both Gouy ion exchange and Stern ion exchange should probably be considered. Moreover, the equilibrium constant often changes with the ion ratio on the clay, which may be due to the presence of various specific adsorption sites on the surface of a different character. Therefore, the exchange equilibrium is governed by many factors which are difficult to determine (16).

2. Monovalent-Divalent Ion Exchange

From the uncorrected Gouy theory, it can be derived that divalent cations are concentrated in the double layer, so the ratio of divalent to monovalent ion is much higher in the double layer than in the equilibrium liquid.

For a single double layer of constant potential, the ion ratio in the double layer as a function of the ratio in the equilibrium solution was computed by van Os (13). A graphical representation of the results of his computations is given in the cited reference.

For interacting double layers of constant charge, Eriksson (14) has derived the following expression:

$$\frac{\sigma_1}{\sigma} = \frac{[c_1/2(c_2Z)^{1/2}] \sinh^{-1} 2(c_2Z)^{1/2}}{(c_1 + c_2 \cosh u)} \qquad \text{(III-22)}$$

in which $\sigma_1 = \sigma_+ + \sigma_-$, $\sigma_2 = \sigma_{2+} + \sigma_{2-}$, and $\sigma = \sigma_1 + \sigma_2$. (It is assumed for simplicity that 1–1 and 2–2 valent electrolytes are used.) c_1 and c_2 are the concentrations of mono- and divalent salts in the equilibrium liquid in moles/dm^3.

$$Z^{1/2} = \frac{4\pi\sigma e}{\beta\epsilon kT} \quad \text{and} \quad \beta = \left(\frac{16\pi e^2 N\, 10^{-3}}{\epsilon kT}\right)^{1/2}$$

When this formula is applied, the negative adsorption of the anions must also be measured. Bolt (15) has measured the Na-Ca exchange equilibrium with illite clay. The experimental points appear to fit the relation of Eriksson quite well, as shown in the reference by a plot of σ_1/σ versus $c_1/\sqrt{c_2}$. The theoretical curve was drawn from the equation, and cosh u was assumed to be unity (weak interaction).

F. SUMMARY OF GOUY DOUBLE-LAYER FORMULAS

1. Symbols and Values

n	local ion concentrations in number of ions/cm^3; n = molarity $\times 10^{-3} \times$ Avogadro's number (6.02×10^{23})
n_+	local cation concentration in number of ions/cm^3
n_-	local anion concentration in number of ions/cm^3
σ	total surface charge = total diffuse charge in esu/cm^2 surface
σ_+	total excess of cation charge opposite negative surface in esu/cm^2 surface
σ_-	total deficit of anion charge opposite negative surface in esu/cm^2 surface
v	valence of ion (number)
e	elementary charge = 4.77×10^{-10} esu
kT	Boltzmann constant $\times$ absolute temperature = 0.4×10^{-13} erg at room temperature
ϵ	the dielectric constant of the medium
κ	$= (8\pi n e^2 v^2/\epsilon kT)^{1/2}$
x	distance from the surface or distance between surfaces
d	half-distance between surfaces
ξ	$= \kappa x$ or κd
Φ	local electric potential; $y = ve\Phi/kT$
Φ_0	surface potential; $z = ve\Phi_0/kT$
Φ_d	midway potential; $u = ve\Phi_d/kT$
$ve\Phi/kT$	$= 1$ for $\Phi = 25$ mV at room temperature and $v = 1$

2. Single Gouy Double Layer

A. Electric-Potential Distribution. The local potentials in the double layer are given by

$$e^{y/2} = \frac{e^{z/2} + 1 + (e^{z/2} - 1)e^{-\xi}}{e^{z/2} + 1 - (e^{z/2} - 1)e^{-\xi}} \tag{III-5}$$

and, for small potentials at the surface,

$$\Phi = \Phi_0 e^{-\kappa x} \qquad \text{(III-5}'\text{)}$$

The initial slope of the potential-distance curve is given by the surface charge:

$$\sigma = \frac{-\epsilon}{4\pi}\left(\frac{d\Phi}{dx}\right)_0$$

Effect of electrolyte addition: The double layer is compressed with increasing electrolyte content, κ increases, and the decay according to equation (III-5′) is more rapid.

For a constant-potential surface, Φ_0 is constant, but the surface charge increases as reflected by the increase in initial slope indicated in Figure 10*a* (Chapter Three).

For a constant-charge surface, the initial slope remains the same, and the value of Φ_0 decreases, as indicated in Figure 10*b* (Chapter Three). The surface potential decreases, and $n^{1/2} \times \sinh ve\Phi_0/2kT$ is constant.

B. Electric-Charge Distribution. The local ion concentrations in the double layer are given by

$$n_+ = n \exp\left(\frac{-ve\Phi}{kT}\right) = ne^{-y} \qquad n_- = n \exp\left(\frac{ve\Phi}{kT}\right) = ne^{y} \qquad \text{(III-1)}$$

The total charge of the double layer is

$$\sigma = \left(\frac{2n\epsilon kT}{\pi}\right)^{1/2} \sinh\left(\frac{z}{2}\right) \qquad \text{(III-6)}$$

or, for small potentials,

$$\sigma = \left(\frac{\epsilon\kappa}{4\pi}\right)\Phi_0 = C\Phi_0 \qquad (C = \text{capacity}) \qquad \text{(III-6}'\text{)}$$

The total charge deficit of ions of the same sign as the surface charge is

$$\sigma_- = \left(\frac{\epsilon nkT}{2\pi}\right)^{1/2}(e^{z/2} - 1) \qquad \text{(III-19)}$$

and the total excess of positive charge is $\sigma_+ = \sigma - \sigma_-$.

The total deficit is the following fraction of the total charge:

$$\frac{\sigma_-}{\sigma} = -\frac{(e^{z/2} - 1)}{(e^{-z/2} - e^{z/2})} \qquad \text{(III-19}a\text{)}$$

Effect of electrolyte addition: Surface with constant potential:

$$\sigma : \sigma' = \sqrt{n} : \sqrt{n'} \qquad \frac{\sigma_-}{\sigma} = \frac{\sigma_-'}{\sigma'}$$

Surface of constant charge:

$$\sigma = \sigma' \qquad \frac{\sigma_-}{\sigma} < \frac{\sigma'_-}{\sigma'} \ (n' > n)$$

Ion-exchange equilibrium: See graph in van Os (13) for monovalent exchange for constant-potential double layer.

3. Interacting Gouy Double Layers

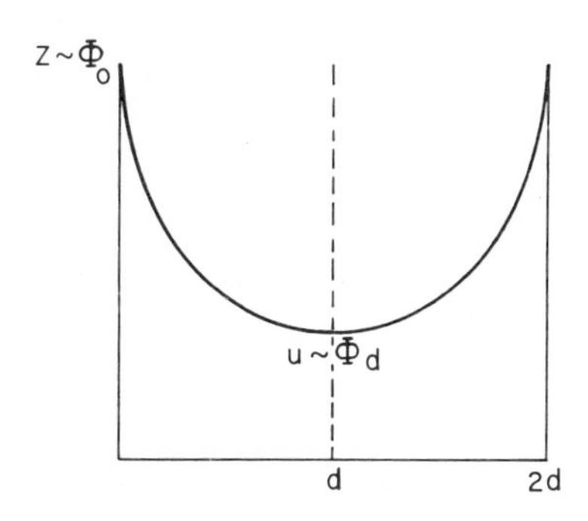

A. Electric-Potential Distribution. The potential distribution is given by

$$\frac{dy}{d\xi} = -(2 \cosh y - 2 \cosh u)^{1/2} \tag{III-11}$$

and

$$\int_z^u dy(2 \cosh y - 2 \cosh u)^{-1/2} = -\kappa d \tag{III-12}$$

For surfaces of constant potential, pairs of z and u values are read from Table IX in Verwey and Overbeek.

For surfaces of constant charge, pairs of values of z and u are read from Table III-2. If u is small and $\cosh u \approx 1$, Verwey and Overbeek's table may be used.

For weak interaction (κd large) and simple additivity of the local potentials of both double layers,

$$u = 8\gamma e^{-\kappa d} \qquad \text{and} \qquad \gamma = \frac{(e^{z/2} - 1)}{(e^{z/2} + 1)} \tag{III-14}$$

may be assumed.

For surfaces of constant potential, the surface charge decreases with increasing interaction; for surfaces of constant charge, the surface potential increases with increasing interaction.

B. ELECTRIC-CHARGE DISTRIBUTION. The total charge of an interacting double layer is

$$\sigma = \left(\frac{\epsilon nkT}{2\pi}\right)^{1/2}(2\cosh z - 2\cosh u)^{1/2} \quad \text{(III-13)}$$

The computation of the total deficit of charge of the same sign as the surface charge leads to elliptic integrals. An explicit expression is obtained if u is small (weak interaction). [See equation (III-20).]

Ion-exchange equilibrium: See equation (III-22) [Eriksson (14)] for monovalent-divalent exchange for constant-charge double layer.

4. Interaction Force and Energy

For a constant-potential surface, the energy of interaction is read from Table XI in Verwey and Overbeek's book.

The interaction force between two double layers is given by Langmuir's formula

$$p = 2nkT(\cosh u - 1) \quad \text{(III-17)}$$

which is valid for both a constant-charge and a constant-potential surface if u is computed according to Table III-2 of this Appendix and Table IX of Verwey and Overbeek, respectively.

For weak interaction (large κd), the following equation can be applied:

$$V_R = \frac{64nkT}{\kappa}\gamma^2 e^{-2\kappa d} \quad \text{(III-18)}$$

G. SHORT-RANGE ELECTROSTATIC INTERACTION BETWEEN LAYERS OF A 2:1 LAYER CLAY

In the region of layer separations in which diffuse double layers have not been developed, the counter ions may either occupy midway positions between the layers, or they may be segregated and associated with the silica sheets of the layers, for example, in hole positions. The electrostatic interaction between the layers is attractive when the counter ions are in midway positions, and repulsive when they are segregated.

1. Counter Ions Associated with Silica Sheets—Calculation of the Repulsive Electrostatic Energy

The counter ions form dipoles with the centers of negative charge in the interior of the layer. The electrostatic interaction may be treated as the repulsion between two plates of dipoles in opposite layers.

The force between two such parallel plates, each a square, a centimeter on edge can be computed by replacing each dipole layer by an equivalent current floating along the edge of the square with the current strength equal to the dipole moment μ per unit area. By applying the law of Biot and Savart (A. S. Ginzbarg), the energy per unit area as a function of distance x between the centers of the closest poles of the dipoles is then given by the following equation:

$$E(x) = -\frac{8\mu^2}{a\epsilon}\left(0.773 - \ln a - \frac{s}{a} - 1 - \frac{x}{a} - \tfrac{1}{2} + \frac{x^2}{2s^2}\ln\frac{(x+2s)x}{(x+s)^2} - \frac{2x}{s}\ln\frac{(x+2s)}{(x+s)} + \ln\frac{(x+2s)^2}{(x+s)}\right)$$

in which s is the pole distance of the dipole.

Example: Assuming that the counter ions are located in the tetrahedral holes in the silica sheets, and that the negative charge in the layer is concentrated in the octahedral hydroxyl groups adjacent to the cations in the holes, then for monovalent cations, the pole distance s will vary between 2.03 A for Li and 3.02 A for Cs. Taking the pole distance for Rb as an example, the dipole moment per unit area will be

$$e \times s \times \frac{10^{16}}{144} = 4.77 \times 10^{-10} \times 2.85 \times 10^{-8} \times \frac{10^{16}}{144}$$

since there is one cation for every 144 A of layer surface.

Taking $\epsilon = 6$ for the interlayer water between 1 and water layers, the factor in the above equation, outside the brackets, will be

$$\frac{8\mu^2}{a\epsilon} = \begin{array}{ll} 0.0118 & \text{for } a = 10{,}000 \text{ A} \\ 0.118 & \text{for } a = 1{,}000 \text{ A} \\ 1.18 & \text{for } a = 100 \text{ A} \end{array}$$

In the following table the values of $E(x)$ for different values of x and a are shown expressed in mJ/m^2.

a, edge length of particle in Å	x, distance between the closest poles of the dipoles of two plates with 1, 2, 3, or 4 water layers between the plates in Å					
	4.0	6.6	9.2	11.0	(23)	Σ
10,000	0.077	0.0735	0.070	0.068^5	(0.061)	0.001
1,000	0.50	0.46	0.43	0.41^5	(0.34)	0.01
100	2.36	1.90	1.73	1.59	(0.96)	0.14
	1	2	3	4	(5)	6

For two clay layers with a dipole layer at each side of each layer, two pairs of dipole layers attract and two pairs repel each other.

Assuming that the dielectric constant of the clay layer is the same as that of the water in between ($\epsilon \approx 6$) when separated by one water layer, then the total repulsive energy is found by taking the energy listed in column 1 plus that in column 5 minus twice the value in column 4. The results are shown in column 6. Apparently, the net repulsion at the most likely plate sizes (1,000–10,000 A) is rather low, that is, smaller than 0.01 mJ/m² (erg/cm²).

2. Counter Ions in Midway Position between the Layers—Computation of the Attractive Electrostatic Energy

The assembly of two negatively charged layers and a layer of cations midway between the layers may be considered condensors. Hence, the change of attractive energy ΔE with a change in interlayer spacing $\Delta x = 2.6$ A accompanying the removal of one water layer is given by:

$$E = \frac{2\pi\sigma^2\Delta x}{\epsilon}$$

For a dielectric constant of 6 and a charge density of 4.5×10^4 esu/cm², ΔE amounts to about 30 mJ/m² (erg/cm²).

References

DOUBLE LAYER THEORY—REFINEMENTS AND COMPUTATIONS

1. Bolt, G. H. (1955), Analysis of the validity of the Gouy-Chapman theory of the electric double layer, *J. Colloid Sci.*, **10,** 206–218.
2. Levine, S. (1953), Interaction of two parallel colloidal plates using a modified Poisson-Boltzmann equation, *Proc. Phys. Soc. A*, **66,** 365–371.
3. Brodowsky, H., and Strehlow, H. (1959), Zur Struktur der elektrochemischen Doppelschicht, *Z. Elektrochem.*, **63,** 262–269.
4. Sparnaay, M. J. (1958), Correction on the theory of the flat diffuse double layer, *Rec. Trav. Chim.*, **77,** 872–888.
5. Loeb, A. L., Overbeek, J. Th. G., and Wiersema, P. H. (1960), *The Electrical Double Layer Around a Spherical Colloid Particle,* Computation of the potential, charge density, and free energy of the electrical double layer around a spherical colloid particle, The M.I.T. Press, Cambridge, Mass.
6. Sweeton, F. H., *Calculation of Suspension Peptization,* Oak Ridge National Laboratory, ORNL-2791, Office of Technical Services, Department of Commerce, Washington, D.C. (extensive tables for the computation of coagulation velocities from the potential energy curves).
7. Devereux, O. F., and de Bruyn, P. L. (1963), *Interaction of Plane Parallel Double Layers,* The M.I.T. Press, Cambridge, Mass.
8. Honig, E. P., and Mul, P. M. (1971), Tables and equations of the diffuse double layer repulsion at constant potential and at constant charge, *J. Colloid Interface Sci.*, **36,** 258–272.
9. Sparnaay, M. J. (1972), *The Electrical Double Layer,* Pergamon, New York.

COUNTER ION EXCHANGE AND NEGATIVE ADSORPTION OF CO-IONS

10. Klaarenbeek, F. W. (1946), Over Donnan-evenwichten bij solen van arabische gom, Thesis, Utrecht; Reference in Kruyt, H. R. (1952), *Irreversible Systems,* Volume I of *Colloid Science,* Elsevier, New York, 192–193.
11. Schofield, R. K., and Talibuddin, O. (1948), Measurement of internal surface by negative adsorption, *Discussions Faraday Soc.,* **3,** 51–56.
12. Bolt, G. H., and Warkentin, B. P. (1958), The negative adsorption of anions by clay suspensions, *Kolloid-Z.,* **156,** 41–46.
13. Van Os, G. A. J. (1943), Ionenuitwisseling en geleidingsvermogen van het zilverjodidesol, Thesis, Utrecht; Reference in Kruyt, H. R. (1952), *Irreversible Systems,* Volume I of *Colloid Science,* Elsevier, New York, 177–180.
14. Eriksson, E. (1952), Cation exchange equilibria on clay minerals, *Soil Sci.,* **74,** 103–113.
15. Bolt, G. H. (1955), Ion adsorption by clays, *Soil Sci.,* **79,** 267–276.
16. Bolt, G. H. (1960), Cations in relation to clay surfaces, *Trans. Intern. Congr. Soil Sci., 7th,* Madison, Wis.
17. Chaussidon, J. (1958), Adsorption négative des anions dans les suspensions d'argile, *Bull. Groupe Franç. Argiles,* **10,** 31–35.
18. de Haan, F. A. M., and Bolt, G. H. (1963), Determination of anion adsorption by clays, *Soil Sci. Soc. Am. Proc.,* **27,** 636–640.
19. Edwards, D. G., Posner, A. M., and Quirk, J. P. (1965), Repulsion of chloride ions by negatively charged clay surfaces. I: Monovalent cation Fithian illite; II: Monovalent cation montmorillonite; III: Di- and trivalent cation clays; *Trans. Faraday Soc.,* **61,** 2808–2815 (I); 2816–2819 (II); 2820–2823 (III).
20. van den Hul, H. J., and Lyklema, J. (1968), Determination of specific surface areas of dispersed materials. Comparison of the negative adsorption method with some other methods, *J. Am. Chem. Soc.,* **90,** 3010–3015.

APPENDIX IV

Van der Waals Attraction Energy between Two Layers

The van der Waals attraction energy between two layers of a 2:1 layer clay is computed according to the following formula:

$$V_A = -\frac{A}{48\pi}\left\{\frac{1}{d^2} + \frac{1}{(d+\Delta)^2} - \frac{2}{(d+\Delta/2)^2}\right\}$$

in which d is the half-distance between the plates measured between the planes of the centers of the oxygen atoms of the tetrahedral sheet, and Δ is the thickness of the unit layer measured between the same planes ($\Delta = 6.60$ A).

The constant A $\sim 10^{-12}$.*

In the following figure, V_A is plotted as a function of the half-distance.†

* See the experiments with mica by Tabor, D., and Winterton, R. H. S. (1969), *Proc. Roy. Soc.*, **A312,** 435–450, and the table on page 445.

† At short distances, at layer separations of 2.6, 3.9, 5.2, and 6.5 A, corresponding to the presence of 1, 2, 3, and 4 monolayers of water between the layers, application of the above formula yields attractive energies of 6.7, 2.4, 1.1, and 0.6 mJ/m² (not shown in the figure).

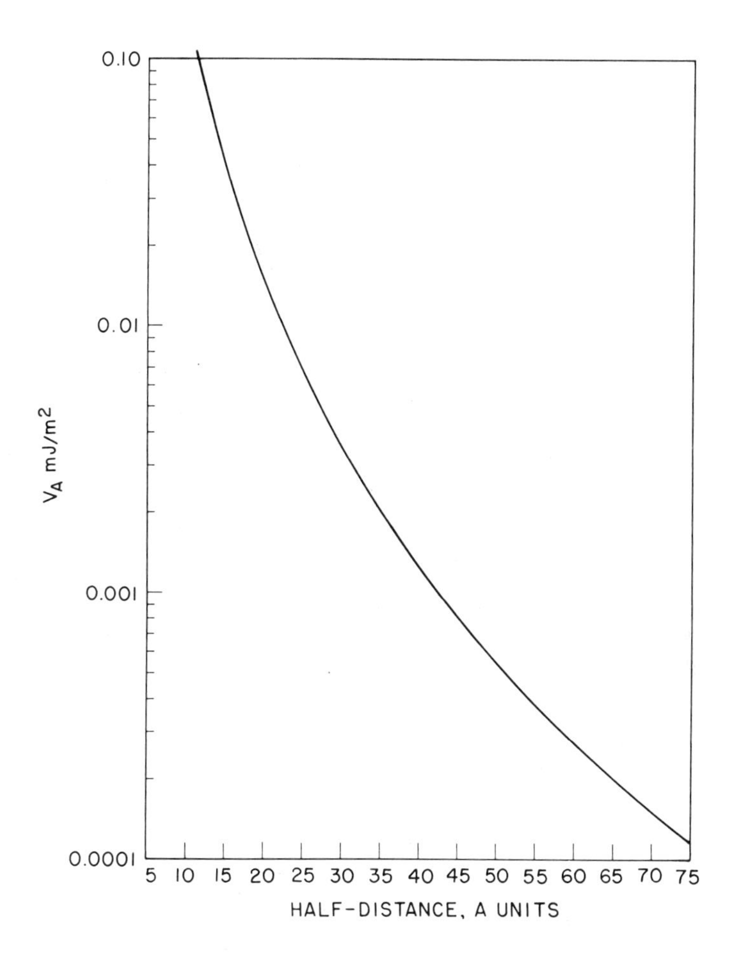
0.10
0.01
0.001
0.0001
V_A mJ/m^2
5 10 15 20 25 30 35 40 45 50 55 60 65 70 75
HALF-DISTANCE, A UNITS

APPENDIX V

Clay Literature

CURRENT

The current clay literature may be traced by consulting

a. Chemical Titles (American Chemical Society) under the heading "Clay."

b. Petroleum Abstracts (The University of Tulsa, 1136 North Lewis Avenue, Tulsa 10, Oklahoma) under the heading "Geochemistry."

TERMINOLOGY, SYMBOLS, UNITS

Manual of Symbols and Terminology for Physicochemical Quantities and Units, 1975, IUPAC, Butterworths, London.

Appendix II: *Definitions, Terminology and Symbols in Colloid and Surface Chemistry,* Part I, 1972, IUPAC, Butterworths, London (*Pure and Applied Chemistry,* **31,** 579–638, 1972).

BOOKS, MONOGRAPHS, REVIEWS

Colloid Chemistry of Clays

B-1 Marshall, C. E. (1949), *The Colloid Chemistry of the Silicate Minerals,* Academic, New York.

B-2 Iler, R. K. (1955), *The Colloid Chemistry of Silica and Silicates,* Cornell University Press, Ithaca, New York.

B-3 *The Colloid Chemistry of Palygorskite* (1964), F. D. Ovcharenko, Editor, Israel Program for Scientific Translations, Jerusalem.

B-4 Swartzen-Allen, S. Lee, and Matijevic, E. (1974), *Surface and Colloid Chemistry of Clays, Chem. Rev.,* **74,** 385–400.

Colloid Chemistry and Surface Chemistry in General

B-5 Kruyt, H. R. (1949 and 1952), *Colloid Science,* Volume I—*Irreversible systems,* 1952; Volume II—*Reversible systems,* 1949, Elsevier, Amsterdam, Houston, New York, London.

B-6 Verwey, E. J. W., and Overbeek, J. Th. G. (1948), *Theory of the Stability of Lyophobic Colloids,* Elsevier, New York, Amsterdam, London, Brussels.

B-7 Freundlich, H. (1922), *Colloid and Capillary Chemistry,* Translation by H. S. Hatfield, Dutton Co., New York.

B-8 Mysels, K. J. (1959), *Introduction to Colloid Chemistry,* Interscience, New York.

B-9 Bikerman, J. J. (1958), *Surface Chemistry,* Second Edition, Academic, New York.

B-10 Adamson, A. W. (1960), *Physical Chemistry of Surfaces,* Interscience, New York.

B-11 Davies, J. T., and Rideal, E. K. (1961), *Interfacial Phenomena,* Academic, New York.

B-12 *Physical Chemistry: Enriching Topics from Colloid and Surface Science* (1975), H. van Olphen and Karol J. Mysels, Editors, Theorex, 8327 La Jolla Scenic Drive, La Jolla, Ca. 92037.

Clay Mineralogy, Chemistry

B-13 *X-ray Identification and Crystal Structure of Clay Minerals* (1951 and 1961), First Edition (G. W. Brindley, Ed.), 1951; Second Edition (G. Brown, Ed.), 1961, Miner. Soc., London.

B-14 *The Differential Thermal Investigation of Clays* (1957), R. C. Mackenzie, Editor, Miner. Soc., London.

B-15 *The Electro-Optical Investigation of Clays* (1971), J. A. Gard, Editor, Miner. Soc., London.

B-16 *The Infrared Spectra of Minerals* (1974), V. C. Farmer, Editor, Miner. Soc. (41 Queen's Gate, London SW7 5HR).

B-17 Weaver, C. E., and Pollard, L. D. (1973), *The Chemistry of Clay Minerals,* Elsevier, Amsterdam.

B-18 Grim, R. E., (1968), *Clay Mineralogy,* Second Edition, McGraw, New York.

B-19 Bates, T. F., Selected electron micrographs of clays and other fine-grained minerals, *Pennsylvania State Univ. Circ. 51,* Min. Ind. Expt. Sta., Coll. Min. Ind.

B-20 *Soil Clay Mineralogy* (1962), C. I. Rich and G. W. Kunze, Editors, The University of Carolina Press, Chapel Hill, N.C.

B-21 Millot, G. (1964), *Geologie des Argiles. Alterations, Sedimentologie, Geochimie,* Masson et Cie., Paris.

B-22 Caillère, S., and Hénin, M. S. (1964), *Minèralogie des Argiles,* Masson et Cie., Paris.

Soils

B-23 Bear, F. E. (1955), *Chemistry of the Soil,* Reinhold, New York.

B-24 Baver, L. D. (1948), *Soil Physics,* Second Edition, Wiley, New York; Chapman, London.

B-25 Tschapek, M. (1949), *Quimica Coloidal del Suelo,* Imprenta y casa editoria Coni, Buenos Aires.

B-26 Jackson, M. L. (1960), *Soil Chemical Analysis,* Prentice-Hall Inc., Englewood Cliffs, New Jersey.

B-27 Marshall, C. E. (1964), *The Physical Chemistry and Mineralogy of Soils,* Volume 1 of *Soil Materials,* Wiley, New York.

B-28 *Metody Badan Gruntow Spoistych* (Methods of Studying Clay Soils) (1975), Barbara Grabowska-Olszewska, Editor, Wydawnictwa Geologiczne, Warsaw, Poland.

B-29 *Soil Components,* Volume 1,—*Organic Components*; Volume 2,—*Inorganic Components,* J. E. Gieseking, Editor, Springer, Berlin.

Silicates, Zeolites

B-30 Eitel, W. (1975), *Silicate Science,* Volume 6: *Silicate Structure and Dispersoid Systems,* Academic, New York.

B-31 Breck, Donald W. (1974), *Zeolite Molecular Sieves: Structure, Chemistry and Use,* Wiley, New York.

Clay-Organic Interactions

B-32 Theng, B. K. G. (1974), *The Chemistry of Clay-Organic Reactions,* Wiley, New York.

B-33 Cairns-Smith, A. G. (1971), *The Life Puzzle—On Crystals and Organisms and on the Possibility of a Crystal as an Ancestor,* Oliver and Boyd, Edinburgh.

Data Collections

B-34 *Reference Clay Minerals,* American Petroleum Institute, Project 49, Columbia University, New York, 1951.

B-35 Scifax D.T.A. Data Index, a punched card index of data and references for differential thermal analysis. Compiled by R. C. Mackenzie. Cleaver-Hume Press Ltd., 31 Wright's Lane, Kensington, London W 8 (1962).

B-36 van der Marel, H. E., and Beutelspacher, H. (1968), *Atlas of Electron Microscopy of Clay Minerals and their Admixtures,* Elsevier, Amsterdam, New York.

B-37 van der Marel, H. W., and Beutelspacher, H. (1976), *Atlas of Infrared Spectroscopy of Clay Minerals and their Admixtures,* Elsevier, Amsterdam, New York.

B-38 Thorez, J. (1975), *Phyllosilicates and Clay Minerals—A Laboratory Handbook for their X-Ray Diffraction Analysis,* G. Lelotte, rue Pisseroule 109, B-4820 DISON, Belgium.
B-39 van Olphen, H., and Fripiat, J. J., *Clay materials Data Handbook,* Pergamon, Oxford (in preparation).

Ceramics

B-40 *Ceramic Fabrication Processes* (1958) (W. D. Kingery, Ed.), Technology Press of Massachusetts Inst. of Technology, Wiley, New York; Chapman, London.
B-41 Kingery, W. D. (1961), *Introduction to Ceramics,* Wiley, New York.
B-42 Searle, A. B., and Grimshaw, R. W. (1959), *The Chemistry and Physics of Clays and Other Ceramic Materials,* Third Edition, Interscience, New York.

REFERENCE CLAYS

The purpose of establishing collections of reference clays is to enable the comparison of research results obtained on identical samples in different laboratories. Also, the collection of data on these samples, as published in the open literature or contributed by participating laboratories, will provide well-characterized materials for the practitioner.

R-1 The first project of this kind was carried out under the auspices of the American Petroleum Institute, and it was entitled "Reference Clay Minerals, Project 49." Data collected on the samples were published in a series of eight reports by Columbia University, New York, 1951. Original batches on which the data were collected are exhausted, however, substitute batches have been collected at the same locations. Samples of these are still available from Wards.

R-2 Recently, the clay Minerals Society in the United States initiated a new collection of so-called "Source Clays," consisting of homogenized batches of hand-picked samples of a number of typical clays. The Source Clay Program is described in the Proceedings of the International Clay Conference, Mexico City, 1975. Samples may be ordered from the repository, c/o Professor W. D. Johns, University of Missouri, Columbia, Mo. 65201.

R-3 An analogous program was initiated in Europe under the auspices of the OECD. A "Bank of non-metallic minerals" (primarily clay minerals) was established. Samples are available through the "Laboratoire de Minéralogie du Museum National d'Histoire Naturelle," 61 rue de Buffon, Paris 5e, France.

For both sample collections data have been obtained from participating laboratories. A compilation and evaluation of the data for the two collections are in preparation.

For all projects, the hand-picked clays were powdered and homogenized. They have not been submitted to any further treatments. In addition to the principal clay mineral indicated by the name of the sample, other minerals are present in the samples usually to a small extent, but occasionally larger amounts of such impurities occur. In research studies, the samples are often purified and sometimes brought into a homoionic condition. It is urged that in publishing research results on the standard samples, the treatment procedures be described in detail.

R-4 The Standard Reference materials collection of the National Bureau of Standards (NBS) comprises two clays of certified chemical composition: Flint Clay (#97a) and Plastic Clay (#98a). In addition, NBS distributes Attapulgus Clay (#GM-2007) having adsorptive characteristics as specified by ASTM D-2007. For catalog write National Bureau of Standards, Washington D.C. 20034.

PERIODIC PUBLICATIONS

Since clay research covers such a variety of fields, publications on clays appear in many scientific and technological journals and conference proceedings. The following periodic publications deal exclusively with clay research:

P-1 *Proceedings of the National Conference on Clays and Clay Minerals,* Sponsored by the Clay Min. Comm. of the Natl. Acad. Sci.-Natl. Res. Coun., Washington, D.C., and from vol. 13 by the Clay Minerals Society.

(1) *Clays and Clay Technology* (1955), Proc. of the First Conference, Bull. 169, Div. Mines, Dept. Natl. Resources, State of California, Ferry Bldg., San Francisco: Univ. California, Berkeley, California, 1952.

(2) *Clays and Clay Minerals* (1954), Proc. of the Second Conference, Natl. Acad. Sci.-Natl. Res. Coun., Pub. 327, Washington, D.C.; Univ. Missouri, Columbia, Missouri, 1953.

(3) *Clays and Clay Minerals* (1955), Proc. of the Third Conference, Natl. Acad. Sci.-Natl. Res. Coun., Pub. 395, Washington, D.C.; Rice Inst., Houston, Texas, 1954.

(4) *Clays and Clay Minerals* (1956), Proc. of the Fourth Conference, Natl. Acad. Sci.-Natl. Res. Coun., Pub. 456, Washington, D.C.; Pennsylvania State Univ., University Park, Pennsylvania, 1955.

(5) *Clays and Clay Minerals* (1958), Proc. of the Fifth Conference,

Natl. Acad. Sci.-Natl. Res. Coun., Pub. 566, Washington, D.C.; Univ. Illinois, Urbana, Illinois, 1956.

(6) *Clays and Clay Minerals* (1959), Proc. of the Sixth Conference, Pergamon Press, New York; Univ. California, Berkeley, California, 1957.

(7) *Clays and Clay Minerals* (1960), Proc. of the Seventh Conference, Pergamon Press, New York; Washington, D.C., 1958.

(8) *Clays and Clay Minerals* (1960), Proc. of the Eighth Conference, Pergamon Press, New York; Univ. Oklahoma, Norman, Oklahoma, 1959.

(9) *Clays and Clay Minerals* (1962), Proc. of the Ninth Conference, Pergamon Press, New York; Purdue Univ., Lafayette, Indiana, 1960 (Features symposia on clay-water relationships with respect to engineering properties of soils, and on clay-organic complexes).

(10) *Clays and Clay Minerals* (1963), Proc. of the Tenth Conference, Pergamon Press, New York: University of Texas, Austin, Texas, 1961 (features symposium on clay-organic complexes).

(11) *Clays and Clay Minerals* (1963), Proc. of the Eleventh Conference, Pergamon Press, New York; National Research Council, Ottawa, Ontario, Canada, 1962 (Features a symposium on clay mineral transformations).

(12) *Clays and Clay Minerals* (1964), Proc. of the Twelfth Conference, Pergamon Press, New York; University of Georgia, Atlanta, Ga., 1963 (Features symposia on high temperature transformations, on mechanism of emplacement (formation) of clay minerals, and on kaolinite.

(13) *Clays and Clay Minerals* (1964), Proc. of the Thirteenth Conference, Pergamon Press, New York; University of Wisconsin, Madison, Wis., 1964 (Features a symposium on structural aspects of layer silicates).

(14) *Clays and Clay Minerals* (1966), Proc. of the Fourteenth Conference, Pergamon Press, New York; University of California, Berkeley, Cal., 1965 (Features symposia on structure and quantitative analysis, on surface reactivity, and on genesis and synthesis of clays).

(15) *Clays and Clay Minerals* (1967), Proc. of the Fifteenth Conference, Pergamon Press, New York; University of Pittsburgh, Pa., 1966 (Features symposia on electron-optical study of smectites, on X-ray diffraction techniques, on mixed-layer minerals, on high temperature reactions, on spectroscopic techniques, on vermiculite studies, and on zeolites).

With the sixteenth issue, the annual proceedings were converted to a journal, entitled *Clays and Clay Minerals*, the first one being numbered volume **16,** 1968. Citations to previous annual proceedings should refer to volume numbers 1–15, with the date of issue of each volume, for example, *Clays and Clay Minerals,* **14,** 53, 1966. The journal is the official journal of the Clay Minerals Society, and it contains papers from the annual conferences, as well as other research papers.

P-2 *Clay Minerals Bulletin* (volumes 1–5, 1949–1964) and its successor *Clay Minerals* (volume 6, 1965–), a publication of the Clay Minerals Group of the Mineralogical Society of Great Britain and Ireland.

P-3 *Clay Science,* a journal published by the Clay Science Society of Japan since 1960. Papers are in English. Japan Publications Trading Co., Ltd., P.O. Box 5030, Tokyo International, Tokyo, Japan; or 1255 Howard St., San Francisco, Cal. 94103.

P-4 *Bulletin du Groupe Français des Argiles,* Tome I–V, 1949–1954, stenciled. Annual printed volumes start with Tome VI, Nouvelle Série No. 1, 1955; Papers in French. Centre National de la Recherche Scientifique, Service des Publications, 3ème Bureau; 13, Quai Anatole France, Paris (7°), France.

P-5 *Reports from the Swedish Society for Clay Research,* Meddelanden från Svenska Föreningen för Lerforskning, Tegellab., Stockholm Ö, Biannual printed volumes start with Nr. 1, February 1954, as separate prints from "Geologiska Föreningens i Stockholm Förhandlingar; Papers in English, German, and Swedish.

P-6 *Proceedings of the International Clay Conferences,* Sponsored by the Association ("Comité," prior to 1966) Internationale Pour l'Etude des Argiles (AIPEA, CIPEA prior to 1966).

1963, Stockholm, Sweden, I. Th. Rosenqvist and P. Graff-Peterson, Eds., Pergamon Press, New York (2 volumes).

1966, Jerusalem, Israel, L. Heller and A. Weiss, Eds., Israel Program for Scientific Translations, Jerusalem (2 volumes).

1969, Tokyo, Japan, L. Heller, Ed., Israel University Press, Jerusalem (2 volumes).

1972, Madrid, Spain, J. M. Serratosa, Ed., Division de Ciencias, C.S.I.C., Serrano, 113, Madrid-6, Spain (Including Kaolin Symposium).

1975, Mexico D. F., Mexico, S. W. Bailey, Ed., Applied Publishing Ltd., P.O. Box 261, Wilmette, Illinois 60091.

P-7 Proceedings of symposia organized by several national clay societies.

(a) *Genèse et Synthèse des Argiles* (1962), organized by the Groupe Français des Argiles, M. R. Hocart, Ed., Centre National de Recherche Scientifique, 15 Quai Anatole France, Paris 7e, France.

The following papers were presented:

Wey, R., and Siffert, B., *Réactions de la silice monomoléculaire en solution avec les ions* Al^{3+} *et* Mg^{2+}.
Gillis, E., and Dekeyser, W., *Expériences avec des gels de silice et d'alumine.*
Caillère. S., and Hénin, S., *Vue d'ensemble sur le problème de la synthèse des minéraux phylliteux à basse température.*
Oberlin, A., Tchoubar, C., Schiller, C., Pézerat, H., and Kovacevic, S., *Etude du fireclay produit par altération de la kaolinite et de quelques fireclays naturels.*
Gastuche, M. C., Fripiat, J. J., and de Kimpe, C., *La genèse des minéraux argileux de la famille du kaolin, I—Aspect colloidal.*
Gastuche, M. C., and de Kimpe, C., *La genèse des minéraux argileux de la famille du kaolin, II—Aspect cristallin.*
Roy, Rustum, *The preparation and properties of synthetic clay minerals.*
Pedro, G., *Genèse des minéraux argileux par lessivage des roches cristallines au laboratoire.*
Correns, C. W., *Beobachtungen über die Bildung und Umbildung von Tonmineralen bei der Zersetzung von Basalten.*
Konta, J., *Crystallization temperature of clay minerals in the molybdenite and cassiterite wolframite ore sands of Northern Bohemia.*
Tchoukhrov, F. V., *Sur la genèse des minéraux argileux dans la zone d'altération superficielle des gites métallifères.*
Mitchell, B. D., *The influence of soil forming factors on clay genesis.*
Nesteroff, W. D., and Sabatier, G., *"Apport" et "néogenèse" dans la formation des argiles des grands fonds marins.*
Millot, G., *Silicifications et néoformations argileuses: problèmes de genèse.*
Lucas, J., *Remarques sur les minéraux argileux interstratifiés et leur genèse.*
Gallitelli, P., *Remarques sur la genèse de quelques minéraux argileux des Apennins de l'Italie du Nord.*
Nicolas, J., *Sur la présence de "glauconie" en Bretagne Centrale.*

(b) *Reunion Hispano-Belga de Minerales de la Arcilla* (1970), C.S.I.C., Distribucion de Publicaciones, Vitrubio, 16, Madrid-6, Spain.

(c) *Atti del I° Congresso Nazionale* (1971), organized by the Gruppo Italiano dell' A.I.P.E.A., F. Veniale and C. Palmonari, Eds., Cooperativa Libraria Universitaria Bologna, P.zza G. Verdi 2/A, Bologna, Italy.

AUTHOR INDEX

SUBJECT INDEX